Fuzzy Sets and Triangular Norms

This book aims to serve as a comprehensive resource that equips readers with the knowledge and practical skills needed to navigate the intricacies of fuzzy set theory, t-norms, and their integration into decision-aided intelligent systems. It provides a comprehensive understanding of aggregation operators and their role in data fusion, risk analysis, and expert opinion aggregation.

- New aggregation operators, entropy measures, t-norm, and t-conorm structures are developed across multiple fuzzy set extensions to better model uncertainty and hesitation in decision-making.
- A wide range of real-world applications, including, tourism planning, smart cities, urban mobility, water security, smart campus automation, energy facility siting, and firefighting helicopter selection, are addressed using advanced multi-criteria decision making methods.
- The chapters collectively emphasize sustainable, data-driven, and uncertainty-aware decision support, contributing solutions in areas such as environmental protection, resource optimization, public services, and technological infrastructure.
- Innovative techniques, such as Lambert W–based aggregation operators, Choquet integral–based entropy, confidence-level aggregation, and fuzzy–machine learning hybrid models, improve the representation of interaction, ambiguity, and complexity in multi-criteria decision problems.

The text is primarily written for senior undergraduates, graduate students, and academic researchers in diverse fields including engineering mathematics, industrial engineering, supply chain management, operations research, manufacturing engineering, production engineering, and applied mathematics.

Intelligent Data-Driven Systems and Artificial Intelligence

Series Editor: Harish Garg

Cognitive Machine Intelligence
Applications, Challenges, and Related Technologies
Inam Ullah Khan, Salma El Hajjami, Mariya Ouaissa, Salwa Belqziz, and Tarandeep Kaur Bhatia

Artificial Intelligence and Internet of Things based Augmented Trends for Data Driven Systems
Anshu Singla, Sarvesh Tanwar, Pao-Ann Hsiung

Modelling of Virtual Worlds Using the Internet of Things
Edited by Simar Preet Singh and Arun Solanki

Data-Driven Technologies and Artificial Intelligence in Supply Chain
Tools and Techniques*Mahesh Chand, Vineet Jain, and Puneeta Ajmera*

Data driven Decision Making and Soft Computing
Applications and Advances
D. Nagarajan, Dragan Pamucar, Zarife Zararsız

Artificial Intelligence Powered Sustainable Development of Business 4.0
Edited by Hemant Kumar Saini, Sita Rani, Vikram Puri, and Noor Zaman Jhanjhi

Artificial Intelligence and Fuzzy Logic for Next Generation Intelligent Transportation Systems
Mariya Ouaissa, Mariyam Ouaissa, Najib Ben Aoun, Anass Cherrafi, Rahul Kar

Blockchain and Software Defined Networking: Opportunities, Challenges, and Future Trends
Sarvesh Tanwar, Sumit Badotra, and Harish Garg

Intelligent Systems: Emerging Trends and Challenges in Deep Neuro-Fuzzy Models and Explainable Artificial Intelligence
Edited by Siddhartha Roy, Soumya Sen, and Agostino Cortesi

Artificial Intelligence Techniques in Mathematical Modeling and Optimization
Edited by Mukesh Kumar Awasthi, Sanoj Kumar, Deepika Saini

Robotics and Intelligent Machines in Smart Agriculture: Emerging Systems and Applications
Edited by Mohammad Shabaz

Fuzzy Sets and Triangular Norms: Aggregation in Decision-Aided Intelligent Systems
Edited by Mehmet Ünver and Gökhan Özçelik

For more information about this series, please visit: www.routledge.com/Intelligent-Data-Driven-Systems-and-Artificial-Intelligence/book-series/CRCIDDSAAI

Fuzzy Sets and Triangular Norms

Aggregation in Decision-Aided Intelligent Systems

Edited by Mehmet Ünver and Gökhan Özçelik

CRC Press
Taylor & Francis Group
Boca Raton London New York

CRC Press is an imprint of the
Taylor & Francis Group, an **informa** business

Designed cover image: Shutterstock

First edition published 2027
by CRC Press
2385 NW Executive Center Drive, Suite 320, Boca Raton FL 33431

and by CRC Press
4 Park Square, Milton Park, Abingdon, Oxon, OX14 4RN

CRC Press is an imprint of Taylor & Francis Group, LLC

ISBN: 978-1-032-86767-0 (hbk)
ISBN: 978-1-032-86766-3 (pbk)
ISBN: 978-1-003-52908-8 (ebk)

DOI: 10.1201/9781003529088

Typeset in Times
by Deanta Global Publishing Services, Chennai, India

Contents

Preface

This volume presents a focused compilation of recent advancements in fuzzy set theory and its integration with triangular norms, with a particular emphasis on intelligent decision-making systems. The contributions in this book reflect both foundational theoretical developments and applied methodologies that demonstrate the versatility of fuzzy logic in addressing complex, real-world problems.

The chapters span a wide array of applications, including multi-criteria decision-making, group decision aggregation, and intelligent systems, in domains such as tourism marketing, sustainable urban logistics, water security, smart cities, and IoT infrastructures. Advanced fuzzy modeling tools—such as interval-valued intuitionistic fuzzy sets, hesitant fuzzy sets, and q-rung orthopair fuzzy frameworks—are extensively employed across these chapters.

In addition to aggregation operator development using t-norms and t-conorms, this volume incorporates entropy measures, fuzzy AHP techniques, and hybrid models combining fuzzy logic with machine learning and goal programming. The integration of these models into intelligent systems highlights the interdisciplinary nature of fuzzy logic and its adaptability to various decision support environments.

This compilation aims to serve as a reference for researchers, practitioners, and graduate students engaged in the development and implementation of fuzzy logic-based decision systems.

The Editors
Mehmet Ünver
Gökhan Özçelik

About the Editors

Mehmet Ünver completed his undergraduate studies in the Department of Mathematics at Ankara University (Türkiye) in 2007, followed by his master's degree at the same institution in 2009. During his master's studies, he focused on summability theory and Korovkin-type approximation theory. He earned his PhD in Mathematics from Ankara University in 2013, continuing his research in summability theory and Korovkin-type approximation theory. He served as a research assistant at the Department of Mathematics, Ankara University, from 2009 to 2015, and later as an assistant professor from 2015 to 2017. Since 2017, he has held the position of Associate Professor at the Department of Mathematics, at Ankara University. His current research pursuits encompass fuzzy measure and fuzzy set theory, multi-criteria decision-making, Korovkin-type approximation theory, and summability theory, resulting in numerous published papers in highly cited journals on topics including summability theory and the applications of fuzzy set theory in multi-criteria decision-making, pattern recognition, aggregation operators, and classification.

Gökhan Özçelik obtained his bachelor's degree from the Department of Industrial Engineering at Gazi University (Türkiye) in 2009. He completed his master's degree in the same department in 2011 and subsequently received his PhD in Industrial Engineering from Gazi University in 2016. Presently, he serves as Associate Professor at Karadeniz Technical University (Türkiye). His research primarily focuses on optimization, multi-criteria decision-making, and fuzzy sets and systems. He has an extensive publication record in reputable journals such as the *European Journal of Operational Research*, *Transportation Research Part E*, *Expert Systems with Applications*, *International Journal of Production Research*, and *Journal of Cleaner Production*, among others.

Contributors

Abdullah, Lazim
Faculty of Science and Technology
University of Malaysia Terengganu
K. Terengganu, Malaysia

Azman, Wan Nur Amira Wan
Faculty of Informatics and Computing
Universiti Sultan Zainal Abidin
Terengganu, Malaysia

Bozyiğit, Mahmut Can
Department of Mathematics
Ankara Yıldırım Beyazıt University
Ankara, Türkiye

Burnaz, Ersin
Department of Marketing and Advertising
Trabzon University
Trabzon, Türkiye

Chatterjee, Prayosi
Department of Mathematics
Kazi Nazrul University
Asansol, India

Dağıstanlı, Hakan Ayhan
Department of Industrial and System Engineering
Turkish Military Academy, National Defense University
Ankara, Türkiye

Dogan, Onur
Department of Management Information Systems
Izmir Bakircay University
Izmir, Türkiye

Elomiya, Akram
Department of Transport Management, Marketing, and Logistics
University of Pardubice
Pardubice, Czech Republic

Erdogan, Ali Mert
Department of Management Information Systems
Izmir Bakircay University
Izmir, Türkiye

Eroğlu, Özer
Graduate School of Natural and Applied Sciences
Gazi University
Ankara, Türkiye

Gürsoy Yılmaz, Beren
Department of Industrial Engineering
Karadeniz Technical University
Trabzon, Türkiye

Jovčić, Stefan
Kayhan, Behice Meltem
Department of Business Administration
Manisa Celal Bayar University
Manisa, Türkiye

Kose, Bayram
Department of Electrical and Electronic Engineering
Izmir Bakircay University
Izmir, Türkiye

Křupka, Jiří
Department of Transport Management, Marketing, and Logistics
University of Pardubice
Pardubice, Czech Republic

Mandal, Utpal
Department of Mathematics & Statistics
Banasthali Vidyapith
Rajasthan, India

Olgun, Murat
Department of Mathematics
Ankara University
Ankara, Türkiye

Özçelik, Gökhan
Department of Industrial Engineering
Karadeniz Technical University
Trabzon, Türkiye

Pairan, Mohamad Ammar
Faculty of Informatics and Computing
Universiti Sultan Zainal Abidin
Terengganu, Malaysia

Radovanović, Marko
Military Academy
University of Defense
Belgrade, Serbia

Seikh, Mijanur Rahaman
Department of Mathematics
Kazi Nazrul University
Asansol, India

Smarandache, Florentin
Department of Mathematics
University of New Mexico
Gallup, NM

Sukono
Department of Mathematics
Universitas Padjadjaran
West Java, Indonesia

Švadlenka, Libor
Department of Transport Management, Marketing, and Logistics
University of Pardubice
Pardubice, Czech Republic

Temel Gencer, Cevriye
Department of Industrial Engineering
Gazi University
Ankara, Türkiye

Ünver, Mehmet
Department of Mathematics
Ankara University
Ankara, Türkiye

Yeni, Fatma Betül
Department of Industrial Engineering
Karadeniz Technical University
Trabzon, Türkiye

Zamri, Nurnadiah
Faculty of Informatics and Computing
Universiti Sultan Zainal Abidin
Besut Terengganu, Malaysia

1 Lambert *W* Aggregation Operators for Intuitionistic Fuzzy Multi-Criteria Decision-Making

Mehmet Ünver, Florentin Smarandache, and Murat Olgun

1.1 INTRODUCTION

Schweizer and Sklar (1983) proposed triangular norms and triangular conorms (t-norms and t-conorms, respectively), which are extensions of metric spaces and build upon the idea of probabilistic metric spaces initially introduced by Menger (1942). In the field of fuzzy sets (FSs), t-norms and t-conorms are important tools used to represent conjunction and disjunction operations, respectively. T-norms are used to model the logical "and" operation between fuzzy propositions, while t-conorms are used to model the logical "or" operation. Various t-norms and t-conorms have been proposed and studied in the literature. Some of the well-known t-norms include the minimum t-norm, product t-norm, Hamacher t-norm, Einstein t-norm, Lukasiewicz t-norm, Dombi t-norm, and drastic t-norm. In the realm of fuzzy logic, various t-conorm operators play a pivotal role in aggregation and decision-making processes. Notably, Bobillo and Straccia (2007) introduced a fuzzy description logic with the product t-norm. Dimuro et al. (2018) proposed the CMin-Integral, a Choquet-like aggregation function based on the minimum t-norm, particularly applicable in fuzzy rule-based classification systems. Dombi (1982) contributed significantly by defining a general class of fuzzy operators and introducing fuzziness measures induced by such operators. Garg (2017) developed generalized Pythagorean fuzzy geometric aggregation operators utilizing the Einstein t-norm and t-conorm for multi-criteria decision-making processes. Gödel (1958) laid the foundation for exploring extensions of the finite standpoint, making vital contributions to fuzzy logic. Hamacher (1978) contributed to the field by investigating logical combinations of vague statements and their associated evaluation functions. Lukasiewicz and Straccia (2008) were instrumental in managing uncertainty and vagueness in description logics, particularly for the Semantic Web. Klement and Mesiar (2005) have contributed significantly to the field of fuzzy logic with a comprehensive study, offering valuable insights into

DOI: 10.1201/9781003529088-1

the multifaceted aspects of triangular norms. These individual contributions have enriched the landscape of t-conorm operators, enhancing their applicability in various domains of fuzzy logic and decision-making. The citation and publication trends for t-norms are illustrated in Figure 1.1. The figure clearly indicates the increasing popularity of this subject over time.

T-norms and t-conorms have become essential tools in fuzzy set theory for various reasons, including the development of aggregation operators. Aggregation operators are used to combine multiple inputs to generate a single output in decision-making processes. The graph depicting the citation and publication trends for aggregation operators is presented in Figure 1.2. It is evident from this figure that the field is gaining popularity and attention over time. Moreover, t-norms and t-conorms have found extensive applications across diverse fields such as decision-making, pattern recognition, image processing, and machine learning. Beliakov et al. (2011) delved into the study of averaging operators for intuitionistic fuzzy sets (IFSs), shedding light on their significance in various computational applications. Keller et al. (1994) made significant strides in the realm of fuzzy integration for pattern recognition. In the domain of image processing, Lopez-Molina et al. (2009) presented an innovative t-norm-based approach to edge detection. Recent research by Olgun et al. (2022) introduced single and interval-valued hybrid enthalpy fuzzy sets, coupled with a TOPSIS based on aggregation designed with t-norms and t-conorms for multi-criteria decision-making (MCDM). Additionally, Ünver et al. (2022) pioneered intuitionistic fuzzy-valued neutrosophic multi-sets and their numerical applications in classification with the help of t-norms and t-conorms. These diverse contributions highlight the broad utility of t-norms and t-conorms in addressing multifaceted challenges across various domain.

Since Zadeh's introduction of fuzzy sets (FSs) (Zadeh, 1965), they have become widely recognized and applied for handling uncertainty in diverse real-world

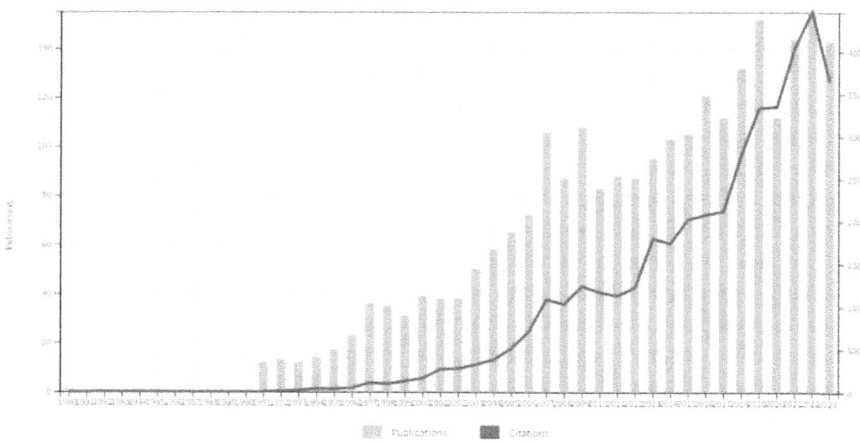

FIGURE 1.1 Citation and publication trends for t-norms in the literature

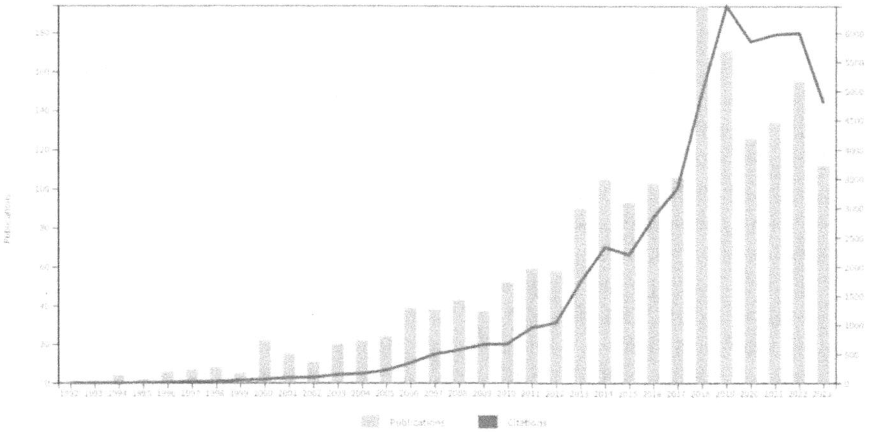

FIGURE 1.2 Citation and publication trends for aggregation operators in the literature

scenarios. FSs are defined by membership functions that map elements from a universal set X to the interval $[0,1]$. Atanassov further expanded this concept by introducing IFSs and intuitionistic fuzzy values (IFVs) (Atanassov, 1999), which utilize membership and non-membership functions satisfying a certain condition. The theory of IFS has garnered widespread attention due to its versatile applications across various domains. Feng et al. (2020) explored Minkowski weighted score functions of IFVs, contributing to the mathematical foundation of IFS. Mao et al. (2013) introduced novel cross-entropy and entropy measures for IFSs, with applications in knowledge-based systems. Song et al. (2019) introduced divergence-based cross-entropy and uncertainty measures for IFSs, applied effectively in decision-making scenarios. Wu and Wang (2021) presented a distance-based knowledge measure for IFSs, offering insights into decision-making processes. Xu (2007) and Xu and Yager (2006) explored intuitionistic fuzzy aggregation operators and geometric aggregation operators, respectively, expanding the toolkit for handling IFSs. Ye (2009) introduced a fuzzy MCDM method based on intuitionistic fuzzy cross-entropy, demonstrating its effectiveness in human–machine systems and cybernetics applications. Feng et al. (2018) provided another perspective on generalized intuitionistic fuzzy soft sets and their relevance in MCDM methods. These individual contributions have enriched the understanding and practical applications of IFSs across diverse fields.

MCDM is an area where t-norms and t-conorms play a crucial role. In MCDM, a set of alternatives is evaluated based on multiple conflicting criteria, where each alternative is represented by a fuzzy set. Aggregation operators are used to combine the individual evaluations into an overall evaluation of each alternative. The effectiveness of a MCDM approach depends on the choice of aggregation operator. Therefore, the development of new aggregation operators contributes to the improvement of MCDM approaches. FSs have emerged as valuable tools in MCDM, facilitating the resolution of complex real-world challenges across various domains.

Akbaş and Dalkılıç (2019) contributed to the field by employing fuzzy multi-criteria linear programming in portfolio selection, offering a robust decision-making framework. Akram et al. (2021) presented a group decision-making approach based on the complex spherical fuzzy VIKOR method, enhancing the collective decision-making process. Kayaturan et al. (2023) proposed a decision-aided evaluation approach for routing on multi-attribute computer networks under uncertainty, showcasing the applicability of FSs in network optimization. Olgun et al. (2021a) introduced a cosine similarity measure based on the Choquet integral for IFSs, opening new avenues for pattern recognition applications. Olgun et al. (2021b) extended the concept of Pythagorean fuzzy points, demonstrating their utility in pattern recognition and topology. Özçelik (2022) compared MCDM approaches against optimization models for finding the safest shortest path on fuzzy networks, shedding light on their relative advantages. Pehlivan et al. (2019) employed a fuzzy analytic hierarchy process and a fuzzy TOPSIS to select players for a national football team, exemplifying the practicality of FS-based decision-making. Yolcu et al. (2021) introduced intuitionistic fuzzy hypersoft sets, adding to the ever-expanding repertoire of FS-based methodologies. Also topological properties of fuzzy sets were examined in several works (Chang, 1968; Çoker, 1997; Lowen, 1976; Olgun et al., 2019; Türkarslan et al., 2021).

The Lambert W function, also known as the omega function or product logarithm, holds a significant place in mathematics, science, and engineering due to its versatile applications. Lambert (1758) first introduced this special function, which serves as the inverse of the function defined by $f(x) = xe^x$, where $x \geq -\frac{1}{e}$. Euler (1921) further delved into the properties of the Lambert W function. Its practical significance extends to physics, as highlighted by Valluri et al. (2000) in their applications in physics. The Lambert W function also finds relevance in inequalities, as evidenced by Hoorfar and Hassani's work (Hoorfar and Hassani, 2008) on inequalities involving the Lambert W function. For analytical approximations related to the Lambert W function, Barry et al. provided valuable insights in Barry et al. (2000). Moreover, Corless et al. (1996) offered a comprehensive analysis, detailing efficient algorithms for its computation. In addition, Jain and Kapoor (2005) explored applications of the Lambert W function in the study of organic solar cells. Lastly, in the field of mathematics, Cayley (1889) and Janson et al. (1993) have benefited from the insights provided by the Lambert W function. The Lambert W function serves as a valuable tool in mathematical modeling and optimization problems, offering a means to express equations in closed form and providing efficient algorithms for computation, as discussed by Lambert (1772).

Despite its wide applicability and importance in mathematical modeling and optimization, the Lambert W function has not been used to define a t-norm in the fuzzy logic literature so far. The construction of t-norms using the additive generators of t-norms is a well-established approach in fuzzy set theory and has led to the development of many families of t-norms with different properties and applications. This chapter presents a novel contribution in the field of t-norms and t-conorms by introducing the Lambert W t-norm and Lambert W t-conorm. These operators are constructed using specific additive generators, and their names are derived from the inverse functions of these generators represented via Lambert W function. Furthermore, we leverage these newly defined operators to establish algebraic operations between IFVs and introduce Lambert W weighted aggregation operators. By applying these operators to MCDM problems involving IFVs, we provide empirical evidence of their efficacy through a numerical example. Our results indicate promising performance of these operators compared to existing aggregation methods, showcasing comparable levels of accuracy and computational efficiency.

The main contributions of this chapter are given as follows:

- This chapter introduces a novel t-norm and t-conorm derived using specific additive generators. These operators provide an approach for handling uncertainty and vagueness in MCDM problems.
- This chapter defines weighted aggregation operators for families of IFVs via Lambert W t-norm and t-conorm. These operators offer a flexible and powerful framework for aggregating IFVs in decision-making processes.
- This chapter highlights the practical implications of using the Lambert W aggregation operators with a comparative and a sensitivity analysis.

The chapter is structured as follows: Section 1.2 provides a brief overview of the necessary background information. In Section 1.3, we introduce the Lambert W t-norm and t-conorm and examine their properties. We visualize the operators' graphs and contours and define the Lambert W weighted aggregation operators for IFV families. Section 1.4 presents a numerical example demonstrating the effectiveness of the proposed operators, including sensitivity and comparative analyses. Lastly, Section 1.5 concludes the chapter, offering insights into potential future research directions.

1.2 PRELIMINARIES

According to Atanassov (1999), an IFS in a set X is represented by a collection of ordered pairs $\{\langle x, \mu_A(x), \nu_A(x)\rangle : x \in X\}$. In the context of intuitionistic fuzzy sets, the membership function μ_A and the non-membership function ν_A are non-negative functions defined on a set X. These functions quantify the degree of membership and non-membership, respectively, for each element $x \in X$. It is required that $\mu_A(x) + \nu_A(x) \leq 1$ holds for every $x \in X$, ensuring that the total degree of membership and non-membership for any element does not exceed one. This condition reflects the inherent uncertainty and vagueness in the assignment of membership to elements of the set. The proposed definition of IFS allows for more accurate representation of vague or uncertain information in real-world problems.

According to Klement and Mesiar (2005) and Schweizer and Sklar (1983), a t-norm is defined as a function $\psi : \mathcal{J}^2 \rightarrow \mathcal{J}$ that satisfies four conditions where $\mathcal{J} = [0,1]$.

- The first condition, called the border condition, states that $\psi(x, 1) = x$ for any $x \in \mathcal{J}$.

- The second condition requires commutativity, which means that $\psi(x, y) = \psi(y, x)$ for any $x, y \in \mathcal{J}$.
- The third condition, called associativity, requires that $\psi(x, \psi(y, z)) = \psi(\psi(x, y), z)$ for any $x, y, z \in \mathcal{J}$.
- The fourth condition is monotonicity, which requires that $\psi(x, y) \leq \psi(x', y')$ whenever $x \leq x'$ and $y \leq y'$ for any $x, x', y, y' \in \mathcal{J}$.

The definition of a t-conorm, as given in Klement and Mesiar (2005) and Schweizer and Sklar (1983), states that it is a function $S : \mathcal{J}^2 \to \mathcal{J}$ that satisfies four conditions.

- The first condition, known as the border condition, states that $\phi(x, 0) = x$ for any $x \in \mathcal{J}$.
- The second condition, called commutativity, states that $\phi(x, y) = \phi(y, x)$ for any $x, y \in \mathcal{J}$.
- The third condition, known as associativity, states that $\phi(x, \phi(y, z)) = \phi(\phi(x, y), z)$ for any $x, y, z \in \mathcal{J}$.
- The fourth condition, referred to as monotonicity, states that $\phi(x, y) \leq \phi(x', y')$ whenever $x \leq x'$ and $y \leq y'$ for any $x, x', y, y' \in \mathcal{J}$.

A t-norm is considered continuous if it exhibits continuity as a bivariate function, and continuous t-norm ψ is referred to as Archimedean if $\psi(x, x) \leq x$ holds true for all $x \in (0,1)$. This property ensures their compatibility with some operations in fuzzy set theory. Specifically, it guarantees that the resulting fuzzy implications and negations maintain important properties like monotonicity and involutiveness, which are desirable in many applications of fuzzy logic.

The following theorem, based on the work of Klement et al. (2004), establishes the equivalence between different conditions for a t-norm T.

Theorem 1. *(Klement et al., 2004) For a t-norm ψ on $\mathcal{J}$, the following conditions are equivalent: (i) ψ is a continuous Archimedean t-norm. (ii) ψ possesses a continuous additive generator, denoted by $g : \mathcal{J} \to [0, \infty]$, with $g(1) = 0$. This means that for all $(x, y) \in \mathcal{J}^2$, the t-norm $\psi(x, y)$ can be expressed as $\psi(x, y) = g^{-1}(g(x) + g(y))$.*

Remark 1. *The dual t-conorm ϕ of a continuous Archimedean t-norm ψ, as mentioned in Klement et al. (2004), is also a continuous Archimedean t-conorm, i.e., $\phi(x, x) > x$ for all $x \in (0, 1)$, and has a representation through an additive generator $h : \mathcal{J} \to [0, \infty]$. Actually, ϕ is expressed using an additive generator $h : \mathcal{J} \to [0, \infty]$ defined as $h(t) = g(1 - t)$.*

1.3 LAMBERT *W* T-NORM AND T-CONORM AND ALGEBRAIC OPERATIONS BETWEEN IFVS

This section presents a novel Archimedean t-norm and t-conorm, highlighting their properties. Furthermore, we establish specific algebraic operations between IFVs.

1.3.1 Introduction of Lambert W t-Norm and t-Conorm

The t-norm theory gains a new additive generator with the introduction of the following theorem.

Theorem 2. *Function g defined on $[0,1]$ by $g(t) = -\frac{\ln t}{t}$ is an additive operator of a t-norm.*

Proof. Observe that $g'(t) = \frac{-1+\ln t}{t^2} < 0$ on the interval $(0,1]$, implying that g is strictly decreasing. Furthermore, we have $g(1) = 0$ and since $\lim_{t \to 0^+} \frac{-\ln t}{t} = \infty$, it follows that $g : \mathcal{J} \to [0, \infty]$. □

Throughout this chapter, we denote by g the function given in Theorem 2, and we define $h(t) = g(1-t)$. Using these functions, we introduce the concepts of the Lambert W t-norm and Lambert W t-conorm.

We require the following proposition beforehand.

Proposition 1. i) *The inverse of $g : [0,1] \to [0, \infty]$ is given by*

$$g^{-1}(t) = \frac{W(t)}{t},$$

where W is the Lambert W function.

ii) *The inverse of $h : [0,1] \to [0, \infty]$ is given by*

$$h^{-1}(t) = \frac{t - W(t)}{t}.$$

Proof. **i)** Since $W(x)e^{W(x)} = x$, we have $W(x) = xe^{-W(x)}$, which yields that $-\ln \frac{W(x)}{x} = W(x)$. Thus, we have $-\frac{\ln \frac{W(x)}{x}}{\frac{W(x)}{x}} = x$. Finally, we have

$$g\left(\frac{W(x)}{x}\right) = x.$$

On the other hand, since $\ln xe^{-\ln x} = \frac{\ln x}{x}$, we obtain $W\left(-\frac{\ln x}{x}\right) = -\ln x$. Therefore, we get

$$\frac{W\left(-\frac{\ln x}{x}\right)}{-\frac{\ln x}{x}} = x.$$

ii) It is clear that $h^{-1} = 1 - g^{-1}$. So, the proof is completed. □

Figures 1.3 and 1.4 depict the additive generators g and h and their inverse, which are represented using the Lambert W function.

The Lambert W function provides a powerful tool for constructing t-norms and t-conorms. In particular, using the additive generator functions g and h, introduced earlier, we can define the Lambert W t-norm and t-conorm as follows:

Definition 1. *The Lambert W t-norm is the Archimedean t-norm generated by the function g and is denoted by LWN. Similarly, the Lambert W t-conorm is the Archimedean t-conorm generated by the function h and is denoted by LWC.*

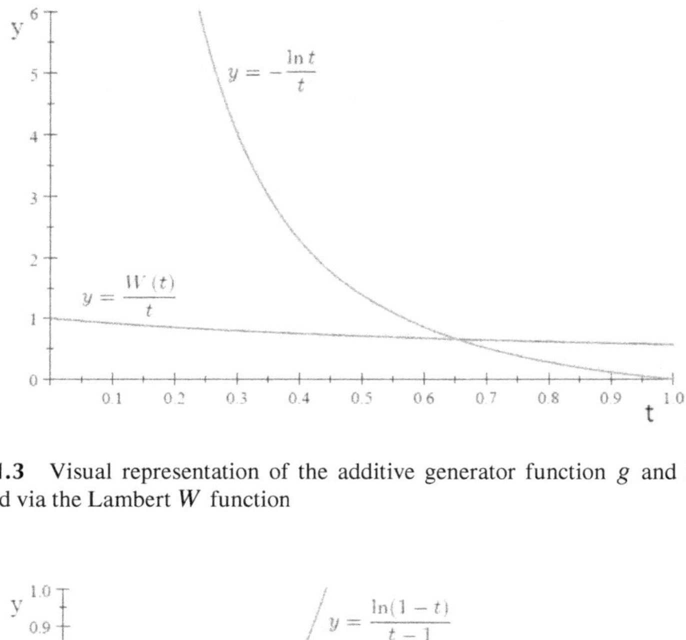

FIGURE 1.3 Visual representation of the additive generator function g and its inverse, represented via the Lambert W function

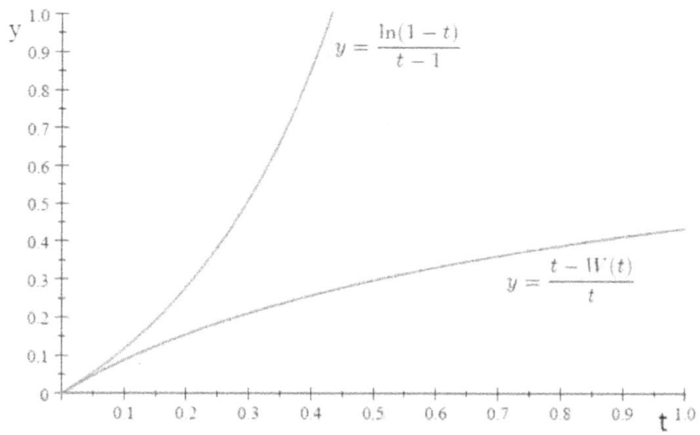

FIGURE 1.4 Visual representation of the additive generator function h and its inverse, represented via the Lambert W function

Remark 2. i) *As stated in Definition 1, we can express the Lambert W t-norm, LWN, and the Lambert W t-conorm, LWC, as follows:*

$$LWN(x,y) = g^{-1}(g(x)+g(y))$$

and

$$LWC(x,y) = h^{-1}(h(x)+h(y)).$$

ii) *To avoid complications, we assume that* $0^{\frac{1}{0}} = 0$*, which is equivalent to* $\lim_{x \to 0^+} x^{\frac{1}{x}} = 0$.

The following theorem provides the expressions for LWN and LWC.

Theorem 3. *For x, y in the domain of the functions, we have*

$$LWN(x,y) = e^{-W\left(\ln\left(x^{\frac{-1}{x}} y^{\frac{-1}{y}}\right)\right)}$$

and

$$LWC(x,y) = 1 - e^{-W\left(\ln\left((1-x)^{\frac{-1}{1-x}}(1-y)^{\frac{-1}{1-y}}\right)\right)}.$$

Proof. By employing Proposition 1 and the relationship $\frac{W(\ln t)}{\ln t} = e^{-W(\ln t)}$, we obtain

$$\begin{aligned} LWN(x,y) &= \sigma^{-1}(\sigma(x)+\sigma(y)) \\ &= \sigma^{-1}\left(-\frac{\ln x}{x} - \frac{\ln y}{y}\right) \\ &= \sigma^{-1}\left(\ln\left(x^{\frac{-1}{x}} y^{\frac{-1}{y}}\right)\right) \\ &= \frac{W\left(\ln\left(x^{\frac{-1}{x}} y^{\frac{-1}{y}}\right)\right)}{\ln\left(x^{\frac{-1}{x}} y^{\frac{-1}{y}}\right)} \\ &= e^{-W\left(\ln\left(x^{\frac{-1}{x}} y^{\frac{-1}{y}}\right)\right)}. \end{aligned}$$

Similarly, we obtain

$$LWC(x,y) = 1 - e^{W\left(\ln\left((1-x)^{\frac{-1}{1-x}}(1-y)^{\frac{-1}{1-y}}\right)\right)}.$$

1.3.2 Visualization of Lambert W t-Norm and t-Conorm

Visualizing t-norms and t-conorms provides valuable insights into their behavior and properties. The Lambert W t-norm and t-conorm are novel operators derived from specific additive generators. The contour plots of these operators provide a visual representation of their behavior over the range of input values. In the Lambert W t-norm plot, the yellow region indicates higher values of the t-norm operator, while the blue region indicates lower values. As the input values increase, the output of the t-norm operator increases, as indicated by the lightening of the yellow region. The contour curves represent the constant values of the t-norm operator, with the darker curves indicating lower values and the lighter curves indicating higher values (see Figure 1.5).

In the Lambert W t-conorm plot, the green region indicates higher values of the t-conorm operator, while the red region indicates lower values. As the input values increase, the output of the t-conorm operator decreases, as indicated by the darkening of the blue region. The contour curves represent the constant values of the t-conorm operator, with the lighter curves indicating higher values and the darker curves indicating lower values (see Figure 1.6).

1.3.3 Algebraic operations between IFVs and Lambert W Weighted Aggregation Operators

Defining algebraic operations between fuzzy sets or values is crucial in many applications of fuzzy logic, especially in decision-making and control systems.

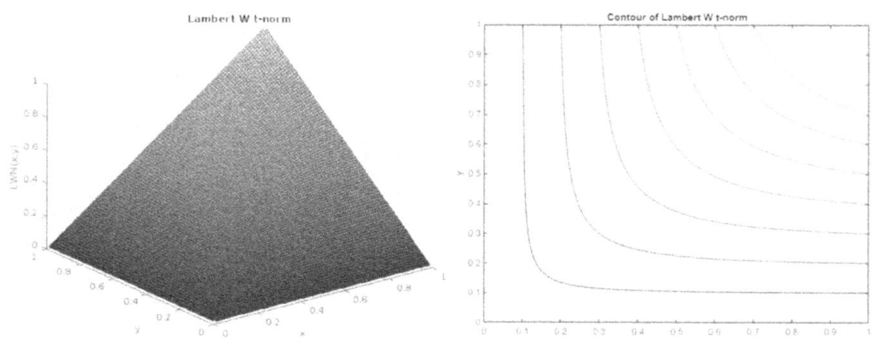

FIGURE 1.5 Visualization of the Lambert W t-norm (LWN)

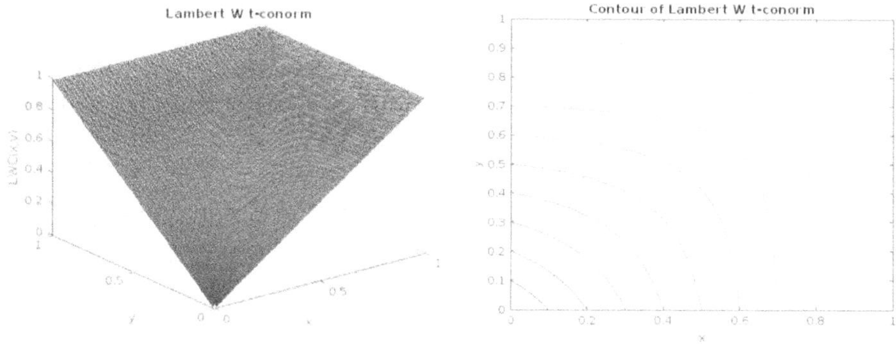

FIGURE 1.6 Visualization of the Lambert W t-conorm (LWC)

In traditional Boolean logic, operations such as AND, OR, and NOT are used to manipulate and combine crisp values, which allows for a clear and unambiguous representation of information. However, when dealing with imprecise or uncertain information, traditional logic fails to provide a suitable framework for manipulation and combination of values. Fuzzy logic, on the other hand, provides a flexible and powerful mathematical framework for handling uncertainty and vagueness in data. The use of algebraic operations in fuzzy logic enables the manipulation and combination of fuzzy values and sets in a systematic and consistent way, allowing for the development of accurate and efficient decision-making and control systems. Therefore, the development of new t-norms and t-conorms and the definition of algebraic operations between IFVs are important steps toward the advancement of fuzzy logic theory and its practical applications. In the literature, the algebraic operations for IFVs have been widely studied (Xu, 2007; Xu and Yager, 2006; Beliakov et al., 2011; Zhao et al., 2010). We define certain algebraic operations between IFVs by utilizing LWN and LWC.

Definition 2. *Consider two IFVs represented by* $\alpha = \langle \mu_\alpha, \nu_\alpha \rangle$ *and* $\beta = \langle \mu_\beta, \nu_\beta \rangle$, *and let* $\lambda \geq 0$ *be a scalar. Then*

i) $\alpha \oplus \beta = 1 - e^{-W\left(\ln\left((1-\mu_\alpha)^{\frac{-1}{1-\mu_\alpha}}(1-\mu_\beta)^{\frac{-1}{1-\mu_\beta}}\right)\right)}, e^{-W\left(\ln\left(\nu_\alpha^{\frac{-1}{\nu_\alpha}}\nu_\beta^{\frac{-1}{\nu_\beta}}\right)\right)},$

ii) $\alpha \otimes \beta = e^{-W\left(\ln\left(\mu_\alpha^{\frac{-1}{\mu_\alpha}}\mu_\beta^{\frac{-1}{\mu_\beta}}\right)\right)}, 1 - e^{-W\left(\ln\left((1-\nu_\alpha)^{\frac{-1}{1-\nu_\alpha}}(1-\nu_\beta)^{\frac{-1}{1-\nu_\beta}}\right)\right)},$

iii) $\lambda\alpha = 1 - e^{-W\left(\ln(1-\mu_\alpha)^{\frac{-\lambda}{1-\mu_\alpha}}\right)}, e^{-W\left(\ln \nu_\alpha^{\frac{-\lambda}{\nu_\alpha}}\right)},$

iv) $\alpha^\lambda = e^{-W\left(\ln \mu_\alpha^{\frac{-\lambda}{\mu_\alpha}}\right)}, 1 - e^{-W\left(\ln(1-\nu_\alpha)^{\frac{-\lambda}{1-\nu_\alpha}}\right)}.$

Now, we introduce new operators for weighted arithmetic and geometric aggregation, based on the Lambert W function. These operators, called the Lambert W

weighted arithmetic aggregation operator ($LWWA$) and the Lambert W weighted geometric aggregation operator ($LWWG$), extend the traditional weighted average aggregation operators by taking into account the nonlinearity of the Lambert W function, which allows for more flexible aggregation of fuzzy values. We start with $LWWA$.

Definition 3. *Let $\{\alpha_i = \langle \mu_i, \nu_i \rangle : i = 1,\dots,n\}$ be a collection of IFVs. The Lambert W weighted arithmetic aggregation operator LWWA is defined by*

$$LWWA(\alpha_1,\dots,\alpha_n) := \bigoplus_{i=1}^{n} \omega_i \alpha_i .$$

Here, $\omega = (\omega_1,\dots,\omega_n)^T$ represents a weight vector, with each component belonging to $\mathcal{J}$, and the sum of the components is equal to one.

Beliakov et al. (2011) presented a method for obtaining an aggregation operator based on the additive generators. Specifically, this method involves using Equality 7 from Beliakov et al. (2011). By using this equality, we give the following theorems.

Theorem 4. *Let $\{\alpha_i = \langle \mu_i, \nu_i \rangle : i = 1,\dots,n\}$ be a collection of IFVs. Then we have*

$$LWWA(\alpha_1,\dots,\alpha_n) = 1 - e^{-W\left(\ln \prod_{i=1}^{n}(1-\mu_i)^{-\frac{\omega_i}{1-\mu_i}}\right)}, e^{-W\left(\ln \prod_{i=1}^{n}\nu_i^{-\frac{\omega_i}{\nu_i}}\right)}.$$

Proof. Using the additive generators of LWN and LWC in Equality 7 of Beliakov et al. (2011), we can obtain

$$\begin{aligned}
\mu_{LWWA(\alpha_1,\dots,\alpha_n)} &= h^{-1}\left(\sum_{i=1}^{n}\omega_i h(\mu_i)\right)\\
&= h^{-1}\left(\sum_{i=1}^{n}\omega_i \frac{-\ln(1-\mu_i)}{1-\mu_i}\right)\\
&= h^{-1}\left(\ln\left(\prod_{i=1}^{n}(1-\mu_i)^{-\frac{\omega_i}{1-\mu_i}}\right)\right)\\
&= 1 - \frac{W\left(\ln\left(\prod_{i=1}^{n}(1-\mu_i)^{-\frac{\omega_i}{1-\mu_i}}\right)\right)}{\ln\left(\prod_{i=1}^{n}(1-\mu_i)^{-\frac{\omega_i}{1-\mu_i}}\right)}.
\end{aligned}$$

Since $\dfrac{W(\ln t)}{\ln t} = e^{-W(\ln t)}$, we get

$$\mu_{LWWA(\alpha_1,\ldots,\alpha_n)} = 1 - \frac{W\left(\ln\left(\prod_{i=1}^{n}(1-\mu_i)^{-\frac{\omega_i}{1-\mu_i}}\right)\right)}{\ln\left(\prod_{i=1}^{n}(1-\mu_i)^{-\frac{\omega_i}{1-\mu_i}}\right)}$$

$$= 1 - e^{-W\left(\ln\left(\prod_{i=1}^{n}(1-\mu_i)^{-\frac{\omega_i}{1-\mu_i}}\right)\right)}.$$

Using the same argument in Equality 7 of Beliakov et al. (2011), we obtain

$$\nu_{LWWA(\alpha_1,\ldots,\alpha_n)} = g^{-1}\left(\sum_{i=1}^{n}\omega_i g(\nu_i)\right)$$

$$= g^{-1}\left(\sum_{i=1}^{n}\omega_i \frac{-\ln\nu_i}{\nu_i}\right)$$

$$= g^{-1}\left(\ln\left(\prod_{i=1}^{n}\nu_i^{-\frac{\omega_i}{\nu_i}}\right)\right)$$

$$= \frac{W\left(\ln\left(\prod_{i=1}^{n}\nu_i^{-\frac{\omega_i}{\nu_i}}\right)\right)}{\ln\left(\prod_{i=1}^{n}\nu_i^{-\frac{\omega_i}{\nu_i}}\right)}$$

$$= e^{-W\left(\ln\left(\prod_{i=1}^{n}\nu_i^{-\frac{\omega_i}{\nu_i}}\right)\right)}.$$

Next, we proceed with the definition of $LWWG$.

Definition 4. *Let* $\{\alpha_i = \langle\mu_i,\nu_i\rangle : i = 1,\ldots,n\}$ *be a collection of IFVs. The Lambert W weighted geometric aggregation operator* $LWWG$ *is defined by*

$$LWWG(\alpha_1,\ldots,\alpha_n) := \bigotimes_{i=1}^{n}\alpha_i^{\omega_i},$$

where $\omega = (\omega_1,\ldots,\omega_n)^T$ *is a weight vector satisfying* $0 \le \omega_i \le 1$ *for any* $i = 1,\ldots,n$ *and* $\sum_{i=1}^{n}\omega_i = 1.$

We present a similar theorem for $LWWG$ as $LWWA$, which we omit the proof since it can be conducted similarly to the proof of Theorem 4.

Theorem 5. *Let* $\{\alpha_i = \langle\mu_i,\nu_i\rangle : i = 1,\ldots,n\}$ *be a collection of IFVs. Then we have*

$$LWWG(\alpha_1,\ldots,\alpha_n)=e^{-W\left(\ln\prod_{i=1}^{n}\mu_i^{-\frac{\omega_i}{\mu_i}}\right)},1-e^{-W\left(\ln\prod_{i=1}^{n}(1-\nu_i)^{-\frac{\omega_i}{1-\nu_i}}\right)}.$$

1.4 APPLICATION OF LAMBERT *W* WEIGHTED AGGREGATION OPERATORS IN MCDM

In this section, we provide a MCDM approach via Lambert W weighted aggregation operators proposed in Section 1.3.3. Then we apply this approach to a MCDM problem. Specifically, we consider a problem adapted from Wei (2008), where an investment company needs to select the best investment option from a set of alternatives based on multiple criteria. Our approach builds upon the existing work in Wei (2008) by using the powerful mathematical tools offered by the Lambert W function to provide more accurate and robust results. We also include a sensitivity analysis and a comparative analysis.

1.4.1 MCDM Application

We first outline the steps involved in the promised MCDM approach:

- **Step 1**: Consider a MCDM problem with m alternatives denoted by $A=\{\mho_1,\ldots,\mho_m\}$ and k criteria denoted by $C=\{C_1,\ldots,C_k\}$.
- **Step 2**: Construct a decision matrix,

$$D=\begin{bmatrix} d_{11} & \cdots & d_{1k} \\ \vdots & \ddots & \vdots \\ d_{m1} & \cdots & d_{mk} \end{bmatrix}$$

where $d_{ij}=\langle\mu_{d_{ij}},\nu_{d_{ij}}\rangle$ is an IFV for each $i=1,\ldots,m$ and $j=1,\ldots,k$.

- **Step 3**: Determine the normalized decision matrix N, where

$$N=\begin{bmatrix} n_{11} & \cdots & n_{1k} \\ \vdots & \ddots & \vdots \\ n_{m1} & \cdots & n_{mk} \end{bmatrix}$$

and each element n_{ij} is defined as

$$n_{ij}=\begin{cases} d_{ij}, & \text{if } d_{ij} \text{ is a benifit criterion} \\ 1-d_{ij}, & \text{if } d_{ij} \text{ s a cost criterion} \end{cases}$$

- **Step 4**: Determine the weights of the criteria and using either the *LWWA* or *LWWG* operators, aggregate each alternative to a single IFV α_j^A or α_j^G, respectively, i.e.,

$$\alpha_j^A = LWWA(n_1,\ldots,n_k)$$

and

$$\alpha_j^G = LWWG(n_1,\ldots,n_k)$$

for any $j = 1,\ldots,m$.

- **Step 5**: Apply a score function S to defuzzify each α_j^A or α_j^G for $j = 1,\ldots,m$, which results in the scores S_j^A or S_j^G, respectively. Thus, we have

$$S_j^A = S(\alpha_j^A)$$

or

$$S_j^G = S(\alpha_j^G)$$

- **Step 6:** The selection of the best alternative is determined using the following:

$$\arg\max_{j=1,\ldots,m} \{S_j^A, S_j^G\}.$$

It means the alternative with the higher score is considered to be the better choice. In other words, the larger the score, the more preferable the alternative is according to the decision criteria and the aggregation method used.

We now apply both *LWWA* and *LWWG* to solve the aforementioned MCDM problem presented in Wei (2008).

- **Step 1–3:** We have utilized the normalized matrix presented in Table 1.1 from the work of Wei (2008), leading us to complete Steps 1–3.

TABLE 1.1
Normalized decision matrix

	C_1	C_2	C_3	C_4
$\mho_1$	$\langle 0.5, 0.4\rangle$	$\langle 0.6, 0.3\rangle$	$\langle 0.3, 0.6\rangle$	$\langle 0.2, 0.7\rangle$
$\mho_2$	$\langle 0.7, 0.3\rangle$	$\langle 0.7, 0.2\rangle$	$\langle 0.7, 0.2\rangle$	$\langle 0.4, 0.5\rangle$
$\mho_3$	$\langle 0.6, 0.4\rangle$	$\langle 0.5, 0.4\rangle$	$\langle 0.5, 0.3\rangle$	$\langle 0.6, 0.3\rangle$
$\mho_4$	$\langle 0.8, 0.1\rangle$	$\langle 0.6, 0.3\rangle$	$\langle 0.3, 0.4\rangle$	$\langle 0.2, 0.6\rangle$
$\mho_5$	$\langle 0.6, 0.2\rangle$	$\langle 0.4, 0.3\rangle$	$\langle 0.7, 0.1\rangle$	$\langle 0.5, 0.3\rangle$

- **Step 4**: Let $\rho_i = \sum_{j=1}^{5} \frac{\mu_{n_{ji}} - \nu_{n_{ji}} + 1}{2}$ for any $i = 1,\ldots,4$. A linear optimization problem can be formulated as:

Solving the optimization problem gives us $w_1 = 0.30$, $w_2 = 0.11$, $w_3 = 0.34$, and $w_4 = 0.25$.

Remark 3. *The inequalities for w_1, w_2, w_3, and w_4 provide partial information about the weights. These inequalities are essentially upper and lower bounds on the possible values of the weights, which are based on some external information or constraints. However, to fully determine the weights, we need to solve an optimization problem that maximizes the overall score while satisfying these constraints on the weights. In this case, we use the ρ_i values as coefficients of the objective function f. The coefficients ρ_i represent the degree of optimism for each criterion, based on the IFVs of the alternatives. These values are used as the coefficients of the objective function f, because the objective is to maximize the overall degree of optimism in the decision process. In other words, the higher the degree of optimism for a particular criterion, the more important it is to the decision maker and thus the higher its weight should be. By using the ρ_i values as coefficients of f, we are effectively giving more weight to the criteria that have higher levels of optimism. The optimal weights obtained from solving the linear optimization problem then reflect the decision maker's overall priorities and preferences, as expressed through the degree of optimism in the IFVs.*

Now we calculate *LWWA* of alternatives. MATLAB code is provided below for alternative $\mho_1$:

Listing 1: MATLAB code to calculate $LWWA(\mho_1)$

```
% Set the values of the weights
w1 = 0.3;
w2 = 0.11;
w3 = 0.34;
w4 = 0.25;
we = [w1, w2, w3, w4];
% Alternative 1
mu1 = [0.5, 0.6, 0.3, 0.2];
nu1 = [0.4, 0.3, 0.6, 0.7];
pm1 = (1 - mu1).^(-we./(1 - mu1));
pn1 = nu1.^(-we./nu1);
% Calculate the product of each row
product_mu1 = prod(pm1);
product_nu1 = prod(pn1);
% Calculate the LWWA using the Lambert W function
lwwamu1 = 1 - exp(-lambertw(log(product_mu1)));
lwwanu1 = exp(-lambertw(log(product_nu1)));
end
```

Table 1.2 shows the aggregated IFVs for all alternatives.

TABLE 1.2
Aggregated IFVs

$\mho_1$	$\langle 0.41374, 0.47783 \rangle$
$\mho_2$	$\langle 0.66274, 0.26027 \rangle$
$\mho_3$	$\langle 0.66274, 0.33274 \rangle$
$\mho_4$	$\langle 0.64449, 0.19562 \rangle$
$\mho_5$	$\langle 0.61596, 0.15792 \rangle$

- **Step 5–6:** We utilize a score function S (see Definition 3 (v) in Chen (2011)) to aggregate the decision matrix. The function is defined as

where $0 \le \gamma \le 1$. When $\gamma = 0.5$, the scores for the alternatives are obtained as $S(\mho_1) = 0.46795$, $S(\mho_2) = 0.70124$, $S(\mho_3) = 0.665$, $S(\mho_4) = 0.72444$, and $S(\mho_5) = 0.72902$. Based on these scores, we can rank the alternatives as $\mho_5 \succ \mho_4 \succ \mho_2 \succ \mho_3 \succ \mho_1$.

Remark 4. *After applying LWWG and conducting similar calculations as LWWA, we obtained the ranking* $\mho_5 \succ \mho_2 \succ \mho_3 \succ \mho_4 \succ \mho_1$. *The MATLAB code used for calculating the aggregated IFV of alternative* $\mho_1$ *is shown below:*

Listing 2: MATLAB code to calculate $LWWG(\mho_1)$

```
% Set the values of the weights
w1 = 0.3;
w2 = 0.11;
w3 = 0.34;
w4 = 0.25;
we = [w1, w2, w3, w4];
% Alternative 1
mu1 = [0.5, 0.6, 0.3, 0.2];
nu1 = [0.4, 0.3, 0.6, 0.7];
pm1 = mu1.^(-we./mu1);
pn1 = (1 - nu1).^(-we./(1 - nu1));
% Calculate the product of each row
product_mu1 = prod(pm1);
product_nu1 = prod(pn1);
% Calculate the LWWA using the Lambert W function
lwwamu1 = exp(-lambertw(log(product_mu1)));
lwwanu1 =1 - exp(-lambertw(log(product_nu1)));;
end
```

Remark 5. *It is worth noting that the best and worst alternatives remain the same in both methods, i.e.,* $\mho_5$ *is the best alternative and* $\mho_1$ *is the worst alternative* *in both LWWA and LWWG. However, the rankings of the other alternatives change between these two aggregation operators. In LWWA, the ranking of the alternatives from best to worst is* $\mho_5 \succ \mho_4 \succ \mho_2 \succ \mho_3 \succ \mho_1$*, whereas in LWWG the ranking is* $\mho_5 \succ \mho_2 \succ \mho_3 \succ \mho_4 \succ \mho_1$*. This difference in ranking can be attributed to the different ways in which the two methods handle the interaction between the criteria. In LWWA, the interaction between criteria is based on the arithmetic mean, whereas in LWWG, it is based on the geometric mean.*

1.4.2 Sensitivity Analysis

In this section, we perform a sensitivity analysis on the MCDM solution by varying the parameter γ between zero and one in the score function S. The score function is a key component of the MCDM framework and is used to rank the alternatives based on their performance on different criteria. Actually, γ plays a crucial role in balancing the importance of membership and non-membership degrees. Specifically, it determines the degree of penalization for non-membership degrees when computing the overall score. Therefore, selecting an appropriate value of γ is essential for obtaining meaningful results in MCDM problems, as it directly affects the final ranking of alternatives. By changing the value of γ, we can explore how sensitive the MCDM solution is to change in this parameter. This analysis is important because it allows us to gain insights into the robustness of the MCDM solution and to identify the range of values for γ that produce the most stable and reliable results. The rankings given in Table 1.3 are obtained by varying γ between zero and one:

The visualization of the rankings for *LWWA* can be seen in Figure 1.7, while Figure 1.8 illustrates the rankings for *LWWG*.

Overall, the rankings for LWWA and LWWG seem to be largely consistent across different values of γ, with some changes in the rankings for certain alternatives when the value of γ is increased or decreased. For example, when $\gamma = 0.7$ for LWWA, the ranking for $\mho_5$ drops to third place, while $\mho_4$ moves up to first place. Despite these variations, it appears that when γ is around 0.5, the rankings obtained by both

TABLE 1.3
Rankings for different γ values

γ	*LWWA*	*LWWG*
0.1	$\mho_5 \succ \mho_4 \succ \mho_2 \succ \mho_3 \succ \mho_1$	$\mho_5 \succ \mho_2 \succ \mho_3 \succ \mho_4 \succ \mho_1$
0.3	$\mho_5 \succ \mho_4 \succ \mho_2 \succ \mho_3 \succ \mho_1$	$\mho_5 \succ \mho_2 \succ \mho_3 \succ \mho_4 \succ \mho_1$
0.5	$\mho_5 \succ \mho_4 \succ \mho_2 \succ \mho_3 \succ \mho_1$	$\mho_5 \succ \mho_2 \succ \mho_3 \succ \mho_4 \succ \mho_1$
0.7	$\mho_4 \succ \mho_2 \succ \mho_5 \succ \mho_3 \succ \mho_1$	$\mho_5 \succ \mho_2 \succ \mho_3 \succ \mho_4 \succ \mho_1$
0.9	$\mho_2 \succ \mho_3 \succ \mho_4 \succ \mho_5 \succ \mho_1$	$\mho_2 \succ \mho_5 \succ \mho_3 \succ \mho_4 \succ \mho_1$

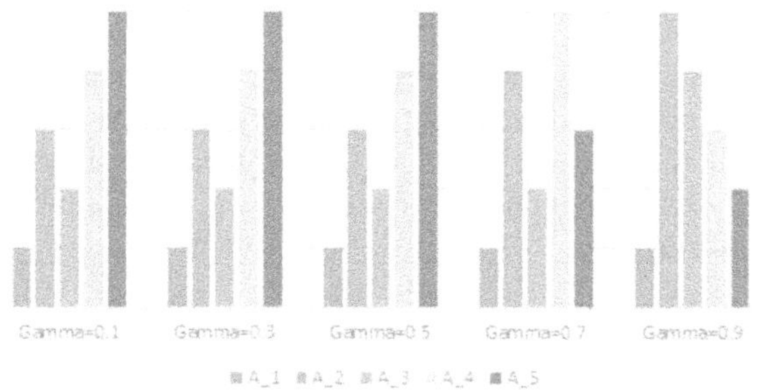

FIGURE 1.7 Rankings with respect to *LWWA*

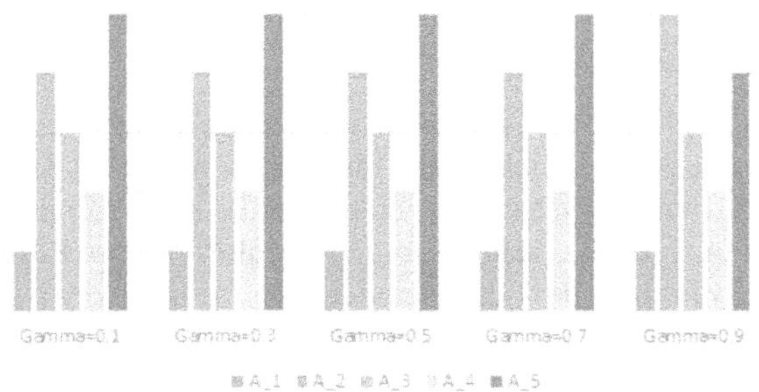

FIGURE 1.8 Rankings with respect to *LWWG*

LWWA and LWWG are quite stable and consistent with each other. This suggests that the weighting of membership and non-membership degrees is balanced, and neither is given too much importance over the other. Therefore, staying around 0.5 may be a reasonable choice for this particular problem.

1.4.3 Comparative Analysis

In this section, we conduct a comparative analysis of the proposed approach by considering various aggregation operators. Specifically, we compare the results obtained using algebraic weighted aggregation operators (Xu, 2007), Einstein weighted aggregation operators (Wang and Liu, 2012), Hamacher weighted aggregation operators (Huang, 2014), and Dombi weighted aggregation operators (Seikh and Mandal, 2021). By examining the outcomes produced by these different aggregation operators, we gain insights into their respective performance and evaluate their suitability for the given problem.

For each solution, we set γ to 0.5. Usage of the same value of γ for each solution ensures a fair and uniform comparison between the different aggregation operators. It allows for a more direct and meaningful comparison of the results obtained from each approach, as they are evaluated under the same parameter setting.

Upon applying various aggregation operators, we have obtained the following rankings for the alternatives:

- Using the algebraic weighted arithmetic and geometric aggregation operators: $\mho_5 \succ \mho_2 \succ \mho_3 \succ \mho_4 \succ \mho_1$.
- Employing the Einstein weighted arithmetic and geometric aggregation operators: $\mho_5 \succ \mho_2 \succ \mho_3 \succ \mho_4 \succ \mho_1$.
- Utilizing the Hamacher weighted arithmetic aggregation operators with the Hamacher product: $\mho_5 \succ \mho_2 \succ \mho_3 \succ \mho_4 \succ \mho_1$ (arithmetic mean) and $\mho_2 \succ \mho_5 \succ \mho_4 \succ \mho_3 \succ \mho_1$ (geometric mean).
- Employing the Dombi weighted arithmetic aggregation operators with $p = 2$: $\mho_5 \succ \mho_2 \succ \mho_3 \succ \mho_4 \succ \mho_1$ (arithmetic mean) and $\mho_2 \succ \mho_5 \succ \mho_4 \succ \mho_1 \succ \mho_3$ (geometric mean).

The consistent rankings obtained across different aggregation operators in our results indicate the robustness and reliability of the proposed approach. The similarity in rankings suggests that the selected operators effectively capture the underlying criteria and their interactions, leading to consistent evaluations of the alternatives. This agreement in rankings provides confidence in the accuracy and stability of the decision-making process, reinforcing the validity of the proposed approach for the given problem. The differences in rankings between the geometric means of the Hamacher and Dombi aggregations can be attributed to the varying treatment of the membership and non-membership degrees in the aggregation process.

In conclusion, the comparative analysis of the proposed approach using different aggregation operators highlights the impact of operator selection on the obtained rankings. While some operators yield consistent rankings across arithmetic and geometric means, others exhibit variations, emphasizing the importance of carefully considering the aggregation method. The choice of operator should be based on the specific characteristics of the problem and the desired interpretation of the results.

1.5 CONCLUSION

This chapter has addressed the challenges of handling uncertainty and vagueness in multi-criteria decision-making problems. By introducing the Lambert W t-norm and t-conorm, derived from specific additive generators, we have provided valuable tools for capturing and managing vague or uncertain decision data. These operators have been further utilized to define Lambert W weighted aggregation operators for families of IFVs. Through a numerical example and the application of these operators to

MCDM problems, we have demonstrated their effectiveness in achieving accurate and reliable results.

Additionally, a comparative analysis of different aggregation operators has shown that the consistent rankings obtained align with the findings presented in the chapter. This reaffirms the superiority of the Lambert *W* aggregation operators in terms of both accuracy and computational efficiency.

Furthermore, a sensitivity analysis conducted by varying the parameter values has provided insights into the impact of different settings on the rankings. It has been observed that certain parameter values in the score function, such as $\gamma \approx 0.5$, lead to stable and consistent rankings, suggesting a balanced weighting between membership and non-membership degrees.

Future research can explore the application of the Lambert *W* aggregation operators to larger and more complex decision-making problems, considering real-world scenarios and datasets to further validate their effectiveness and practicality.

Data Availability

The data has been used here is cited in the manuscript.

Conflicts of Interest

The authors declare that they have no conflicts of interest.

REFERENCES

S. Akbaş and T. E. Dalkılıç. Decision making for portfolio selection by fuzzy multi criteria linear programming. *Communications Faculty of Sciences University of Ankara Series A1 Mathematics and Statistics*, 68(2):2238–2257, 2019.

M. Akram, C. Kahraman, and K. Zahid. Group decision-making based on complex spherical fuzzy VIKOR approach. *Knowledge-Based Systems*, 216:106793, 2021.

K. T. Atanassov. *Intuitionistic Fuzzy Sets*. Springer-Verlag Berlin Heidelberg, 1999.

D. A. Barry, J. Y. Parlange, L. Li, H. Prommer, C. J. Cunningham, and F. Stagnitti. Analytical approximations for real values of the Lambert W-function. *Mathematics and Computers in Simulation*, 53(1–2):95–103, 2000.

G. Beliakov, H. Bustince, D. P. Goswami, U. K. Mukherjee, and N. R. Pal. On averaging operators for Atanassov's intuitionistic fuzzy sets. *Information Sciences*, 181(6):1116–1124, 2011.

F. Bobillo and U. Straccia. A fuzzy description logic with product t-norm. In *2007 IEEE International Fuzzy Systems Conference*, 2007, London, UK.

A. Cayley. A theorem on trees. *Quarterly Journal of Mathematics*, Oxford Series 23:376–378, 1889.

C. L. Chang. Fuzzy topological spaces. *Journal of Mathematical Analysis and Applications*, 24(1):182–190, 1968.

T. Y. Chen. A comparative analysis of score functions for multiple criteria decision making in intuitionistic fuzzy settings. *Information Sciences*, 181(17):3652–3676, 2011.

D. Çoker. An introduction to intuitionistic fuzzy topological spaces. *Fuzzy Sets and Systems*, 88(1):81–89, 1997.

R. M. Corless, G. H. Gonnet, D. E. Hare, D. J. Jeffrey, and D. E. Knuth. On the Lambert W function. *Advances in Computational Mathematics*, 5:329–359, 1996.

G. P. Dimuro, G. Lucca, J. A. Sanz, H. Bustince, and B. Bedregal. Cmin-integral: A Choquet-like aggregation function based on the minimum t-norm for applications to fuzzy rule-based classification systems. In Vicenç Torra, Radko Mesiar, Bernard De Baets (Eds.), *Aggregation Functions in Theory and in Practice*, pages 83–95, 2018, Springer Nature.

J. Dombi. A general class of fuzzy operators, the DeMorgan class of fuzzy operators and fuzziness measures induced by fuzzy operators. *Fuzzy Sets and Systems*, 8(2):149–163, 1982.

L. Euler. De serie lambertina plurimisque eius insignibus proprietatibus. *Leonhardi Euleri Opera Omnia, Ser. 1, Opera Mathematica*, 6:350–369, 1921.

F. Feng, H. Fujita, M. I. Ali, R. R. Yager, and X. Liu. Another view on generalized intuitionistic fuzzy soft sets and related multiattribute decision making methods. *IEEE Transactions on Fuzzy Systems*, 27(3):474–488, 2018.

F. Feng, Y. Zheng, J. C. R. Alcantud, and Q. Wang. Minkowski weighted score functions of intuitionistic fuzzy values. *Mathematics*, 8(7):1143, 2020.

H. Garg. Generalized Pythagorean fuzzy geometric aggregation operators using Einstein t-norm and t-conorm for multicriteria decision-making process. *International Journal of Intelligent Systems*, 32(6):597–630, 2017.

V. K. Gödel. Über eine bisher noch nicht benützte erweiterung des finiten standpunktes. *Dialectica*, 12(3–4):280–287, 1958.

H. Hamacher. Uber logische verknunpfungenn unssharfer aussagen undderen zugenhorige bewertungsfunktione. *Progress in Cybernetics and Systems Research*, 3:267–288, 1978.

A. Hoorfar and M. Hassani. Inequalities on the Lambert W function and hyperpower function. *Journal of Inequalities in Pure and Applied Mathematics*, 9(2):5–9, 2008.

J. Y. Huang. Intuitionistic fuzzy Hamacher aggregation operators and their application to multiple attribute decision making. *Journal of Intelligent & Fuzzy Systems*, 27(1):505–513, 2014.

A. Jain and A. Kapoor. A new approach to study organic solar cell using Lambert W-function. *Solar Energy Materials and Solar Cells*, 86(2):197–205, 2005.

S. Janson, D. E. Knuth, T. Łuczak, and B. Pittel. The birth of the giant component. *Random Structures & Algorithms*, 4(3):233–358, 1993.

G. Ç. Kayaturan, G. Özçelik, and A. Gökçe. Decision-aided evaluation of paths for routing on multi-attribute computer network consisting of encoded paths under uncertainty. *Expert Systems with Applications*, 223:119881, 2023.

J. M. Keller, P. Gader, H. Tahani, J. H. Chiang, and M. Mohamed. Advances in fuzzy integration for pattern recognition. *Fuzzy Sets and Systems*, 65(2–3):273–283, 1994.

E. P. Klement and R. Mesiar. *Logical, Algebraic, Analytic and Probabilistic Aspects of Triangular Norms*. Elsevier, 2005.

E. P. Klement, R. Mesiar, and E. Pap. Triangular norms. Position paper iii: Continuous t-norms. *Fuzzy Sets and Systems*, 145(3):439–454, 2004.

J. H. Lambert. Observationes variae in mathesin puram. *Acta Helvetica*, 3(1):128–168, 1758.

J. H. Lambert. *Observations analytiques*. In Nouveaux mémoires de l'Académie royale dessciences et belles-lettres, 1, Berlin, 1772.

C. Lopez-Molina, H. Bustince, J. Fernández, E. Barrenechea, P. Couto, and B. De Baets. A t-norm based approach to edge detection. In *Bio-Inspired Systems: Computational and Ambient Intelligence: 10th International Work-Conference on Artificial Neural Networks, IWANN 2009, Salamanca, Spain, June 10–12, 2009. Proceedings, Part I 10*, pages 302–309. Springer, 2009.

R. Lowen. Fuzzy topological spaces and fuzzy compactness. *Journal of Mathematical Analysis and Applications*, 56(3):621–633, 1976.

T. Lukasiewicz and U. Straccia. Managing uncertainty and vagueness in description logics for the semantic web. *Journal of Web Semantics*, 6(4):291–308, 2008.

J. Mao, D. Yao, and C. Wang. A novel cross-entropy and entropy measures of IFSs and their applications. *Knowledge-Based Systems*, 48:37–45, 2013.

K. Menger. Statistical metrics. *Proceedings of the National Academy of Sciences of the United States of America*, 28(12):535, 1942.

M. Olgun, M. Ünver, and Ş. Yardımcı. Pythagorean fuzzy topological spaces. *Complex & Intelligent Systems*, 5(2):177–183, 2019.

M. Olgun, E. Türkarslan, M. Ünver, and J. Ye. A cosine similarity measure based on the Choquet integral for intuitionistic fuzzy sets and its applications to pattern recognition. *Informatica*, 32(4):849–864, 2021a.

M. Olgun, M. Ünver, and Ş. Yardımcı. Pythagorean fuzzy points and applications in pattern recognition and Pythagorean fuzzy topologies. *Soft Computing*, 25(7):5225–5232, 2021b.

M. Olgun, E. Türkarslan, J. Ye, and M. Ünver. Single and interval-valued hybrid enthalpy fuzzy sets and a TOPSIS approach for multicriteria group decision making. *Mathematical Problems in Engineering*, 2022(1):2501321, 2022.

G. Özçelik. The attitude of mcdm approaches versus the optimization model in finding the safest shortest path on a fuzzy network. *Expert Systems with Applications*, 203:117472, 2022.

N. Y. Pehlivan, Y. Ünal, and C. Kahraman. Player selection for a national football team using fuzzy AHP and fuzzy TOPSIS. *Journal of Multiple-Valued Logic & Soft Computing*, 32(5–6):369–405, 2019.

B. Schweizer and A. Sklar. *Probabilistic Metric Spaces*. North-Holland, New York, 1983.

M. R. Seikh and U. Mandal. Intuitionistic fuzzy Dombi aggregation operators and their application to multiple attribute decision-making. *Granular Computing*, 6:473–488, 2021.

Y. Song, Q. Fu, Y. F. Wang, and X. Wang. Divergence-based cross entropy and uncertainty measures of Atanassov's intuitionistic fuzzy sets with their application in decision making. *Applied Soft Computing*, 84:105703, 2019.

E. Türkarslan, M. Ünver, and M. Olgun. q-rung orthopair fuzzy topological spaces. *Lobachevskii Journal of Mathematics*, 42(2):470–478, 2021.

M. Ünver, E. Türkarslan, N. Çelik, M. Olgun, and J. Ye. Intuitionistic fuzzy-valued neutrosophic multi-sets and numerical applications to classification. *Complex & Intelligent Systems*, 8(2):1703–1721, 2022.

S. R. Valluri, D. J. Jeffrey, and R. M. Corless. Some applications of the Lambert W function to physics. *Canadian Journal of Physics*, 78(9):823–831, 2000.

W. Wang and X. Liu. Intuitionistic fuzzy information aggregation using Einstein operations. *IEEE Transactions on Fuzzy Systems*, 20(5):923–938, 2012.

G. W. Wei. Maximizing deviation method for multiple attribute decision making in intuitionistic fuzzy setting. *Knowledge-Based Systems*, 21(8):833–836, 2008.

X. Wu, Y. Song, and Y. Wang. Distance-based knowledge measure for intuitionistic fuzzy sets with its application in decision making. *Entropy*, 23(9):1119, 2021.

Z. Xu. Intuitionistic fuzzy aggregation operators. *IEEE Transactions on Fuzzy Systems*, 15(6):1179–1187, 2007.

Z. Xu and R. R. Yager. Some geometric aggregation operators based on intuitionistic fuzzy sets. *International Journal of General Systems*, 35(4):417–433, 2006.

J. Ye. Multicriteria fuzzy decision-making method based on the intuitionistic fuzzy cross-entropy. In *2009 International Conference on Intelligent Human-Machine Systems and Cybernetics*, pages 59–61. IEEE, 2009.

A. Yolcu, F. Smarandache, and T. Y. Öztürk. Intuitionistic fuzzy hypersoft sets. *Communications Faculty of Sciences University of Ankara Series A1 Mathematics and Statistics*, 70(1):443–455, 2021.

L. A. Zadeh. Fuzzy sets. *Information and Control*, 8(3):338–353, 1965.

H. Zhao, Z. Xu, M. Ni, and S. Liu. Generalized aggregation operators for intuitionistic fuzzy sets. *International Journal of Intelligent Systems*, 25(1):1–30, 2010.

2 A Group Decision Aggregation-Based IVIF-MARCOS and Goal Programming Approach for Intelligent Decision-Making in Tourism Marketing

Ersin Burnaz and Gökhan Özçelik

2.1 INTRODUCTION

The income obtained from the tourism sector is highly important for countries. In this context, countries compete to expand their tourism economies and gain a larger market share. As is well known, 2020 was a year of economic contraction across the world due to the COVID-19 pandemic, and the tourism sector was one of the industries most severely affected by this contraction. The last year without a global pandemic, 2019, provides remarkable data to understand the economic importance of tourism. Although growth was below expectations compared to previous years, the tourism sector continued to expand strongly in 2019. Uncertainty due to Brexit, geopolitical and trade tensions, and the global economic slowdown put pressure on growth. Despite all these negativities, 1.46 billion international tourists traveled around the world in 2019, and 1.48 trillion dollars were obtained from these tourists. The tourism sector grew by 7% compared to 2017 and by 6% compared to 2018. The real growth in international tourism revenues between 2009 and 2019 was 54%. In 2019, the total exports gained from international tourism reached 1.7 trillion dollars. As of 2019, the tourism sector constituted the third largest export category in the world after fuels and chemicals. Over the past ten years, tourism has been one of the world's largest and fastest-growing sectors (World Tourism Organization, 2021). In light of these data, under the assumption that there would be no global crisis like the COVID-19 pandemic, the tourism sector is seen as very important for the world

 DOI: 10.1201/9781003529088-2

economy. When Turkey is analyzed specifically, according to 2019 data, 51,747,198 tourists visited Turkey, and a tourism income of 34,520,332,000 dollars was obtained. In 2019, the ratio of Turkey's tourism income to GDP was 4.6% (RTMCT, 2021). Considering the number of visitors coming to the country globally, Turkey ranks among the top 10 countries in terms of tourist arrivals but cannot enter the top 10 in terms of tourism income (World Tourism Organization, 2021). In other words, Turkey cannot obtain the expected income from its visitors. To attract more high-income visitors and to gain higher tourism revenue, diversification in tourism should be increased. In addition to sea-sand-sun-based mass tourism, other types of tourism should also be offered to visitors. According to the classification by the Republic of Turkey Ministry of Culture and Tourism, there are many types of tourism: river tourism, hunting tourism, mountaineering, golf tourism, air sports, religious tourism, Silk Road tourism, winter sports tourism, congress tourism, birdwatching tourism, cave tourism, health and thermal tourism, underwater diving tourism, marine tourism, and highland tourism. Besides, destinations should diversify their touristic products to attract more visitors. In this way, it can be ensured that visitors with different characteristics are attracted to the country. The Eastern Black Sea Region is one of the regions of Turkey with important touristic resources. In recent years, interest from Arab tourists in the region has increased considerably. The Eastern Black Sea Development Agency has carried out many studies to increase the number and diversity of tourists coming to this region.

The provinces (TR90 region) supported by this agency in terms of economic development by supporting projects are Ordu, Giresun, Trabzon, Rize, Artvin, and Gümüşhane. It is expected that these six provinces, by acting cooperatively rather than competitively, will generate more income from regional tourism. This idea constitutes the main motivation of the study.

To this end, a two-stage research methodology is developed. In the first stage, the aim is to reveal the general performance map in terms of tourism types for present and future situations. In this regard, the performances of the provinces are evaluated considering 15 types of tourism determined by the Republic of Turkey Ministry of Culture and Tourism for both the present and future. To evaluate the performances, the interval-valued intuitionistic fuzzy measurement alternatives and ranking according to the compromise solution (IVIF-MARCOS) method, which handles multi-criteria group decision-making (MCGDM) processes, is employed for the first time in the context of tourism management. Furthermore, the stability and validity of the IVIF-MARCOS method are tested by employing the IVIF-TOPSIS and IVIF-SAW methods through a set of scenarios constructed by varying criteria weights. In the second stage, the aim is to better manage the tourism potential at operational, tactical, and strategic levels. In this context, a goal programming model is developed, enabling the provinces to act as pioneers for other cities in one or more tourism types for the present and future situations. Besides, according to the computational results, managerial insights are provided by emphasizing the key findings.

The main accomplishments of the study can be outlined from three perspectives:

- **Theoretical perspective:** To the best of our knowledge, this is the first study to propose an integrated methodology combining the IVIF-MARCOS method and a goal programming model specifically developed for strategic tourism management and provincial assignment planning.
- **Practical perspective:** The proposed IVIF-MARCOS method, requiring relatively less effort for the evaluation process, can be employed in various application areas from service to production systems when addressing MCGDM problems. Moreover, the developed goal programming model based on the IVIF decision matrix can readily be applied for selection or assignment processes.
- **Managerial perspective:** Thanks to the proposed two-stage research methodology, extensive analyses are performed through a case study with a real-world background for the provinces of the TR90 region in Turkey. In addition, vital insights are provided for operational, tactical, and strategic planning periods. Therefore, this study intends to create a framework for governments to develop sustainable strategies in tourism marketing.

The remainder of the study is structured as follows. Section 2.2 presents a literature review, focusing on studies employing fuzzy-based multi-criteria decision-making (MCDM) methods in tourism marketing and management and those referring to the measurement alternatives and ranking according to the compromise solution (MARCOS) method and its fuzzy extensions in MCDM problems. Section 2.3 summarizes the main definitions regarding interval-valued intuitionistic fuzzy sets (IVIFSs) and introduces the IVIF-MARCOS method and the goal programming model. Section 2.4 conducts a case study with a real background. Additionally, to test the stability and validity of the IVIF-MARCOS method, its results are compared with those obtained from IVIF-TOPSIS and IVIF-SAW methods under randomly generated criteria weights. Managerial insights based on the computational results are also discussed. Finally, Section 2.5 concludes the study and highlights directions for future research.

2.2 LITERATURE REVIEW

After Zadeh introduced fuzzy set theory, Bellman and Zadeh adapted this theory to multi-criteria decision-making (MCDM) problems under uncertainty (Zadeh, 1965; Bellman and Zadeh, 1970). Since then, the handling of uncertainty using fuzzy sets has accelerated studies related to this concept (Pelissari et al., 2021). Currently, numerous studies have employed fuzzy-based MCDM methods across various problem domains (Ocampo et al., 2018; Yilmaz et al., 2017; Martín et al., 2019; Özçelik and Nalkıran, 2021; Türkarslan et al., 2021; Lin et al., 2021; Boyacı and Şişman, 2022). In this section, recent and noteworthy studies are presented under two main headings: (i) studies employing fuzzy-based MCDM methods in tourism management and marketing and (ii) studies referring to the MARCOS method and its fuzzy extensions in MCDM problems. According to the literature review, it is observed that many studies utilizing fuzzy-based MCDM techniques have been conducted in different application areas within tourism management. Most of these studies consider triangular fuzzy numbers in their approaches. Mijajlović et al. (2020) and Bire et al. (2021) focused on competitiveness in tourism management; Li et al. (2013) and Yu et al. (2017) studied hotel selection problems; Chang (2015) and Vatankhah et al. (2021) investigated issues related to tourism and hospitality employees; Fu and Tzeng (2016), Qin (2017), and Krishnan et al. (2021) addressed general tourism applications; Liao et al. (2019) explored travel agency selection; Peng et al. (2019) and Tian and Peng (2020) focused on tourism attractions; Yang and Wang (2020) and Mishra et al. (2022) studied low-carbon tourism; Abrishamchi et al. (2021) worked on sustainable tourism; and García Mestanza and Bakhat (2021) addressed overtourism. Tan et al. (2025) investigated hotel performance evaluation using a hybrid SOCP-MCDM approach; Kang et al. (2024) prioritized sustainable post-pandemic tourism policies through a hybrid spherical fuzzy MCDM methodology. Overall, the notable recent studies are comparatively presented in Table 2.1 in terms of subject, methodology, and the types of fuzzy numbers and aggregation operators employed. After the MARCOS method was introduced (Stević et al., 2020), it quickly attracted the attention of researchers and has been applied in various fields. Moreover, many fuzzy extensions of the MARCOS method have been proposed. From the literature, it is seen that in most studies utilizing fuzzy-based MARCOS methods, other fuzzy-based MCDM methods are also incorporated. Generally, triangular fuzzy numbers are employed in these studies. It is further revealed that many studies applying the fuzzy-based MARCOS method focus on transportation-related problems (Pamucar et al., 2021a; Bouraima et al., 2021; Vrtagić et al., 2021; Tadić et al., 2021; Majidi et al., 2021; Blagojević et al., 2021; Stević et al., 2021a; Simić et al., 2020; Stanković et al., 2020). Additionally, the fuzzy-based MARCOS method has been utilized in supply chain management (Badi and Pamucar, 2020; Stević et al., 2020; Chakraborty et al., 2020; Ulutaş et al., 2020), construction (Anysz et al., 2021; Celik and Gul, 2021), sustainable production (Stević et al., 2021b; Hasheminasab et al., 2022; Boral and Chakraborty, 2021), failure mode and effects analysis (FMEA) (Boral et al., 2021), information and communication technology (ICT) (Torkayesh and Torkayesh, 2021), and transportation technologies such as hydrogen buses and battery electric vehicles (Pamucar et al., 2021b; Ecer, 2021). It has also been applied in resource selection problems (Pramanik et al., 2021) and tourism-related studies (Mijajlović et al., 2020). The notable studies related to the MARCOS method are comparatively presented in Table 2.2, categorized by subject, methodology, and the fuzzy numbers and aggregation operators used. When the literature is examined, it is revealed that, although several fuzzy extensions of the MARCOS method exist, studies employing the MARCOS method in an interval-valued intuitionistic fuzzy (IVIF) environment are extremely limited. Therefore, this study aims to contribute to the literature by employing an IVIF-MARCOS- and goal programming-based methodology specifically developed for strategic tourism management. In addition to addressing this gap, extensive analyses are conducted on the provinces of the TR90 region in Turkey in the context of tourism marketing.

TABLE 2.1
Some remarkable studies employing fuzzy-based MCDM methods in tourism management and marketing

Author(s)	Subject	Methodology	Fuzzy numbers	Aggregation operators
Tan et al. (2025)	Hotel performance evaluation	Hybrid SOCP-MCDM (second-order cone programming + MCDM)	—	—
Kang et al. (2024)	Sustainable post-pandemic tourism policy prioritization	Hybrid spherical fuzzy AHP + spherical fuzzy WASPAS	Spherical fuzzy numbers	Weighted averaging
Mishra et al. (2022)	Evaluation of low-carbon tourism strategy	ARAS	Interval-valued intuitionistic fuzzy (IVIF) numbers	IVIF-weighted averaging (IVIFWA), IVIF-weighted geometric operator (IVIFWG)
Krishnan et al. (2021)	Evaluation of the smart e-tourism applications	VIKOR	Interval type-2 trapezoidal fuzzy numbers	Interval type-2 trapezoidal-fuzzy weighted with zero inconsistency (IT2TR-FWZIC)
Erdenejargal et al. (2021)	Evaluation of the natural landscape aesthetic	AHP	Boolean and fuzzy logic	—
Bire et al. (2021)	Regional destination competitiveness	DEMATEL, fuzzy TOPSIS	Triangular fuzzy numbers	—
Tang et al. (2021)	Tourism quality evaluation	Hesitant fuzzy Taguchi method	Hesitant fuzzy numbers	—
Abrishamchi et al. (2021)	Sustainable island-based tourism development	Fuzzy DELPHI, fuzzy TODIM	Triangular fuzzy numbers	—
García Mestanza and Bakhat (2021)	Overtourism	Fuzzy AHP, fuzzy MAIRCA	Triangular fuzzy numbers	—
Yang and Wang (2020)	Evaluating performances of low-carbon tourism scenic spots	AHP	Intuitionistic fuzzy values	IFV aggregation operator

(Continued)

TABLE 2.1 (CONTINUED)
Some remarkable studies employing fuzzy-based MCDM methods in tourism management and marketing

Author(s)	Subject	Methodology	Fuzzy numbers	Aggregation operators
Tian and Peng (2020)	Tourism attraction recommendation	ANP, TODIM	Picture fuzzy numbers	—
Vatankhah et al. (2021)	Stress among tourism and hospitality employees	AHP	Triangular fuzzy numbers, reciprocal fuzzy numbers	—
Fu et al. (2020)	The selection model for horizontal alliances between hotels and airlines	Nominal group technique (NGT), fuzzy TOPSIS, multi-choice goal programming (MCGP)	Triangular fuzzy numbers	—
Mijajlović et al. (2020)	Competitiveness of spa centers	FUCOM, fuzzy MARCOS	Triangular fuzzy numbers	—
Peng et al. (2019)	Selection of tourism attraction	Mathematical programming model	Single-valued neutrosophic numbers (SVNNs)	Generalized weighted single-valued neutrosophic power Shapley Choquet average (GWSVNPSCA), generalized weighted single-valued neutrosophic power Shapley Choquet geometric (GWSVNPSCG)
Lo and Liou (2019)	Medical tourism	DEMATEL, fuzzy TOPSIS	Triangular fuzzy numbers	—
Martín et al. (2019)	Determining satisfaction of international tourist	TOPSIS	Triangular fuzzy numbers	—
Liao et al. (2019)	Optimal online travel agencies (OTAs) selection	Fuzzy DELPHI, DEMATEL, ANP	Triangular fuzzy numbers	—

(Continued)

TABLE 2.1 (CONTINUED)
Some remarkable studies employing fuzzy-based MCDM methods in tourism management and marketing

Author(s)	Subject	Methodology	Fuzzy numbers	Aggregation operators
Ocampo et al. (2018)	Modeling the causal relationships of the antecedents of organizational citizenship behavior in the hospitality industry	Fuzzy DEMATEL	Triangular fuzzy numbers	Arithmetic mean
Qin (2017)	Tourism recommender system	AHP	Symmetric triangular interval type-2 fuzzy numbers	Symmetric triangular interval type-2 fuzzy Hamy mean (STIT2FHM) operator, weighted symmetric triangular interval type-2 fuzzy Hamy mean (WSTIT2FHM) operator
Yu et al. (2017)	Selecting hotels on a tourism website	Multi-attributive border approximation area comparison (MABAC)	Interval type-2 trapezoidal fuzzy numbers (IT2TrFNs)	—
Mardani et al. (2016)	Tourism and hospitality industry: A systematic review			—
Fu and Tzeng (2016)	Establish safety management system for hot spring hotels	AHP	Triangular fuzzy numbers	—
Chang (2015)	Public relations personnel selection	Fuzzy DELPHI, ANP, TOPSIS	Triangular fuzzy numbers	—
Chen (2014)	Suitable festival activities	Preference relation analytic hierarchy process (PR AHP)	Reciprocal fuzzy PR	—

(Continued)

TABLE 2.1 (CONTINUED)
Some remarkable studies employing fuzzy-based MCDM methods in tourism management and marketing

Author(s)	Subject	Methodology	Fuzzy numbers	Aggregation operators
Horng et al. (2014)	Industrial environment analysis	Fuzzy DELPHI, ANP, DEMATEL	Triangular fuzzy numbers	—
Li et al. (2013)	Hotel selection preferences	Choquet integral	—	Weighted arithmetic mean (WAM), ordered weighted averaging (OWA)

TABLE 2.2
Sme remarkable studies referring to MARCOS method and its fuzzy extensions

Authors	Subject	Methodology	Fuzzy numbers	Aggregation operators
Aydin et al. (2025)	Survival craft performance assessment on cargo ships	IT2F-MARCOS	Interval type-2 fuzzy numbers	IT2F aggregation and center of area defuzzification
Wang et al. (2024)	Risk prioritization in sustainable maritime logistics	IVPF-AHP–EDAS	Interval-valued Pythagorean fuzzy sets	Weighted averaging aggregation for IVPF sets
Akram et al. (2023)	E-government adoption decision-making in Pakistan	SF-AHP and q-ROF-VIKOR	Spherical fuzzy and q-rung orthopair fuzzy sets	Spherical weighted aggregation, q-ROF distance-based aggregation
Salimian et al. (2022)	Supplier selection	VIKOR, MARCOS	Interval-valued intuitionistic fuzzy numbers	Weighted geometric average approach
Pamucar et al. (2021a)	Sustainable road transportation	Fuzzy FUCOM, MARCOS	Neutrosophic fuzzy numbers	Single-valued neutrosophic fuzzy number Dombi weighted arithmetic averaging (SVNFNDWAA)
Bouraima et al. (2021)	The performance of sub-Saharan African (SSA) railways	Entropy, MARCOS	—	—
Vrtagić et al. (2021)	Ranking road sections	IMF SWARA, fuzzy MARCOS, the data envelopment analysis (DEA)	Triangular fuzzy numbers	Dombi aggregator, Bonferroni aggregator
Tadić et al. (2021)	The selection of intermodal transport system scenarios	Fuzzy DELPHI, fuzzy Factor Relationship (FARE), fuzzy MARCOS	Triangular fuzzy numbers	—

(Continued)

TABLE 2.2 (CONTINUED)
Sme remarkable studies referring to MARCOS method and its fuzzy extensions

Authors	Subject	Methodology	Fuzzy numbers	Aggregation operators
Majidi et al. (2021)	Sustainability of the maritime supply chain (MSC)	SWARA, MARCOS, CoCoSo	—	—
Anysz et al. (2021)	Construction cost	TOPSIS, AHP, FUCOM, MARCOS	—	—
Celik and Gul (2021)	Dam construction safety	BWM, MARCOS	Interval type-2 fuzzy numbers	—
Boral ve Chakraborty (2021)	Failure analysis of CNC machines	IT2F-DEMATEL, IT2F-AHP, IT2F-MARCOS	Interval type-2 fuzzy numbers	—
Blagojević et al. (2021)	Sustainable traffic management	Fuzzy FUCOM, fuzzy PIPRECIA, fuzzy MARCOS	Triangular fuzzy numbers	Fuzzy power Heronian mean operator
Torkayesh and Torkayesh (2021)	Information and communication technology (ICT)	Level-based weight assessment (LBWA) MARCOS	—	—
Stević et al. (2021b)	Sustainable production	Novel Rough PIPRECIA, fuzzy MARCOS	Triangular fuzzy numbers	Rough Dombi weighted geometric averaging (RNDWGA)
Stević et al. (2021a)	Traffic safety evaluation	CRITIC, DEA, MARCOS	—	—
Boral et al. (2021)	Failure modes, effects, and criticality analysis (FMECA)	IT2F-DEMATEL, IT2F-MAIRCA, IT2F-MARCOS, IT2F-TOPSIS	Interval type-2 fuzzy numbers	Simple weighted average approach
Pamucar et al. (2021b)	The hydrogen bus development	BWM, MARCOS	Triangular fuzzy numbers	Bonferroni aggregator

(Continued)

TABLE 2.2 (CONTINUED)
Sme remarkable studies referring to MARCOS method and its fuzzy extensions

Authors	Subject	Methodology	Fuzzy numbers	Aggregation operators
Ecer (2021)	Performance assessment of battery electric vehicles	SECA, MARCOS, MAIRCA, COCOSO, ARAS, COPRAS	—	—
Pramanik et al. (2021)	Resource selection in mobile crowd computing	EDAS, ARAS, MABAC, MARCOS, COPRAS	—	—
Hasheminasab et al. (2022)	Fossil fuel sustainable decisions	MARCOS, E-SWARA	—	—
Badi and Pamucar (2020)	Supplier selection for steelmaking company	GREY-MARCOS	—	—
Stević et al. (2020)	Sustainable supplier selection	MARCOS	—	—
Chakraborty et al. (2020)	Supplier selection in iron and steel industry	D-MARCOS	—	—
Ulutaş et al. (2020)	Stackers selection in a logistics system	CCSD method, ITARA MARCOS	—	—
Simić et al. (2020)	Railway infrastructure	Fuzzy MARCOS	Picture fuzzy numbers	The picture fuzzy Hamacher weighted average (PFHWA)
Mijajlović et al. (2020)	The competitiveness of spa centers	FUCOM, fuzzy MARCOS	Triangular fuzzy numbers	—
Stanković et al. (2020)	Road traffic risk analysis	Fuzzy MARCOS, fuzzy PIPRECIA, fuzzy SAW, fuzzy TOPSIS	Triangular fuzzy numbers	—

2.3 METHODOLOGY

In this section, the two-stage research methodology is presented following the introduction of the basic concepts of interval-valued intuitionistic fuzzy sets (IVIFSs). In the first stage, the IVIF-MARCOS method is developed for multi-criteria group decision-making (MCGDM) processes and explained step by step. At this stage, the aim is to reveal a general performance map of the provinces based on different tourism types for both the present and future situations using the proposed IVIF-MARCOS method. Subsequently, in the second stage, a goal programming model is developed to enable the assignment of the provinces based on tourism diversity. This stage is intended to ensure that certain provinces focus on specific tourism types and act as pioneers for other cities, thereby contributing to increased tourism income. A schematic representation of the two-stage research methodology is provided in Figure 2.1.

2.3.1 PRELIMINARIES

Remark 2.1 (Yeni and Özçelik, 2019): In classical fuzzy set theory, every element of the fuzzy set has a single membership value "$\mu_{\tilde{A}}(x)$" between zero and one. However, due to the complexity of certain problems, there may be hesitations regarding preferences, and thus it may not always be accurate that the degree of nonmembership "$v_{\tilde{A}}(x)$" is simply equal to "1-$\mu_{\tilde{A}}(x)$". In such a case, intuitionistic fuzzy set, an extension of the classical fuzzy set, introduced by Atanassov (1986), can be suitable and convenient to express DMs' preferences. Later, Atanassov and Gargov (1989) generalized IFSs to interval-valued intuitionistic fuzzy sets (IVIFSs). The definition of IVIFS is given as follows. Let $\mathcal{H} \subset 2^{[0,1]}$ be the set of all closed subintervals of [0,1] and X be a universe of discourse. An interval-valued intuitionistic fuzzy set in $\tilde{A}$ over X is an object having the form:

$$\tilde{A} = \{\langle x, \tilde{\mu}_{\tilde{A}}(x), \tilde{v}_{\tilde{A}}(x)\rangle \mid. \tag{2.1}$$

where the functions:

$\tilde{\mu}_{\tilde{A}} : X \rightarrow \mathcal{H}$ and $\tilde{v}_{\tilde{A}} : X \rightarrow \mathcal{H}$ with the condition

$$0 \leq \sup \tilde{\mu}_{\tilde{A}}(x) + \sup\ \tilde{v}_{\tilde{A}}(x) \leq 1, \forall x \in X. \tag{2.2}$$

For each $x \in X$, $\tilde{\mu}_{\tilde{A}}(x)$ and $\tilde{v}_{\tilde{A}}(x)$ are closed intervals whose lower and upper end points are denoted by $\tilde{\mu}_{\widetilde{AL}}(x)$, $\tilde{\mu}_{\widetilde{AU}}(x)$, $\tilde{v}_{\widetilde{AL}}(x)$ and $\tilde{v}_{\widetilde{AU}}(x)$seriatim.

Interval-valued intuitionistic fuzzy set $\tilde{A}$ is then denoted by

$$\tilde{A} = \{\langle x, [\tilde{\mu}_{\widetilde{AL}}(x), \tilde{\mu}_{\widetilde{AU}}(x)], [\tilde{v}_{\widetilde{AL}}(x), \tilde{v}_{\widetilde{AU}}(x)]\rangle \mid x \in X\} \tag{2.3}$$

where $0 \leq \tilde{\mu}_{\widetilde{AU}}(x) + \tilde{v}_{\widetilde{AU}}(x) \leq 1$, $\tilde{\mu}_{\widetilde{AL}}(x) \geq 0, \tilde{v}_{\widetilde{AL}}(x) \geq 0$.

For each element x, the hesitancy degree of an interval-valued intuitionistic fuzzy interval of $x \in X$ in $\tilde{A}$ is calculated as follows:

$$\pi_{\tilde{A}}(x) = 1 - \tilde{\mu}_{\tilde{A}}(x) - \tilde{v}_{\tilde{A}}(x) = \left(\left[1 - \tilde{\mu}_{\widetilde{AU}}(x) - \tilde{v}_{\widetilde{AU}}(x)\right], \left[1 - \tilde{\mu}_{\widetilde{AL}}(x) - \tilde{v}_{\widetilde{AL}}\right]\right) \quad (2.4)$$

For convenience, let assume $\tilde{\mu}_{\tilde{A}}(x) = [a,b]$ and $\tilde{v}_{\tilde{A}}(x) = [c,d]$, then an IVIF value is denoted by $\tilde{A} = ([a,b],[c,d])$ and called as an interval-valued intuitionistic fuzzy number (IVAIFN).

Remark 2.2 (Yeni and Özçelik, 2019): The basic algebraic operations between interval-valued intuitionistic fuzzy numbers are given in the following (Xu, 2007):

Let $\tilde{A}_1 = ([a_1,b_1],[c_1,d_1])$ and $\tilde{A}_2 = ([a_2,b_2],[c_2,d_2])$ be two IVAIFNs. Then,

$$\tilde{A}_1 \oplus \tilde{A}_2 = \left(\left[a_1 + a_2 - a_1a_2, b_1 + b_2 - b_1b_2\right], \left[c_1c_2, d_1d_2\right]\right), \quad (2.5)$$

$$\tilde{A}_1 \otimes \tilde{A}_2 = \left(\left[a_1a_2, b_1b_2\right], \left[c_1 + c_2 - c_1c_2, d_1 + d_2 - d_1d_2\right]\right), \quad (2.6)$$

$$\lambda\tilde{A}_1 = \left(\left[1-(1-a_1)^{\lambda}, 1-(1-b_1)^{\lambda}\right], \left[c_1^{\lambda}, d_1^{\lambda}\right]\right), \quad \lambda > 0, \quad (2.7)$$

$$\tilde{A}_1^{\lambda} = \left(\left[a_1^{\lambda}, b_1^{\lambda}\right], \left[1-(1-c_1)^{\lambda}, 1-(1-d_1)^{\lambda}\right]\right) \quad \lambda > 0 \quad (2.8)$$

The subtraction and division operations cannot be directly derived by addition and multiplication operations. They can be extended as follows (Zhao et al., 2016):

$$\tilde{A}_1 \ominus \tilde{A}_2 = \begin{cases} \left(\left[\frac{a_1 - a_2}{1 - a_2}, \frac{b_1 - b_2}{1 - b_2}\right], \left[\frac{c_1}{c_2}, \frac{d_1}{d_2}\right]\right) \textit{if } a_1 \geq a_2, b_1 \geq b_2, c_1 \leq c_2, d_1 \leq d_2, \\ \textit{and } c_2 > 0, d_2 > 0, a_2 \neq 1, b_2 \neq 1 \\ \textit{and } c_1(1-a_2) \leq c_2(1-a_1), \\ d_1(1-b_2) \leq d_2(1-b_1) \\ ([0,0],[1,1]) \textit{ otherwise} \end{cases} \quad (2.9)$$

$$\tilde{A}_1 \oslash \tilde{A}_2 = \begin{cases} \left(\left[\frac{a_1}{a_2}, \frac{b_1}{b_2}\right], \left[\frac{c_1 - c_2}{1 - c_2}, \frac{d_1 - d_2}{1 - d_2}\right]\right) \textit{if } a_1 \leq a_2, b_1 \leq b_2, c_1 \geq c_2, d_1 \geq d_2, \\ \textit{and } a_2 > 0, b_2 > 0, c_2 \neq 1, d_2 \neq 1 \\ \textit{and } a_1(1-c_2) \leq a_2(1-c_1), \\ b_1(1-d_2) \leq b_2(1-d_1) \\ ([0,0],[1,1]) \textit{otherwise} \end{cases} \quad (2.10)$$

In the MCGDM environment, an operator that aggregates individual DM's judgments into a single matrix must be used. In this study, IIFWA, the most basic (and common) operator, is preferred for the aggregation process in IVIF-MARCOS,

IVIF-TOPSIS, and IVAIF-SAW methods due to its simplicity and applicability in studying with different qualified DMs.

Let $\tilde{A}_j = ([a_j, b_j], [c_j, d_j])$, ($j$ = 1, 2, …, n) be a collective of IVIFNs and let IIFWA: $Q^n \rightarrow Q$, if

$$IIFWA_w\left(\tilde{A}_1, \tilde{A}_2, \ldots, \tilde{A}_n\right) = w_1\tilde{A}_j \oplus w_2\tilde{A}_2 \oplus \ldots \oplus w_n\tilde{A}_n, \tag{2.11}$$

where Q is the set of all IVIFNs, $w = (w_1, w_2 \ldots, w_n)$ is the weight vector of the IVIFNs $\tilde{A}_j\left(j = 1, 2, \ldots, n\right)$, and $w_j > 0, \sum_{j=1}^{n} w_j = 1$. Here, the w_j can be considered as the ith DM's opinion importance in decision-making process. The IIFWA operator can be computed as follows (Xu and Chen, 2007):

$$IIFWA_w\left(\tilde{A}_1, \tilde{A}_2, \ldots, \tilde{A}_n\right) = \left(\left[1 - \prod_{j=1}^{n} (1 - a_j)^{w_j}, 1 - \prod_{j=1}^{n} (1 - b_j)^{w_j} \right], \left[\prod_{j=1}^{n} (c_j)^{w_j}, \prod_{j=1}^{n} (d_j)^{w_j} \right] \right) \tag{2.12}$$

If all weights of DMs are equal, the IIFWA operator is reduced to an IIFA operator.

2.3.2 IVIF-MARCOS Method

In this subsection, an extension of MARCOS method, introduced by Stević et al. (2020), is proposed called IVIF-MARCOS for interval-valued intuitionistic fuzzy environment. The proposed IVIF-MARCOS method is explained step by step as follows:

Step 1: Construction of the IVIF decision matrix $\left(\tilde{X}_l\right)$ for each decision maker.

$$\tilde{X}_l = \left[\tilde{x}_{ijl}\right]_{nxm} = \begin{bmatrix} \tilde{x}_{11l} & \cdots & \tilde{x}_{1ml} \\ \tilde{x}_{21l} & \cdots & \tilde{x}_{2ml} \\ \vdots & \vdots & \vdots \\ \tilde{x}_{n1l} & \cdots & \tilde{x}_{xnl} \end{bmatrix} \tag{2.13}$$

where $\tilde{x}_{ijl} = \left(\left[\mu_{Lijl}, \mu_{Uijl}\right], \left[v_{Lijl}, v_{Uijl}\right]\right)$ is the IVIFN of the ith ($i \in \{1, 2, \ldots, n\}$) alternative with respect to jth ($j \in \{1, 2, \ldots, m\}$) criterion and lth ($l \in \{1, 2, \ldots, k\}$) decision maker. To specify the each $\tilde{x}_{ijl}$, the IVIFN scale, given in Table 2.3, is employed. In this step, the weight of each criterion is also determined for each DM.

$$W_l = \left[w_{jl}\right]_{1xm} \tag{2.14}$$

where w_{jl} is crisp value denoting the lth DM's judgment on the importance of the jth criterion. Here, $\sum_{j=1}^{m} w_{jl} = 1$. The weights of the criteria are aggregated based on weighted arithmetic average considering importance weight of each DM.

Step 2: Construction of the aggregated IVIF decision matrix $\left(\tilde{X}\right)$.

$$\tilde{X} = \left[\tilde{x}_{ij}\right]_{nxm} = \begin{bmatrix} \tilde{x}_{11} & \cdots & \tilde{x}_{1m} \\ \tilde{x}_{21} & \cdots & \tilde{x}_{2m} \\ \vdots & \vdots & \vdots \\ \tilde{x}_{n1} & \cdots & \tilde{x}_{xn} \end{bmatrix} \quad (2.15)$$

where $\tilde{x}_{ij} = ([\mu_{Lij}, \mu_{Uij}], [v_{Lij}, v_{Uij}])$. The individual DM's matrices are aggregated in one IVIF decision matrix by employing the IIFWA operator given in Equation (2.12).

Step 3: Calculation of the weighted aggregated IVIF decision matrix.

The weighted IVIFNs $\left(\tilde{r}_{ij}\right)$ are calculated by employing Equation (2.17) as follows:

$$\tilde{R} = \left[\tilde{r}_{ij}\right]_{nxm} \quad (2.16)$$

$$\tilde{r}_{ij} = w_j \otimes \tilde{x}_{ij} \quad (2.17)$$

where w_j represents the IVIF weight of the jth criterion.

Step 4: Creation of the $\tilde{S}_i$ matrix.

The values of the $\tilde{S}_i$ matrix are obtained by employing Equation (2.18) as follows:

$$\tilde{S}_i = \sum_{j}^{m} \tilde{r}_{ij} \quad (2.18)$$

Step 5: Determination of the utility degrees of the alternatives.

TABLE 2.3
An IVIFN scale to rate alternatives on criteria (Zavadskas et al., 2014)

Linguistic term	IVIFNs
Very important (VI)	$([0.9, 0.9], [0.1, 0.1])$
Important (I)	$([0.4, 0.7625], [0, 0.2115])$
Medium (M)	$([0.15, 0.5125], [0.25, 0.4625])$
Unimportant (U)	$([0, 0.3625], [0.4, 0.6125])$
Very unimportant (VU)	$([0.1, 0.1], [0.9, 0.9])$

The utility degrees of alternatives are calculated by employing Equations (2.19) and (2.20) as follows:

$$\tilde{K}_i^+ = \frac{\tilde{S}_i}{\tilde{S}_{is}} \tag{2.19}$$

$$\tilde{K}_i^- = \frac{\tilde{S}_i}{\tilde{S}_{ais}} \tag{2.20}$$

where $\tilde{S}_{is}$ represents ideal solution and corresponds $\left[(1,1),(0,0)\right]$ while $\tilde{S}_{ais}$ represents anti-ideal solution and corresponds $\left[(0,0),(1,1)\right]$.

Step 6: Identification of the utility function of the alternatives.

The utility functions are computed by using Equation (2.21). Here, $f\left(\tilde{K}_i^+\right)$ represents the utility function of the ideal solution, $f\left(\tilde{K}_i^-\right)$ represents the utility function of the anti-ideal solution. $f\left(\tilde{K}_i^+\right)$ and $f\left(\tilde{K}_i^-\right)$ are computed by employing Equations (2.22) and (2.23).

$$f\left(\tilde{K}_i\right) = \frac{\tilde{K}_i^+ + \tilde{K}_i^-}{1 + \dfrac{1 - f\left(\tilde{K}_i^+\right)}{f\left(\tilde{K}_i^+\right)} + \dfrac{1 - f\left(\tilde{K}_i^-\right)}{f\left(\tilde{K}_i^-\right)}} \tag{2.21}$$

$$f\left(\tilde{K}_i^+\right) = \frac{\tilde{K}_i^-}{\tilde{K}_i^- + \tilde{K}_i^+} \tag{2.22}$$

$$f\left(\tilde{K}_i^-\right) \frac{\tilde{K}_i^+}{\tilde{K}_i^- + \tilde{K}_i^+} \tag{2.23}$$

Here, 1 corresponds $\left[(1,1),(0,0)\right]$ in the IVIF number.

Step 7: Rank the alternatives.

In the last step, the final values of the utility functions are ranked in the decreasing order. The most desired alternative is the one with the highest utility value.

2.3.3 A Goal Programming Model

In this section, a goal programming model is developed based on normalized aggregated fuzzy decision matrix to assign the potential provinces in the TR90 region to the existing 15 tourism types. The model ensuring the provinces to focus on one or more tourism types, considering their performance in the relevant tourism diversity, is explained in detail as follows:

Indices:

i : index for the alternative (province) $(i = 1, 2, \ldots, M)$
j : index for the criterion (tourism type) $(j = 1, 2, \ldots, N)$

Parameters:
w_j : the importance weight of the jth criterion
$\ddot{a}_{ij}$: the defuzzified value of the performance value of the ith alternative under the jth criterion in the normalized aggregated fuzzy decision matrix
α : the maximum assignment amount for each tourism type
β : the maximum assignment amount for each province

Decision variables:

$$x_{ij} : \begin{cases} \textit{if } i^{th} \textit{ province is assigned to } j^{th} \textit{ tourism type}, 1 \\ \qquad\qquad \textit{otherwise}, 0 \end{cases}$$

Deviation variables:
d_j^+ : positive deviation from the target value
d_j^- : negative deviation from the target value

Objective function:

$$Min\,Z = \sum_{j}^{m} w_j d_j^- \tag{2.24}$$

Constraints:

$$\sum_{i}^{n} \ddot{a}_{ij} x_{ij} - d_j^+ + d_j^- = 1; \forall j \tag{2.25}$$

$$\sum_{i}^{n} x_{ij} \le \alpha; \forall j \tag{2.26}$$

$$\sum_{j}^{m} x_{ij} \le \beta; \forall i \tag{2.27}$$

$$x_{ij} \in \{0,1\}; \forall i, j \tag{2.28}$$

$$d_j^+, d_j^- \ge 0; \forall j \tag{2.29}$$

The objective function (2.24) minimizes the total weighted negative deviation amounts from the target value. Here, for each criterion, the target value corresponds to the best normalized defuzzified value. Constraint (2.25) relates to the goal constraints for each tourism type.

Constraint (2.26) restricts the number of provinces that can be assigned to a tourism type. Constraint (2.27) restricts the number of tourism types that can be assigned to a province. Constraint (2.28) relates to the binary decision variables. Constraint (2.29) imposes the nonnegativity for positive and negative deviation variables (Figure 2.1).

2.4 A CASE STUDY WITH REAL BACKGROUND

In this section, the two-stage research methodology is applied to a case study with a real-world background for the northeastern region of Turkey, referred to as TR90, as shown in Figure 2.2. The provinces in the TR90 region correspond to the alternatives and are coded as follows: A1: Artvin, A2: Giresun, A3: Gümüşhane, A4: Ordu, A5: Rize, and A6: Trabzon. Similarly, the 15 tourism types are coded as: C1: river tourism, C2: hunting tourism, C3: mountaineering, C4: golf tourism, C5: air sports, C6: religious tourism, C7: Silk Road tourism, C8: winter sports tourism, C9: congress tourism, C10: birdwatching tourism, C11: cave tourism, C12: health and thermal tourism, C13: underwater diving tourism, C14: marine tourism, and C15: highland tourism. To this end, the performances of the provinces under the 15 tourism types

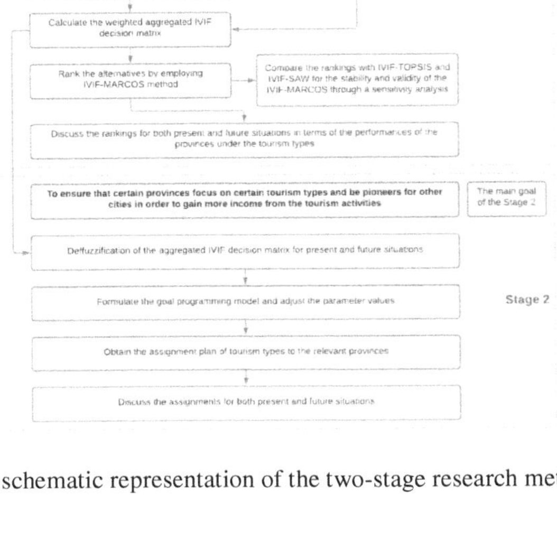

FIGURE 2.1 The schematic representation of the two-stage research methodology

are evaluated by four decision makers (DMs), one of whom is an academician specializing in tourism studies in Trabzon, while the other three are professionals working at the Eastern Black Sea Development Agency. The weights of the DMs are determined based on their years of service. The evaluations and related information about the DMs are provided in Table 2.4. For this region, the importance weights of the tourism types are determined separately for the present and future attractiveness scenarios to evaluate the performances of the provinces across both periods. The importance weights of the tourism types are presented in Figure 2.3.

2.4.1 Computational Results

In the following, the performance rankings of the provinces in the TR90 region for the present, future situations, and the equally weighted scenario, obtained using the proposed IVIF-MARCOS method and the existing IVIF-TOPSIS and IVIF-SAW methods, are presented in Table 2.5. In this study, the IVIF-TOPSIS method proposed by Bai (2013) and the IVIF-SAW method handled by Yeni and Özçelik (2019) are employed. Furthermore, the process steps regarding the proposed IVIF-MARCOS method are provided in Özçelik (2022). To test the validity of the proposed IVIF-MARCOS method, a sensitivity analysis is performed. To this end, the tourism types corresponding to the criteria are weighted randomly, and 15 different cases are generated, as presented in Table 2.6. For these analyses, Spearman's rank correlation coefficient (rs), which evaluates ordinal or ranked variables, is utilized. According to Spearman (1987), rs values are considered highly valid within the range $0.99 \geq rs \geq 0.7$. In this study, $rs1$ and $rs2$ denote the Spearman's rank correlation coefficients between IVIF-MARCOS & IVIF-TOPSIS and IVIF-MARCOS & IVIF-SAW, respectively.

The computational results are visualized in terms of performance rankings in Figure 2.4. According to the results, it is observed that there is a strong correlation between the proposed IVIF-MARCOS method and the existing IVIF-TOPSIS and IVIF-SAW methods. Furthermore, the correlation between IVIF-MARCOS and IVIF-SAW is calculated as 1 across all cases, meaning that the rankings are identical

FIGURE 2.2 The TR90 region and the provinces in this region

TABLE 2.4
The information dealing with the DMs and their evaluations

Decision makers	Criteria	Alternatives						Decision makers	Criteria	Alternatives					
		A1	A2	A3	A4	A5	A6			A1	A2	A3	A4	A5	A6
DM1	C1	VI	U	U	U	I	U	DM3	C1	I	M	M	U	I	M
$w_1 = 0.357$	C2	M	M	M	M	M	M	$w_3 = 0.286$	C2	U	U	U	U	U	U
10-year	C3	I	M	M	U	M	M	8-year expert	C3	I	M	I	M	I	I
academician	C4	VU	VU	VU	VU	VU	VU		C4	U	U	U	M	U	U
	C5	U	U	U	U	U	U		C5	M	U	M	U	M	M
	C6	U	U	M	M	M	I		C6	I	I	VI	VI	I	VI
	C7	M	M	M	M	M	M		C7	I	M	VI	M	VI	VI
	C8	M	M	I	U	U	I		C8	M	M	I	M	M	M
	C9	U	U	U	M	M	I		C9	VU	U	VU	M	U	I
	C10	I	I	I	I	I	I		C10	M	U	M	U	M	U
	C11	U	U	I	U	U	I		C11	U	M	I	U	U	M
	C12	U	U	U	U	I	U		C12	I	M	M	I	I	VI
	C13	U	U	U	U	U	U		C13	U	U	U	U	M	U
	C14	M	I	VU	I	M	M		C14	U	U	VU	M	U	M
	C15	I	VI	M	I	VI	VI		C15	I	I	I	I	VI	VI
DM2	C1	I	M	U	M	VI	M	DM4	C1	I	M	U	U	VI	M
$w_2 = 0.321$	C2	M	U	M	U	M	U	$w_4 = 0.036$	C2	M	M	M	M	M	M
9-year expert	C3	M	U	U	U	M	I	1-year expert	C3	VI	VI	I	I	VI	VI
	C4	VU	VU	VU	U	VU	VU		C4	U	U	U	U	U	U
	C5	U	U	U	I	M	I		C5	M	M	M	I	M	I
	C6	M	M	I	U	U	VI		C6	U	M	I	U	M	VI
	C7	VU	VU	VU	VU	VU	I		C7	U	M	VI	M	M	VI

(Continued)

TABLE 2.4 (CONTINUED)
The information dealing with the DMs and their evaluations

Decision makers	Criteria	Alternatives A1	A2	A3	A4	A5	A6	Decision makers	Criteria	Alternatives A1	A2	A3	A4	A5	A6
	C8	M	M	I	U	M	U		C8	M	I	I	M	I	I
	C9	VU	VU	VU	VU	VU	M		C9	U	M	U	M	U	I
	C10	VI	M	I	M	M	M		C10	M	M	M	M	M	M
	C11	U	U	VI	U	U	I		C11	M	M	I	M	M	M
	C12	M	U	U	U	M	I		C12	M	M	M	M	I	VI
	C13	U	U	VU	M	U	U		C13	U	U	VU	U	U	U
	C14	U	U	VU	U	U	U		C14	M	M	VU	I	M	I
	C15	I	VI	VI	M	VI	I		C15	I	I	I	I	VI	VI

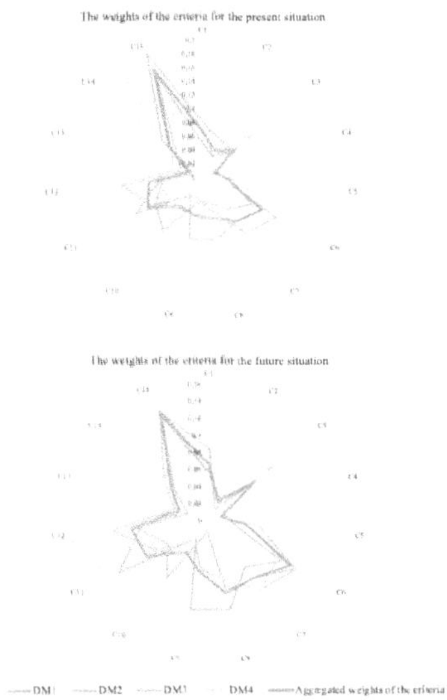

FIGURE 2.3 The visualization of the ratings related to the weights of the criteria according to present and future situations

for both methods in every scenario. Therefore, it can be deduced that the proposed IVIF-MARCOS method is a valid and efficient tool for multi-criteria group decision-making (MCGDM) problems under an interval-valued intuitionistic fuzzy environment.

After evaluating the performance rankings of the provinces using the proposed IVIF-MARCOS method for the present and future situations, the developed goal programming model is employed to ensure the promotion of tourism diversity in the assignment of the provinces. To this end, $\alpha = 1$ and $\beta = 3$ are considered in the goal programming model. Thus, to generate higher income from tourism activities, the aim is to have certain provinces focus on specific types of tourism and act as pioneers for other cities. Subsequently, the assignment plan of tourism types for the relevant provinces is obtained by employing the proposed goal programming model and is presented in Table 2.7.

2.4.2 Discussion

This section is devoted to presenting the managerial insights by highlighting the key findings obtained from the computational results. In the first stage, the provinces are evaluated to reveal the general performance map of the provinces according to tourism types for the present and future situations. These evaluations are conducted considering the present and future conditions for operational-tactical and strategic planning periods, respectively. In addition, for observational purposes, evaluations are also made under a scenario where the importance of tourism types is considered

TABLE 2.5
The performance rankings of the provinces according to the addressed methods for present and future situations

	IVIF-MARCOS			IVIF-TOPSIS			IVIF-SAW		
Provinces	**Equally weighted**	**Present situation**	**Future situation**	**Equally weighted**	**Present situation**	**Future situation**	**Equally weighted**	**Present situation**	**Future situation**
A1: Artvin	**4**	**5**	**4**	4	4	4	4	5	4
A2: Giresun	**5**	**4**	**5**	6	5	5	5	4	5
A3: Gümüşhane	**3**	**3**	**3**	2	2	2	3	3	3
A4: Ordu	**6**	**6**	**6**	5	6	6	6	6	6
A5: Rize	**2**	**2**	**2**	3	3	3	2	2	2
A6: Trabzon	1	1	1	1	1	1	1	1	1
rs1				0.89	0.89	0.94			
rs2							1.00	1.00	1.00

TABLE 2.6
Randomly assigned criteria weights

Criteria	Weights														
	Case 1	Case 2	Case 3	Case 4	Case 5	Case 6	Case 7	Case 8	Case 9	Case 10	Case 11	Case 12	Case 13	Case 14	Case 15
C1	0.109	0.108	0.105	0.083	0.104	0.080	0.019	0.050	0.073	0.110	0.023	0.070	0.015	0.010	0.041
C2	0.041	0.109	0.108	0.105	0.083	0.104	0.080	0.019	0.050	0.073	0.110	0.023	0.070	0.015	0.010
C3	0.010	0.041	0.109	0.108	0.105	0.083	0.104	0.080	0.019	0.050	0.073	0.110	0.023	0.070	0.015
C4	0.015	0.010	0.041	0.109	0.108	0.105	0.083	0.104	0.080	0.019	0.050	0.073	0.110	0.023	0.070
C5	0.070	0.015	0.010	0.041	0.109	0.108	0.105	0.083	0.104	0.080	0.019	0.050	0.073	0.110	0.023
C6	0.023	0.070	0.015	0.010	0.041	0.109	0.108	0.105	0.083	0.104	0.080	0.019	0.050	0.073	0.110
C7	0.110	0.023	0.070	0.015	0.010	0.041	0.109	0.108	0.105	0.083	0.104	0.080	0.019	0.050	0.073
C8	0.073	0.110	0.023	0.070	0.015	0.010	0.041	0.109	0.108	0.105	0.083	0.104	0.080	0.019	0.050
C9	0.050	0.073	0.110	0.023	0.070	0.015	0.010	0.041	0.109	0.108	0.105	0.083	0.104	0.080	0.019
C10	0.019	0.050	0.073	0.110	0.023	0.070	0.015	0.010	0041	0.109	0.108	0.105	0.083	0.104	0.080
C11	0.080	0.019	0.050	0.073	0.110	0.023	0.070	0.015	0.010	0.041	0.109	0.108	0.105	0.083	0.104
C12	0.104	0.080	0.019	0.050	0.073	0.110	0.023	0.070	0.015	0.010	0.041	0.109	0.108	0.105	0.083
C13	0.083	0.104	0.080	0.019	0.050	0.073	0.110	0.023	0.070	0.015	0.010	0.041	0.109	0.108	0.105
C14	0.105	0.083	0.104	0.080	0.019	0.050	0.073	0.110	0.023	0.070	0.015	0.010	0.041	0.109	0.108
C15	0.108	0.105	0.083	0.104	0.080	0.019	0.050	0.073	0.110	0.023	0.070	0.015	0.010	0.041	0.109

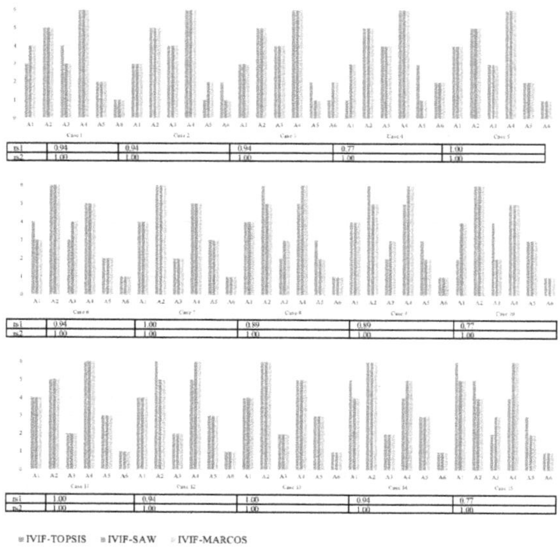

FIGURE 2.4 Visualization of the rankings of the alternatives according to the employed methods

equal. Accordingly, the provinces in the TR90 region are ranked across 15 tourism types using the proposed IVIF-MARCOS method. Analyses are performed for three different criteria weight scenarios, as mentioned above. The rankings obtained by the IVIF-MARCOS method are compared with those obtained by the IVIF-TOPSIS and IVIF-SAW methods, revealing a strong correlation among them. This result confirms that the IVIF-MARCOS method can produce valid and reliable rankings (see Table 2.5). The rankings are obtained for the equally weighted scenario, and for the present and future situations, considering the performances of the provinces across all tourism types. According to the IVIF-MARCOS method, the three provinces with the highest tourism potential in the TR90 region are Trabzon, Rize, and Gümüşhane, respectively, across all three scenarios (see Table 2.5). While Giresun ranks fourth in the present situation, it falls to fifth in the future scenario. Similarly, Artvin, ranked fifth in the present, rises to fourth place in the future. Additionally, Ordu consistently exhibits the lowest tourism potential in the TR90 region across all evaluated scenarios (see Table 2.5). In the second stage, the provinces are assigned to specific tourism types using the goal programming model to ensure that certain provinces specialize and act as pioneers, thereby contributing to increased tourism income. These assignments are designed considering the present and future situations for operational-tactical and strategic planning periods, respectively. It is observed that almost all tourism types can be realized across almost all provinces in the TR90 region. The assignment plan aims to maximize tourism income by enhancing tourism diversity. For example, although river tourism is assigned to Rize, this does not imply that Artvin cannot engage in river tourism. Instead, it indicates that Rize specializes in

TABLE 2.7
The assignment plan of tourism types to the relevant provinces

Provinces	Assignments present situation (for operational-tactical-level planning periods)			Assignments future situation (for strategic planning period)		
A1: Artvin	Mountaineering	Birdwatching tourism	—	River tourism	Mountaineering	Birdwatching tourism
A2: Giresun	Highland tourism	—	—	Highland tourism	—	—
A3: Gümüşhane	Hunting tourism	Winter sports tourism	Cave tourism	Winter sports tourism	Cave tourism	—
A4: Ordu	Golf tourism	Air sports tourism	Marine tourism	Golf tourism	Air sports tourism	Marine tourism
A5: Rize	River tourism	Health and thermal tourism	Underwater diving tourism	Hunting tourism	Health and thermal tourism	Underwater diving tourism
A6: Trabzon	Religious tourism	Silk Road tourism	Congress tourism	Religious tourism	Silk Road tourism	Congress tourism

river tourism and leads other provinces in this type (see Table 2.7). Based on Table 2.7, it is found that Artvin should focus on mountaineering and birdwatching tourism in the present situation. The existence of trails suitable for mountaineering and hiking in the Kaçkar Mountains (TCP, 2022a) shows that Artvin offers sufficient opportunities for mountaineering tourism. Additionally, high hills in Kemalpaşa, Murgul, Borçka, and Artvin, located on bird migration routes, offer favorable conditions for bird and butterfly enthusiasts (TCP, 2022b). Although river sports are currently popular in many streams of Rize, the Çoruh River stands out for Artvin, offering even greater potential. The Çoruh River, which flows through Yusufeli, is recognized as the second-best rafting river in the world after the Zambezi River, and the Fourth World River Cross Country Championship was held there (TCP, 2022c). Giresun is identified as suitable for specializing in highland tourism. Several plateaus, such as Kümbet Plateau, Hanalanı (Kulakkaya) Plateau, Melikli Obası Plateau, and Bektaş Plateau (RTMCT, 2022a), exist in Giresun. Among the TR90 provinces, there are three plateau tourism centers in Giresun, six in Ordu, two in Rize, and four in Trabzon, while Artvin and Gümüşhane have none (RTMCT, 2022b). Highland tourism is assigned to Giresun because other tourism types assigned to Ordu and Trabzon contribute more significantly to the overall tourism efficiency of the region. Gümüşhane is assigned hunting tourism, winter sports tourism, and cave tourism in the present situation. Although hunting opportunities in Gümüşhane are limited and seasonal, the region's geography, consisting of 89% mountains and plains, provides favorable conditions. The diversity of hunting animals and the popularity of hunting sports (TCP, 2022d) support this specialization. Rize also has hunting areas but is expected to become more prominent in the future. Regarding winter sports, Gümüşhane hosts four winter sports centers compared to only one in Rize (RTMCT, 2022c), making it the leading province in this area. For cave tourism, there is only one cave open to tourists each in Trabzon and Gümüşhane, and Gümüşhane's assignment is due to the optimization model's contribution to regional efficiency. Golf tourism, air sports, and marine tourism are assigned to Ordu. Although there are currently no golf facilities in the TR90 region, Ordu is prioritized due to plans for a new golf course. Additionally, Ordu is the only province with two Blue Flag beaches (BFT, 2022) and has numerous beaches suitable for marine tourism. Boztepe and other areas in Ordu also offer opportunities for paragliding and air sports. Health and thermal tourism and underwater diving tourism are assigned to Rize. While river tourism is prominent in Rize today, hunting tourism is projected to gain importance in the future following Artvin's specialization. The Fırtına Stream in Rize, fed by streams from the Kaçkar Mountains, is highly suitable for river sports (TCP, 2022e). Rize also has several thermal springs, including the Ayder Spa, İkizdere Spa, Andon Spring, Şimşirli Spring, and Çayırlı Mineral Water, making it an attractive destination for health and thermal tourism (TCP, 2022f). Although underwater diving is not currently prominent in the TR90 region, this study determines that Rize is the most suitable province to specialize in this area for both present and future periods. Finally, Trabzon is assigned religious tourism, Silk Road tourism, and congress tourism. Trabzon houses many historical churches such as the Sumela Monastery and the Santa Maria Church (RTMCT, 2022d), alongside numerous historical mosques, making it ideal for religious tourism. Trabzon is also a key stop along the Northern Anatolian Silk Road, containing historical inns, baths, and bazaars. Furthermore, Trabzon's geographic location and developed transportation infrastructure promote its potential for congress tourism. Several four- and five-star hotels, university facilities, and public institutions host various congresses, panels, symposiums, and sports competitions (TCP, 2022g).

2.5 CONCLUDING REMARKS

This chapter focuses on tourism marketing and management for Turkey's Eastern Black Sea region, referred to as the TR90 region, which includes several important touristic resources. In this regard, a two-stage research methodology is proposed. In the first stage, the performances of the provinces in the TR90 region are evaluated across various tourism types for both the present and future periods. To this end, the IVIF-MARCOS method, which addresses group decision-making processes, is employed. In the second stage, a goal programming model is developed to enable provinces to specialize and act as pioneers for other cities in one or more tourism types, thereby better managing their tourism potential in the present and future. Based on the present and future assessments, the assignment of provinces is performed separately for operational-tactical and strategic planning periods. Managerial insights are provided by highlighting the key findings. The main contribution of this study is the presentation of a two-stage research methodology that supports sustainable tourism marketing strategies for governments, integrating the newly applied IVIF-MARCOS method for provincial evaluations and a developed goal programming model for the assignment of provinces to tourism types. Additionally, this is the first study to address tourism marketing for the TR90 region while considering both present and future situations. Future research directions include the following: (i) extending the methodology by incorporating different fuzzy-based MCDM methods, (ii) applying other fuzzy number types to represent the uncertainty in evaluations, and (iii) conducting a similar study at the micro level by considering district-level analyses within the TR90 region.

REFERENCES

Abrishamchi, A., Mounesan, A. A., & Maknoon, R. (2021). A fuzzy multi-criteria approach to sustainable island-based tourism development. *Scientia Iranica*, *28*(2), 682–699.

Akram, M., Khan, A., Luqman, A., Senapati, T., & Pamucar, D. (2023). An extended MARCOS method for MCGDM under 2-tuple linguistic q-rung picture fuzzy environment. *Engineering Applications of Artificial Intelligence*, *120*, 105892.

Anysz, H., Nicał, A., Stević, Ž., Grzegorzewski, M., & Sikora, K. (2021). Pareto optimal decisions in multi-criteria decision making explained with construction cost cases. *Symmetry*, *13*(1), 46.

Atanassov, K. T. (1986). Intuitionistic fuzzy sets. *Fuzzy Sets and Systems*, *20*(1), 87–96.

Atanassov, K., & Gargov, G. (1989). Interval valued intuitionistic fuzzy sets. *Fuzzy Sets and Systems*, *31*(3), 343–349.

Aydin, M., Camliyurt, G., Gul, M., Sezer, S. I., Celik, E., & Akyuz, E. (2025). An interval type-2 fuzzy MARCOS modelling to assess performance effectiveness of survival craft on cargo ship. *Ocean Engineering*, *326*, 120899.

Badi, I., & Pamucar, D. (2020). Supplier selection for steelmaking company by using combined Grey-MARCOS methods. *Decision Making: Applications in Management and Engineering*, *3*(2), 37–48.

Bai, Z. Y. (2013). An interval-valued intuitionistic fuzzy TOPSIS method based on an improved score function. *The Scientific World Journal*, *2013*, 879089, 1–6.

Bellman, R. E., & Zadeh, L. A. (1970). Decision-making in a fuzzy environment. *Management Science*, *17*(4), B-141.

Bire, R. B., Conterius, A. L. F., & Nasar, A. (2021). Drivers of regional destination competitiveness: A DEMATEL-Fuzzy TOPSIS approach. *The Indonesian Journal of Geography*, *53*(1), 144–152.

Blagojević, A., Kasalica, S., Stević, Ž., Tričković, G., & Pavelkić, V. (2021). Evaluation of safety degree at railway crossings in order to achieve sustainable traffic management: A novel integrated fuzzy MCDM model. *Sustainability*, *13*(2), 832.

Boral, S., & Chakraborty, S. (2021). Failure analysis of CNC machines due to human errors: An integrated IT2F-MCDM-based FMEA approach. *Engineering Failure Analysis*, 105768.

Boral, S., Chaturvedi, S. K., Howard, I., Naikan, V. N. A., & McKee, K. (2021). An integrated interval type-2 fuzzy sets and multiplicative half quadratic programming-based MCDM framework for calculating aggregated risk ranking results of failure modes in FMECA. *Process Safety and Environmental Protection*, *150*, 194–222.

Bouraima, M. B., Stević, Ž., Tanackov, I., & Qiu, Y. (2021). Assessing the performance of Sub-Saharan African (SSA) railways based on an integrated Entropy-MARCOS approach. *Operational Research in Engineering Sciences: Theory and Applications*, *4*(2), 13–35.

Boyacı, A. Ç., & Şişman, A. (2022). Pandemic hospital site selection: A GIS-based MCDM approach employing Pythagorean fuzzy sets. *Environmental Science and Pollution Research*, *29*(2), 1985–1997.

Celik, E., & Gul, M. (2021). Hazard identification, risk assessment and control for dam construction safety using an integrated BWM and MARCOS approach under interval type-2 fuzzy sets environment. *Automation in Construction*, *127*, 103699.

Chakraborty, S., Chattopadhyay, R., & Chakraborty, S. (2020). An integrated D-MARCOS method for supplier selection in an iron and steel industry. *Decision Making: Applications in Management and Engineering*, *3*(2), 49–69.

Chang, K. L. (2015). The use of a hybrid MCDM model for public relations personnel selection. *Informatica*, *26*(3), 389–406.

Chen, C. A. (2014). Suitable festival activities for Taiwan's tourism and nation branding with the application of the PR AHP program. *Asia Pacific Journal of Tourism Research*, *19*(12), 1381–1398.

Ecer, F. (2021). A consolidated MCDM framework for performance assessment of battery electric vehicles based on ranking strategies. *Renewable and Sustainable Energy Reviews*, *143*, 110916.

Erdenejargal, N., Dorjsuren, B., Choijinjav, L., Doljin, D., Enkhbold, A., Munkhuu, B., ... & Girma, A. (2021). Evaluation of the natural landscape aesthetic: A case study of Uvs province, Mongolia. *Polish Journal of Environmental Studies*, *30*(5), 4497–4509.

Fu, H. H., & Tzeng, S. Y. (2016). Applying fuzzy multiple criteria decision-making approach to establish safety-management system for hot spring hotels. *Asia Pacific Journal of Tourism Research*, *21*(12), 1343–1356.

Fu, Y. K., Huang, W., & Liao, C. N. (2020). The selection model for horizontal alliances between hotels and airlines: An integrated application of NGT, fuzzy TOPSIS and MCGP methods. *Tourism Review*, *75*(4), 681–698.

García Mestanza, J., & Bakhat, R. (2021). A fuzzy AHP-MAIRCA model for overtourism assessment: The case of Malaga Province. *Sustainability*, *13*(11), 6394.

Hasheminasab, H., Hashemkhani Zolfani, S., Kazimieras Zavadskas, E., Kharrazi, M., & Skare, M. (2022). A circular economy model for fossil fuel sustainable decisions based on MADM techniques. *Economic Research-Ekonomska Istraživanja*, *35*(1), 564–582.

Horng, J. S., Liu, C. H., Chou, S. F., Yin, Y. S., & Tsai, C. Y. (2014). Developing a novel hybrid model for industrial environment analysis: A study of the gourmet and tourism industry in Taiwan. *Asia Pacific Journal of Tourism Research*, *19*(9), 1044–1069.

Kang, D., Manirathinam, T., Sandra, M., Pragathi, S., Narayanamoorthy, S., Yazid Bajuri, M., ... & Ahmadian, A. (2024). Measuring priorities of sustainable post-pandemic tourism policy factors using novel fuzzy multi-criteria decision making approach. *Journal of Sustainable Tourism*, *32*(11), 2286–2303.

Krishnan, E., Mohammed, R., Alnoor, A., Albahri, O. S., Zaidan, A. A., Alsattar, H., ... & Alazab, M. (2021). Interval type 2 trapezoidal-fuzzy weighted with zero inconsistency combined with VIKOR for evaluating smart e-tourism applications. *International Journal of Intelligent Systems*, *36*(9), 4723–4774.

Li, G., Law, R., Vu, H. Q., & Rong, J. (2013). Discovering the hotel selection preferences of Hong Kong inbound travelers using the Choquet Integral. *Tourism Management*, *36*, 321–330.

Liao, S. K., Hsu, H. Y., & Chang, K. L. (2019). OTAs selection for hot spring hotels by a hybrid MCDM model. *Mathematical Problems in Engineering*, *2019*, (1), 4251362, 1–9.

Lin, C. P., Chen, S. H., Trac, L. V. T., & Wu, C. F. (2021). An expert-knowledge-based model for evaluating cultural tourism strategies: A case of Tainan City, Taiwan. *Journal of Hospitality and Tourism Management*, *49*, 214–225.

Lo, H. W., & Liou, J. J. (2019). Factors influencing medical tourism adoption in Malaysia: A DEMATEL-Fuzzy TOPSIS approach". *Computers & Industrial Engineering*, *137*, 106005.

Majidi, A., Mirzapour Al-e-Hashem, S. M., & Hashemkhani Zolfani, S. (2021). Sustainability ranking of the Iranian major ports by using MCDM methods. *Mathematics*, *9*(19), 2451.

Mardani, A., Jusoh, A., Zavadskas, E. K., Kazemilari, M., Ungku, N. U. A., & Khalifah, Z. (2016). Application of multiple criteria decision-making techniques in tourism and hospitality industry: A systematic review. *Transformations in Business & Economics*, *15*(1), 192–213.

Martín, J. C., Saayman, M., & du Plessis, E. (2019). Determining satisfaction of international tourist: A different approach. *Journal of Hospitality and Tourism Management*, *40*, 1–10.

Mijajlović, M., Puška, A., Stević, Ž., Marinković, D., Doljanica, D., Jovanović, S. V., Stojanović, I., & Beširović, J. (2020). Determining the competitiveness of spa-centers in order to achieve sustainability using a fuzzy multi-criteria decision-making model. *Sustainability*, *12*(20), 8584.

Mishra, A. R., Chandel, A., & Saeidi, P. (2022). Low-carbon tourism strategy evaluation and selection using interval-valued intuitionistic fuzzy additive ratio assessment approach based on similarity measures. *Environment, Development and Sustainability*, *24*(5), 7236–7282.

Ocampo, L. A., Tan, T. A. G., & Sia, L. A. (2018). Using fuzzy DEMATEL in modeling the causal relationships of the antecedents of organizational citizenship behavior (OCB) in the hospitality industry: A case study in the Philippines. *Journal of Hospitality and Tourism Management*, *34*, 11–29.

Özçelik, G. (2022). A group decision aggregation-based IVIF-MARCOS and goal programming approach for intelligent decision-making in tourism marketing (IVIF_MARCOS_steps), Github-gozcelik87 repository. https://github.com/gozcelik87/IVIF_MARCOS_steps

Özçelik, G., & Nalkıran, M. (2021). An extension of EDAS method equipped with trapezoidal bipolar fuzzy information: An application from healthcare system. *International Journal of Fuzzy Systems*, *23*(7), 2348–2366.

Pamucar, D., Ecer, F., & Deveci, M. (2021a). Assessment of alternative fuel vehicles for sustainable road transportation of United States using integrated fuzzy FUCOM and neutrosophic fuzzy MARCOS methodology. *Science of The Total Environment*, *788*, 147763.

Pamucar, D., Iordache, M., Deveci, M., Schitea, D., & Iordache, I. (2021b). A new hybrid fuzzy multi-criteria decision methodology model for prioritizing the alternatives of the hydrogen bus development: A case study from Romania. *International Journal of Hydrogen Energy*, *46*(57), 29616–29637.

Pelissari, R., Oliveira, M. C., Abackerli, A. J., Ben-Amor, S., & Assumpção, M. R. P. (2021). Techniques to model uncertain input data of multi-criteria decision-making problems: A literature review. *International Transactions in Operational Research*, *28*(2), 523–559.

Peng, J. J., Tian, C., Zhang, Z. Q., Song, H. Y., & Wang, J. Q. (2019). Single-valued neutrosophic power shapley choquet average operators and their applications to multi-criteria decision-making. *Mathematics*, *7*(11), 1081.

Pramanik, P. K. D., Biswas, S., Pal, S., Marinković, D., & Choudhury, P. (2021). A comparative analysis of multi-criteria decision-making methods for resource selection in mobile crowd computing. *Symmetry*, *13*(9), 1713.

Qin, J. (2017). Interval type-2 fuzzy Hamy mean operators and their application in multiple criteria decision making. *Granular Computing*, *2*(4), 249–269.

Salimian, S., Mousavi, S. M., & Antucheviciene, J. (2022). An interval-valued intuitionistic fuzzy model based on extended VIKOR and MARCOS for sustainable supplier selection in organ transplantation networks for healthcare devices. *Sustainability*, *14*(7), 3795.

Simić, V., Soušek, R., & Jovčić, S. (2020). Picture fuzzy MCDM approach for risk assessment of railway infrastructure. *Mathematics*, *8*(12), 2259.

Stanković, M., Stević, Ž., Das, D. K., Subotić, M., & Pamučar, D. (2020). A new fuzzy MARCOS method for road traffic risk analysis. *Mathematics*, *8*(3), 457.

Stević, Ž., Das, D. K., & Kopić, M. (2021a). A novel multiphase model for traffic safety evaluation: A case study of South Africa. *Mathematical Problems in Engineering*, *2021*, 1–22.

Stević, Ž., Karamaşa, Ç., Demir, E., & Korucuk, S. (2021b). Assessing sustainable production under circular economy context using a novel rough-fuzzy MCDM model: A case of the forestry industry in the Eastern Black Sea region. *Journal of Enterprise Information Management*, *38*(1), 261–291.

Stević, Ž., Pamučar, D., Puška, A., & Chatterjee, P. (2020). Sustainable supplier selection in healthcare industries using a new MCDM method: Measurement of alternatives and ranking according to compromise solution (MARCOS). *Computers & Industrial Engineering*, *140*, 106231.

Tadić, S., Kovač, M., Krstić, M., Roso, V., & Brnjac, N. (2021). The selection of intermodal transport system scenarios in the function of Southeastern Europe regional development. *Sustainability*, *13*(10), 5590.

Tan, Y., Park, S., de Medeiros, A. M. A., & Wanke, P. (2025). Cost-benefit analysis in UK hotels: A hybrid SOCP-MCDM approach. *Tourism Management*, *106*, 105034.

Tang, M., Wang, T. D., & Peng, D. H. (2021). An improved Taguchi multi-criteria decision-making method based on the hesitant fuzzy correlation coefficient and its application in quality evaluation. *Journal of Ambient Intelligence and Humanized Computing, 12*(8), 8241–8254.

Tian, C., & Peng, J. (2020). An integrated picture fuzzy ANP-TODIM multi-criteria decision-making approach for tourism attraction recommendation. *Technological and Economic Development of Economy, 26*(2), 331–354.

Torkayesh, A. E., & Torkayesh, S. E. (2021). Evaluation of information and communication technology development in G7 countries: An integrated MCDM approach. *Technology in Society, 66*, 101670.

Türkarslan, E., Ye, J., Ünver, M., & Olgun, M. (2021). Consistency fuzzy sets and a cosine similarity measure in fuzzy multiset setting and application to medical diagnosis. *Mathematical Problems in Engineering, 2021*(1), 9975983, 1–9.

Ulutaş, A., Karabasevic, D., Popovic, G., Stanujkic, D., Nguyen, P. T., & Karaköy, Ç. (2020). Development of a novel integrated CCSD-ITARA-MARCOS decision-making approach for stackers selection in a logistics system. *Mathematics, 8*(10), 1672.

Vatankhah, S., Bouzari, M., & Safavi, H. P. (2021). Unraveling the fuzzy predictors of stress at work. *International Journal of Organizational .Analysis, 29*(2), 277–300.

Vrtagić, S., Softić, E., Subotić, M., Stević, Ž., Dordevic, M., & Ponjavic, M. (2021). Ranking road sections based on MCDM model: New improved fuzzy SWARA (IMF SWARA). *Axioms, 10*(2), 92.

Wang, Y., Wang, W., Wang, Z., Deveci, M., Roy, S. K., & Kadry, S. (2024). Selection of sustainable food suppliers using the Pythagorean fuzzy CRITIC-MARCOS method. *Information Sciences, 664*, 120326.

World Tourism Organization. (2021). *International Tourism Highlights, 2020 Edition.* UNWTO, Madrid.

Xu, Z. S. (2007). Methods for aggregating interval-valued intuitionistic fuzzy information and their application to decision making. *Control and Decision, 22*(2), 215–219.

Xu, Z., & Chen, J. (2007). On geometric aggregation over interval-valued intuitionistic fuzzy information. In *Fourth International Conference on Fuzzy Systems and Knowledge Discovery, 2007 (FSKD 2007)*, vol 2. IEEE, pp. 466–471.

Yang, X., & Wang, Z. J. (2020). Intuitionistic fuzzy hierarchical multi-criteria decision making for evaluating performances of low-carbon tourism scenic spots. *International Journal of Environmental Research and Public Health, 17*(17), 6259.

Yeni, F. B., & Özçelik, G. (2019). Interval-valued Atanassov intuitionistic Fuzzy CODAS method for multi criteria group decision making problems. *Group Decision and Negotiation, 28*(2), 433–452.

Yilmaz, O. F., Oztaysi, B., Durmusoglu, M. B., & Oner, S. C. (2017). Determination of material handling equipment for lean in-plant logistics using fuzzy analytical network process considering risk attitudes of the experts. *International Journal of Industrial Engineering, 24*(1), 81–122.

Yu, S. M., Wang, J., & Wang, J. Q. (2017). An interval type-2 fuzzy likelihood-based MABAC approach and its application in selecting hotels on a tourism website. *International Journal of Fuzzy Systems, 19*(1), 47–61.

Zadeh, L. A. (1965). Fuzzy sets. *Information Control, 8*(3), 338–353.

Zavadskas, E. K., Antucheviciene, J., Hajiagha, S. H. R., & Hashemi, S. S. (2014). Extension of weighted aggregated sum product assessment with interval-valued intuitionistic fuzzy numbers (WASPAS-IVIF). *Applied Soft Computing, 24*, 1013–1021.

Zhao, H., Xu, Z., & Yao, Z. (2016). Interval-valued intuitionistic fuzzy derivative and differential operations. *International Journal of Computational Intelligence Systems, 9*(1), 36–56.

TCP-Turkish Culture Portal. (2022a). https://www.kulturportali.gov.tr/turkiye/artvin/TurizmAktiviteleri/kackar-daglari, Erişim Tarihi: 26.01.2022.

TCP-Turkish Culture Portal. (2022b). https://www.kulturportali.gov.tr/turkiye/artvin/TurizmAktiviteleri/artvin375129, Erişim Tarihi: 26.01.2022.

TCP-Turkish Culture Portal. (2022c). https://www.kulturportali.gov.tr/turkiye/artvin/TurizmAktiviteleri/rafting194999, Erişim Tarihi: 26.01.2022.

RTMCT-Republic of Türkiye Ministry of Culture and Tourism. (2022a). https://yigm.ktb.gov.tr/TR-10123/giresun-yaylalari.html, Erişim Tarihi: 26.01.2022.

RTMCT-Republic of Türkiye Ministry of Culture and Tourism. (2022b). https://yigm.ktb.gov.tr/TR-11533/yayla-turizm-merkezleri.html, Erişim Tarihi: 26.01.2022.

TCP-Turkish Culture Portal. (2022d). https://www.kulturportali.gov.tr/turkiye/gumushane/TurizmAktiviteleri/av-turizm, Erişim Tarihi: 26.01.2022.

RTMCT-Republic of Türkiye Ministry of Culture and Tourism. (2022c). https://yigm.ktb.gov.tr/TR-10177/kis-sporlari-turizm-merkezlerine-iliskin-genel-bilgiler.html, Erişim Tarihi: 26.01.2022.

BFT-Blue Flag Türkiye. (2022). http://www.mavibayrak.org.tr/tr/Default.aspx, Erişim Tarihi:13.01.2022.

TCP-Turkish Culture Portal. (2022e). https://www.kulturportali.gov.tr/turkiye/rize/TurizmAktiviteleri/rafting-1, Erişim Tarihi: 26.01.2022.

TCP-Turkish Culture Portal. (2022f). https://www.kulturportali.gov.tr/turkiye/rize/TurizmAktiviteleri/termal-turizm-1, Erişim Tarihi: 26.01.2022.

RTMCT-Republic of Türkiye Ministry of Culture and Tourism. (2022d). https://yigm.ktb.gov.tr/TR-9993/trabzon.html, Erişim Tarihi: 26.01.2022.

TCP-Turkish Culture Portal. (2022g). https://www.kulturportali.gov.tr/turkiye/trabzon/TurizmAktiviteleri/kongre-turizmi, Erişim Tarihi: 26.01.2022.

RTMCT-Republic of Türkiye Ministry of Culture and Tourism. (2021). https://yigm.ktb.gov.tr/TR-201116/turizm-gelirleri-ve-giderleri.html, Erişim Tarihi:25.10.2021.

3 Sustainable Urban Logistics Evaluation in Smart Cities

A Multi-Criteria Group Decision-Making Approach Using the Hesitant Fuzzy Linguistic ARAS Method

Fatma Betül Yeni, Behice Meltem Kayhan, and Beren Gürsoy Yılmaz

3.1 INTRODUCTION

As rapid urbanization, population growth, and environmental challenges continue to reshape modern cities, the concept of the smart city has emerged as a transformative approach to addressing these complex issues. Smart cities leverage advanced information and communication technologies (ICTs) to enhance urban services, optimize resource use, and improve the overall quality of life for citizens (Ahad et al., 2020). The integration of digital solutions in various domains—including energy management, healthcare, public safety, and environmental sustainability—plays a pivotal role in fostering urban resilience and adaptability (Pandya et al., 2023). Among these domains, urban mobility stands as a critical component, ensuring the efficient and sustainable movement of people and goods within smart city infrastructures.

Urban mobility in smart cities plays a fundamental role in mitigating challenges such as traffic congestion, air pollution, and inefficient transportation networks. The development of sustainable and intelligent mobility solutions enhances accessibility, reduces carbon emissions, and optimizes connectivity across urban areas (Miskolczi et al., 2021; Richter et al., 2022). Efficient mobility systems not only alleviate traffic-related issues but also align with broader sustainability goals, such as promoting low-emission transport alternatives, improving multimodal connectivity, and ensuring equitable access to transportation networks (Standing et al., 2019; Miskolczi

DOI: 10.1201/9781003529088-3

et al., 2021). Given the increasing complexity of mobility planning in smart cities, structured decision-making approaches are essential for evaluating and selecting the most suitable mobility solutions.

To address this complexity, multi-criteria decision-making (MCDM) methods have been widely applied in urban planning for assessing mobility alternatives. These methods provide structured frameworks to evaluate various transportation solutions based on multiple conflicting criteria. However, traditional MCDM approaches often rely on crisp numerical assessments or conventional fuzzy techniques, which may not fully capture the linguistic vagueness and subjective uncertainty inherent in expert evaluations. Since mobility planning involves qualitative assessments alongside quantitative data, fuzzy set theory has been widely employed to manage uncertainty in decision-making. Despite the advantages of fuzzy logic in decision-making, classical fuzzy approaches require decision makers to choose a single linguistic term, limiting their ability to reflect hesitation and ambiguity in evaluations. To overcome this limitation, hesitant fuzzy sets (HFSs) were introduced by Torra (2010) as a generalization of fuzzy sets, allowing for situations where multiple membership functions may be considered simultaneously. Classical fuzzy linguistic approaches often restrict experts to selecting a single linguistic term, limiting their ability to convey nuanced judgments. However, in complex decision-making scenarios, experts may hesitate between multiple linguistic terms when assessing alternatives (Rodríguez et al., 2012). To overcome this limitation, Rodríguez et al. (2013) proposed a decision-making model based on hesitant fuzzy linguistic term sets (HFLTSs), enabling experts to express hesitation among multiple linguistic terms, thereby capturing a broader range of preferences and improving the robustness of group decision-making.

Despite the increasing use of MCDM techniques in urban mobility decision-making, limited research has explored the integration of HFLTS for handling linguistic vagueness in assessing mobility alternatives. Addressing this gap, this study proposes a structured decision-making approach that integrates HFLTS into the Additive Ratio Assessment (ARAS) method, providing a flexible and precise evaluation of mobility solutions under uncertainty. This study applies a group decision-making framework to rank and select well-known urban mobility alternatives within smart cities. The evaluated alternatives include self-driving vehicles, electric public transport, smart bike-sharing systems, and intelligent traffic management systems, each assessed based on multiple criteria. Given the complexity of selecting and ranking these alternatives, expert evaluations and linguistic assessments are incorporated to ensure a comprehensive decision-making process. By integrating HFLTS with the ARAS method, this study presents a systematic approach for evaluating urban mobility solutions in smart city environments. The proposed HFL-ARAS methodology captures linguistic vagueness, enhances the accuracy of expert evaluations, and facilitates a sustainability-driven selection process for urban mobility alternatives.

The rest of this chapter is structured as follows: Section 3.2 reviews the literature on smart cities and multi-criteria decision-making methods. Section 3.3 presents the methodology, detailing the integration of HFLTS with the ARAS method. Section 3.4 illustrates the application of this methodology through a case study in the context of smart cities. Finally, Section 3.5 discusses the results and provides concluding remarks on the implications of the findings for sustainable urban mobility planning in smart cities.

3.2 LITERATURE

In this section, studies from the literature focusing on smart cities and utilizing MCDM methods based on fuzzy set theory are analyzed. This review provides an overview of recent research, with an emphasis on key areas central to smart city development. The selected studies are summarized in Table 3.1, which outlines the main subject, the MCDM method used, and the type of fuzzy set applied.

Vehicle sharing systems have gained significant importance within the context of smart cities due to their efficiency, sustainability, and flexibility. Their ability to integrate with digital platforms and provide real-time data positions them as a crucial component of future urban mobility solutions. For instance, Manirathinam et al. (2024) investigated e-scooter sharing services, emphasizing performance, accessibility, reliability, and responsiveness from the user's perspective. Similarly, Eren and Katanalp (2022) studied bike-sharing station site selection, focusing on various land use types. Lyu et al. (2023) and Rani and Potika (2024) explored risk analysis in smart cities, employing the Analytic Hierarchy Process (AHP) for flood risk assessments in metro systems and AHP-MARCOS for wildfire risk evaluation, respectively. Both studies utilized different fuzzy approaches to address inherent uncertainties.

Furthermore, Otay et al. (2024), İbrahim et al. (2024), Alsattar et al. (2023), Shao et al. (2023), and Demircan and Yetilmezsoy (2023) concentrated on the sustainability of systems, including sustainable energy systems, smart living solutions, sustainable development levels in smart cities, and smart sustainable waste management. On the other hand, Çakır and Demircioğlu (2023), Gündoğdu (2022), Mishra et al. (2022), and Hanine et al. (2021) analyzed the barriers to becoming smart cities, the adoption of smart technologies, and the suitability and dimensions of smart city development. Their research underscores critical challenges and factors influencing the transition toward smart urban environments, offering valuable insights into the complexities of implementing advanced technologies in urban planning and development.

Various fuzzy-based MCDM techniques have been applied in urban mobility decision-making to handle complexity, subjectivity, and uncertainty. Among these, TOPSIS, AHP, and DEMATEL are widely used in evaluating sustainability, risk assessment, and infrastructure planning. Despite their strengths, these fuzzy-based MCDM techniques often fail to account for linguistic vagueness, as they primarily rely on fixed numerical scales rather than flexible linguistic terms.

While previous studies have extensively applied MCDM methods in urban planning, limited research has explored the integration of HFLTS for evaluating urban mobility alternatives. Unlike traditional fuzzy methods, HFLTS enables experts to express hesitation when selecting linguistic terms, allowing for greater flexibility in decision-making under uncertainty.

TABLE 3.1
Selected studies on smart cities and MCDM in the literature

References	MCDM method	Main subject	Fuzzy set
Otay et al. (2024)	BWM and TOPSIS	Sustainable energy systems	Multi-expert Pythagorean fuzzy
İbrahim et al. (2024)	FWZIC and FDOSM	Simulation tools for sustainable MaaS	Interval-valued Fermatean fuzzy
Manirathinam et al. (2024)	APPRESAL	E-scooter sharing service	Triangular fuzzy numbers
Rani and Potika (2024)	AHP-MARCOS	Wildfire risk analysis	Pythagorean fuzzy numbers
Alsattar et al. (2023)	CPOS	Sustainable smart living	C-PFSs
Shao et al. (2023)	DEMATEL and TOPSIS	Sustainable development level of smart cities	Z-fuzzy
Lyu et al. (2023)	AHP	Flood risk assessment of metro systems in smart city development	Interval, triangular, and trapezoidal
Demircan and Yetilmezsoy (2023)	AHP and TOPSIS	Smart sustainable waste management	Triangular
Andreou et al. (2023)	Best-worst method (BWM)	QoS of the available service	Intuitionistic preference relations (IPRs)
Çakır and Demircioğlu (2023)	PROMETHEE	The suitability of seven cities for smart cities	Circular intuitionistic fuzzy set (C-IFS)
Gündoğdu (2022)	DEMATEL	Barriers of becoming smart energy cities	Hesitant fuzzy Z-number
Mishra et al. (2022)	CoCoSo	Adoption of IoT barriers in waste management	Fermatean fuzzy sets
Eren and Katanalp (2022)	AHP, VIKOR,	Bike-sharing station site selection	Linear function or sigmoidal function
Krishankumar et al. (2021)	EDAS	Prioritize zero-carbon measures	Hesitant fuzzy linguistic terms
Hanine et al. (2021)	AHP, DEMATEL	The most influential dimensions and criteria for smart city development	Intuitionistic fuzzy set
Yörükoğlu and Aydın (2021)	AHP	Smart waste management Ssstems	Spherical fuzzy

Notes: APPRESAL: approach for preference, performance, and ranking evaluation with satisfaction level; CoCoSo: the combined compromise solution; C-PFSs: circular-pythagorean fuzzy sets; CPOS: conditional probabilities by opinion scores; EDAS: evaluation based on distance from average solution; FDOSM: fuzzy decision by opinion score method; FWZIC: fuzzy-weighted zero-inconsistency; QoS: Quality of service.

Addressing this gap, this study integrates HFLTS with the ARAS method, offering a systematic approach to evaluating urban mobility alternatives. The HFL-ARAS framework provides several advantages over traditional methods:

- Improved linguistic flexibility—Experts can express hesitation between multiple linguistic terms, enhancing decision-making accuracy.
- Enhanced suitability for urban mobility planning—HFL-ARAS is particularly suited for context-aware mobility decision-making, as it captures linguistic uncertainty more effectively.

This study builds on existing literature by explicitly addressing linguistic vagueness in urban mobility evaluation, bridging the research gap, and offering a more comprehensive decision-making framework.

In summary, the literature on smart cities and MCDM applications highlights a diverse range of approaches to sustainability, risk analysis, and resource efficiency. However, limited research integrates HFL approaches in urban mobility decision-making. By applying HFL-ARAS, this study contributes to structured, flexible, and expert-driven decision-making for urban mobility planning.

3.3 METHODOLOGY

In this study, a structured methodology is employed to evaluate urban mobility alternatives for smart cities using multi-criteria decision-making techniques. The methodology involves the identification of relevant alternatives and criteria, the application of HFLTS, and the implementation of the HFL-ARAS method for ranking and selection. The framework of the methodology is illustrated in Figure 3.1.

FIGURE 3.1 Framework of the proposed methodology

3.3.1 Hesitant Fuzzy Linguistic Term Sets (HFLTSs)

HFLTSs are first introduced by Rodríguez et al. (2012) to provide flexibility to decision makers in expressing assessments when uncertainty exists among multiple linguistic terms. In 2015, a redefinition of HFLTS in mathematical form is provided by Liao et al. (2015a), as follows:

Let $S=\{s_\alpha | s_0 \le s_\alpha \le s_g, \alpha \in [0,g]\}$ be a linguistic term set. An HFLTS, denoted as H_S, over X is expressed as $H_S=\{\langle x, h_S(x)\rangle | x \in X\}$. Here, L represents the number of linguistic terms in $h_S(x)$ and $s_{\varphi_l}(x)(for\, l=1,2,\ldots,L)$ are continuous terms within S. The set $h_S(x)$, representing possible degrees of a linguistic variable x in S, is defined as a hesitant fuzzy linguistic element (HFLE) and given by $h_S(x)=\{s_{\varphi_l}(x) | s_{\varphi_l}(x) \in S, \varphi_l \in [0,g], l=1,2,\ldots,L\}$. Once HFLEs are obtained, the lower and upper bounds for each HFLE can be derived, and subsequently, the HFLE can be encapsulated by its envelope. The envelope of an HFLE h_S is represented in Equation (3.1), where h_S^- and h_S^+ are the smallest and largest terms within h_S, respectively (Rodríguez et al., 2012; Liao et al., 2019).

$$env(h_S)=[h_S^-, h_S^+] \tag{3.1}$$

Table 3.2 provides the rules for converting linguistic terms to HFLEs and their envelopes. The score and variance of an HFLE h_S are given in Equations (3.2) and (3.3), respectively (Liao et al., 2019). Based on these concepts, Liao et al. developed an approach for comparing two different HFLEs, h_S^1 and h_S^2 (Liao et al., 2015b):

$$\rho(h_s)=\frac{1}{L}\sum_{s_{\varphi l}\in h_S} s_{\varphi_l} = s_{\frac{1}{L}\sum_{l=1}^{L}\varphi_l} \tag{3.2}$$

$$\sigma(h_s)=\frac{1}{L}\sqrt{\sum_{s_{\varphi l},\varphi k\in h_S}\left(s_{\varphi_l}-s_{\varphi_k}\right)^2} \tag{3.3}$$

TABLE 3.2
Transformation rules from linguistic terms to HFLEs and their envelopes (Rodríguez et al., 2012)

	Transformation rule (TR)	Envelope of HFLEs
TR_1	$TR_1(s_k)=\{s_k \| s_k \in S\}$	$env(h_S)=[s_k, s_k]$
TR_2	$TR_2(greater\,than\, s_i)=\{s_k \| s_k \in S\, ve\, s_k \ge s_i\}$	$env(h_S)=[s_i, s_\tau]$
TR_3	$TR_3(lower\,than\, s_j)=\{s_k \| s_k \in S\, ve\, s_k \le s_j\}$	$env(h_S)=[s_{-\tau}, s_j]$
TR_4	$TR_4(between\, s_i\, and\, s_j)=\{s_k \| s_k \in S\, ve\, s_i \le s_k \le s_j\}$	$env(h_S)=[s_i, s_j]$

According to these concepts:

- If $\rho\left(h_S^1\right) > \rho\left(h_S^2\right)$, then $h_S^1 > h_S^2$.
- If $\rho\left(h_S^1\right) = \rho\left(h_S^2\right)$ and $\sigma\left(h_S^1\right) > \sigma\left(h_S^2\right)$, then $h_S^1 < h_S^2$.
- If $\left(h_S^1\right) = \sigma\left(h_S^2\right)$, then $h_S^1 = h_S^2$.

Moreover, HFLEs can vary in the number of linguistic terms they contain. To address this, a supplementary adjustment is required (Zhu and Xu, 2014). To achieve uniformity, an additional term, as specified in Equation (3.4), should be added to a briefly expressed HFLE. In Equation (3.4), s^+ represents the maximum linguistic term in the briefly expressed HFLE and s^- the minimum. Additionally, $\zeta\left(0 \le \zeta \le 1\right)$ is a parameter set according to experts' risk preferences (Liao et al., 2019).

$$\bar{s} = \zeta s^+ + \left(1 - \zeta\right) s^- \tag{3.4}$$

3.3.2 Hesitant Fuzzy Linguistic ARAS (HFL-ARAS)

The ARAS method, originally proposed by Zavadskas and Turskis (2010), operates by identifying an ideal solution as a benchmark alternative. This ideal solution serves as a reference point against which all other alternatives are comparatively assessed. In this process, each alternative is evaluated by calculating a benefit value, which reflects its relative performance in relation to the ideal. The ARAS method thus facilitates decision-making by ranking alternatives based on their calculated benefit values. The relative efficiency of each alternative is influenced by the assigned weights of the criteria, indicating the importance of each criterion in the decision problem. By incorporating these weighted criteria, ARAS enables a systematic approach to selecting the most suitable alternative, taking into account the proportional impact of each criterion on the overall decision.

In this section, an extension of the ARAS method for the HFL environment is presented. Specifically, the HFL-ARAS method, as proposed by Liao et al. (2015a), is explained, where the values assigned to alternatives under each criterion are expressed as Hesitant Fuzzy Linguistic Elements (HFLEs).

The steps of the HFL-ARAS method are as follows (Liao et al., 2015a, 2019):

Step 1: *Defining the alternatives, the criteria, and the criteria weights.*
In this step, the criteria (where $i = 1, 2, \ldots, m$) according to which the alternatives will be evaluated are identified. Once the criteria (where $j = 1, 2, \ldots, n$) are selected, the importance of each criterion is determined. Here, w_j represents the weight of the j th criterion.

Step 2: *Constructing the hesitant fuzzy linguistic decision matrices.*

In this step, an HFL decision matrix is created for each expert or decision maker using HFLTS. $H_S^{U_k}$ denotes the HFL decision matrix constructed by the kth expert ($k = 1,2,\ldots,e$), where $h_s^{U_k}(x_{ij})$ is the performance value of the ith alternative under the jth criterion as expressed in HFLEs by the kth expert.

$$H_S^{U_k} : \begin{Bmatrix} h_s^{U_k}(x_{01}) & h_s^{U_k}(x_{02}) & \ldots & h_s^{U_k}(x_{0j}) & \ldots & h_s^{U_k}(x_{0n}) \\ h_s^{U_k}(x_{11}) & h_s^{U_k}(x_{12}) & \ldots & h_s^{U_k}(x_{1j}) & \ldots & h_s^{U_k}(x_{1n}) \\ \vdots & \vdots & \ddots & \vdots & \ddots & \vdots \\ h_s^{U_k}(x_{i1}) & h_s^{U_k}(x_{i2}) & \ldots & h_s^{U_k}(x_{ij}) & \ldots & h_s^{U_k}(x_{in}) \\ \vdots & \vdots & \ddots & \vdots & \ddots & \vdots \\ h_s^{U_k}(x_{m1}) & h_s^{U_k}(x_{m2}) & \ldots & h_s^{U_k}(x_{mj}) & \ldots & h_s^{U_k}(x_{mn}) \end{Bmatrix} \tag{3.5}$$

The ideal reference row (i.e., row $i = 0$) is determined by each decision maker for the best possible performance value under each criterion. If experts do not have specific opinions, Equations (3.6) and (3.7) are used to create the ideal row for benefit and cost criteria, respectively.

$$h_s^{U_k}(x_{0j}) = \max_i h_s^{U_k}(x_{ij}) \tag{3.6}$$

$$h_s^{U_k}(x_{0j}) = \min_i h_s^{U_k}(x_{ij}) \tag{3.7}$$

Step 3: *Normalizing HFL decision matrices.*

In this step, the decision matrices, created with HFLEs by each expert, are normalized, considering whether each criterion is benefit based or cost based. For benefit-based criteria, Equation (3.9) is used for normalization, while for cost-based criteria, Equation (3.10) is applied. Here, $\left(\bar{H}_S^{U_k}\right)$ represents the normalized HFL decision matrix of the kth expert, and $\bar{h}_s^{U_k}(x_{ij})$ denotes the normalized performance value of the ith alternative under the jth criterion.

$$\bar{H}_S^{U_k} : \begin{Bmatrix} \bar{h}_s^{U_k}(x_{01}) & \bar{h}_s^{U_k}(x_{02}) & \ldots & \bar{h}_s^{U_k}(x_{0j}) & \ldots & \bar{h}_s^{U_k}(x_{0n}) \\ \bar{h}_s^{U_k}(x_{11}) & \bar{h}_s^{U_k}(x_{12}) & \ldots & \bar{h}_s^{U_k}(x_{1j}) & \ldots & \bar{h}_s^{U_k}(x_{1n}) \\ \vdots & \vdots & \ddots & \vdots & \ddots & \vdots \\ \bar{h}_s^{U_k}(x_{i1}) & \bar{h}_s^{U_k}(x_{i2}) & \ldots & \bar{h}_s^{U_k}(x_{ij}) & \ldots & \bar{h}_s^{U_k}(x_{in}) \\ \vdots & \vdots & \ddots & \vdots & \ddots & \vdots \\ \bar{h}_s^{U_k}(x_{m1}) & \bar{h}_s^{U_k}(x_{m2}) & \ldots & \bar{h}_s^{U_k}(x_{mj}) & \ldots & \bar{h}_s^{U_k}(x_{mn}) \end{Bmatrix} \tag{3.8}$$

$$\bar{h}_s^{U_k}\left(x_{ij}\right)=\left\{s_{\frac{\varphi_l}{\sum_{i=1}^{m}\varphi_l}}\left(x_{ij}\right)\middle|s_{\varphi_l}\left(x_{ij}\right)\in h_s^{U_k}\left(x_{ij}\right)\right\} \tag{3.9}$$

$$\bar{h}_s^{U_k}\left(x_{ij}\right)=\left\{s_{\frac{1/\varphi_l}{\sum_{i=1}^{m}1/\varphi_l}}\left(x_{ij}\right)\middle|s_{\varphi_l}\left(x_{ij}\right)\in h_s^{U_k}\left(x_{ij}\right)\right\} \tag{3.10}$$

Step 4: *Aggregating normalized HFL decision matrices.*
In this step, the normalized HFL decision matrices from each expert are combined into a single HFL decision matrix $\left(H_S\right)$ using the generalized hesitant fuzzy linguistic weighted averaging (GHFLWA) operator, as proposed by Zhang and Wu (Zhang and Wu, 2014). Here, $h_s\left(x_{ij}\right)$is the aggregated and normalized performance value of the ith alternative under the jth criterion.

$$H_S:\left\{\begin{vmatrix} h_s\left(x_{01}\right) & h_s\left(x_{02}\right) & \cdots & h_s\left(x_{0j}\right) & \cdots & h_s\left(x_{0n}\right) \\ h_s\left(x_{11}\right) & h_s\left(x_{12}\right) & \cdots & h_s\left(x_{1j}\right) & \cdots & h_s\left(x_{1n}\right) \\ \vdots & \vdots & \ddots & \vdots & \ddots & \vdots \\ h_s\left(x_{i1}\right) & h_s\left(x_{i2}\right) & \cdots & h_s\left(x_{ij}\right) & \cdots & h_s\left(x_{in}\right) \\ \vdots & \vdots & \ddots & \vdots & \ddots & \vdots \\ h_s\left(x_{m1}\right) & h_s\left(x_{m2}\right) & \cdots & h_s\left(x_{mj}\right) & \cdots & h_s\left(x_{mn}\right) \end{vmatrix}\right\} \tag{3.11}$$

$$h_s\left(x_{ij}\right)=GHFLWA_{\ddot{e}}\left(\bar{h}_s^{U_1}\left(x_{ij}\right),\bar{h}_s^{U_2}\left(x_{ij}\right),\ldots,\bar{h}_s^{U_e}\left(x_{ij}\right)\right)=\left(\oplus_{k=1}^{e}\mu_{U_k}\bar{h}_s^{U_k}\left(x_{ij}\right)^{\lambda}\right)^{1/\lambda}=$$

$$\bigcup_{s_{\varphi_l}^{U_1}\left(x_{ij}\right)\in\bar{h}_s^{U_1}\left(x_{ij}\right),s_{\varphi_l}^{U_2}\left(x_{ij}\right)\in\bar{h}_s^{U_2}\left(x_{ij}\right),\ldots,s_{\varphi_l}^{U_e}\left(x_{ij}\right)\in\bar{h}_s^{U_e}\left(x_{ij}\right)}\left\{s_{\left(\sum_{k=1}^{e}\mu_{U_k}s_{\varphi_l}^{U_k}\left(x_{ij}\right)^{\lambda}\right)^{1/\lambda}}\right\},\lambda\in\left[0,1\right] \tag{3.12}$$

where μ_{U_k} represents the weight of the kth expert.

Step 5: *Weighting the aggregated HFL decision matrix.*
Using the criteria weights $\left(w_j\right)$ determined in Step 1 and the aggregated HFL decision matrix from Step 4, the weighted normalized HFL decision matrix $\left(\hat{H}_S\right)$ is obtained, as shown in Equation (3.14). Here, $\hat{h}_s\left(x_{ij}\right)$ is the weighted and normalized performance value of the ith alternative under the jth criterion.

$$\hat{H}_S : \begin{Bmatrix} \hat{h}_s(x_{01}) & \hat{h}_s(x_{02}) & \dots & \hat{h}_s(x_{0j}) & \dots & \hat{h}_s(x_{0n}) \\ \hat{h}_s(x_{11}) & \hat{h}_s(x_{12}) & \dots & \hat{h}_s(x_{1j}) & \dots & \hat{h}_s(x_{1n}) \\ \vdots & \vdots & \ddots & \vdots & \ddots & \vdots \\ \hat{h}_s(x_{i1}) & \hat{h}_s(x_{i2}) & \dots & \hat{h}_s(x_{ij}) & \dots & \hat{h}_s(x_{in}) \\ \vdots & \vdots & \ddots & \vdots & \ddots & \vdots \\ \hat{h}_s(x_{m1}) & \hat{h}_s(x_{m2}) & \dots & \hat{h}_s(x_{mj}) & \dots & \hat{h}_s(x_{mn}) \end{Bmatrix} \tag{3.13}$$

$$\hat{h}_s(x_{ij}) = w_j h_s(x_{ij}) \tag{3.14}$$

Step 6: *Calculating the total performance value for each alternative.*

In this step, the performance value $(\tilde{P}_i)$ of each alternative is calculated using Equation (3.15). Subsequently, the benefit degree (Q_i) of each alternative, which indicates its relative proximity to the ideal alternative, is obtained using Equation (3.16). The mean value of P_i is determined by averaging the subscripts of the linguistic terms in $\tilde{P}_i$.

$$\tilde{P}_i = \oplus_{j=1}^{n} \hat{h}_s(x_{ij}) = \left\{ s_{\sum_{l=1}^{L} \varphi_l} \middle| s_{\varphi_l} \in \hat{h}_s(x_{ij}) \right\} \tag{3.15}$$

$$Q_i = \frac{P_i}{P_0} \tag{3.16}$$

Step 7: *Ranking the alternatives.*

The alternatives are ranked in descending order based on their benefit degrees (Q_i). The alternative with the highest benefit degree is considered the most desirable or best-suited alternative for the intended purpose.

3.4 CASE STUDY

In this section, the HFL-ARAS method is applied to assess urban mobility solutions within the context of smart cities. The decision-making process is structured around six key criteria, and the analysis is conducted with the involvement of three field experts. The implementation consists of three main stages: identifying potential alternatives for urban mobility processes, evaluating these alternatives under six different criteria using the HFL-ARAS method, and proposing the most appropriate solutions. Accordingly, the application is carried out step by step following the HFL-ARAS method described in Section 3.3.2.

Step 1: *Defining the alternatives, the criteria, and the criteria weights.*
After examining urban mobility activities in smart cities and discussing the topic with experts in the field, six alternatives, as presented in Table 3.3, are identified. Subsequently, the key criteria for evaluating these alternatives are established, which are also listed in Table 3.4. These criteria serve as the foundation for the assessment process in the following stages.

In determining the criteria weights, the opinions of the three experts are consulted. Each expert ranked the six criteria according to their importance. The average of these expert evaluations is then calculated, and the rankings are normalized to a 0–1 scale. Based on this process, the criteria weights are established as follows:

$$w_1 : 0.206; w_2 : 0.270; w_3 : 0.238; w_4 : 0.111; w_5 : 0.127; \textit{and } w_6 : 0.048$$

Step 2: *Constructing the hesitant fuzzy linguistic decision matrices.*
In this step, HFL decision matrices are created for each expert using the HFLTSs provided in Table 3.5. Initially, the experts' evaluations are collected based on this HFLTS, and these evaluations are presented in Table 3.6.

TABLE 3.3
Urban mobility alternatives and their descriptions

Alternatives	Description
Self-driving vehicles (A1)	Self-driving vehicles (SDVs) are autonomous systems that rely on high-level automation, operating independently of human input by managing critical driving functions, such as steering, acceleration, and braking, and are typically classified at SAE automation level 4 or higher for use on public roadways (Yaqoob, et al., 2019; Poszler et al., 2023; URL-1).
Electric public transport (A2)	Electric public transport refers to public transportation systems that are powered by electricity rather than fossil fuels, offering a more sustainable and environmentally friendly alternative for urban and intercity travel (Glotz-Richter and Koch, 2016; Ribeiro et al., 2024).
Car-sharing systems (A3)	Car-sharing systems refer to technology-driven mobility services that enable multiple users to share vehicles on demand, reducing private car ownership, traffic congestion, and emissions (Das and Ahmed, 2025; Wang et al., 2024).
Smart bike-sharing systems (A4)	Smart bike-sharing systems are app-based services that let users rent and return bikes on demand, promoting eco-friendly and efficient urban transport (Midgley, 2009; Shir et al., 2023).
Autonomous drones (A5)	Autonomous drones are unmanned aerial vehicles (UAVs) that operate without human intervention, using AI and sensors to perform tasks such as surveillance, traffic monitoring, environmental data collection, and delivery services (Nguyen, 2021; Manju, 2021).
Intelligent traffic management systems (A6)	Intelligent traffic management systems are advanced solutions that use real-time data, sensors, and AI to optimize traffic flow, reduce congestion, and enhance road safety in urban areas (Li et al., 2017; Khanna et al., 2019).

TABLE 3.4
Criteria and their descriptions

Criteria	Description	References
Low-cost implementation (C1)	Low implementation and operating costs that are sustainable in the long term	Yucesan et al. (2024); da Silva et al. (2022); UNESCAP (2019)
Sustainability (C2)	The extent to which it reduces carbon emissions, its impact on the environment, and how sustainable its use of resources is	Yucesan et al. (2024); da Silva et al. (2022); Morfoulaki and Papathanasiou (2021); UNESCAP (2019)
User satisfaction (C3)	How well it can be adopted by users, its usability, and the level of comfort it provides	Yucesan et al. (2024); Morfoulaki and Papathanasiou (2021); Munhoz et al. (2020); UNESCAP (2019)
Ease of implementation (C4)	The ease of integration with existing urban infrastructure and its overall feasibility	Yucesan et al. (2024); Morfoulaki and Papathanasiou (2021); Ceder (2021); UNESCAP (2019)
Time efficiency (C5)	The extent to which it speeds up the transportation process and reduces waiting times	Morfoulaki and Papathanasiou (2021); Munhoz et al. (2020); UNESCAP (2019)
Ease of maintenance and operation (C6)	The frequency of maintenance required by the systems and how operational continuity will be ensured	Ceder (2021); Munhoz et al. (2020); Gallo and Marinelli (2020); Ozaki (2018)

TABLE 3.5
The set of linguistic terms considered for the HFL-Aras method (Liu et al., 2014)

Linguistic terms	s_i
Extremely high (EH)	s_8
Very high (VH)	s_7
High (H)	s_6
Slightly high (SH)	s_5
Medium (M)	s_4
Slightly low (SL)	s_3
Low (L)	s_2
Very low (VL)	s_1
Extremely low (EL)	s_0

TABLE 3.6
Decision matrix in terms of HFLTS for experts

Alternatives		Criteria					
		C1	C2	C3	C4	C5	C6
Expert 1	A1	L	M	H	SL	VH	L
	A2	M	At least VH	H	M	M	SL
	A3	Between SH and H	H	Between SH and H	H	SH	M
	A4	VH	VH	SH	H	M	SH
	A5	VL	M	M	Between SL and M	VH	VL
	A6	Between VL and L	H	H	SH	H	Between LandSL
Expert 2	A1	VL	H	SH	M	VH	L
	A2	Between SL and M	EH	H	H	H	SL
	A3	H	Between SH and H	H	VH	M	Between SLandM
	A4	VH	VH	SH	H	M	SH
	A5	Between VL and L	M	Between M and SH	M	EH	VL
	A6	L	H	EH	SH	H	M
Expert 3	A1	Between EL and VL	Between SH and H	H	Between M and SH	At least VH	L
	A2	M	VH	H	Between H and VH	M	M
	A3	VH	At least VH	H	EH	Between M and SH	H
	A4	Between SH and H	Between VH and EH	Between H and VH	VH	M	H
	A5	L	SH	M	L	EH	Between ELandVL
	A6	L	VH	VH	Between M and SH	Between H and VH	SL

Using the linguistic assessments from this table, HFL decision matrices are constructed for each expert. Subsequently, the complementary adjustment approach, described in Equation (3.4), is applied to convert the HFL decision matrices for each expert, with the ζ parameter set to 0.5. The resulting matrices are presented in Table 3.7.

TABLE 3.7
Transformed HFL decision matrix of experts

Alternatives		Criteria					
		C1	**C2**	**C3**	**C4**	**C5**	**C6**
Expert 1 (E1)	**A0**	$\{s_7, s_7, s_7\}$	$\{s_7, s_{7.5}, s_8\}$	$\{s_6, s_6, s_6\}$	$\{s_6, s_6, s_6\}$	$\{s_7, s_7, s_7\}$	$\{s_5, s_5, s_5\}$
	A1	$\{s_2, s_2, s_2\}$	$\{s_4, s_4, s_4\}$	$\{s_6, s_6, s_6\}$	$\{s_3, s_3, s_3\}$	$\{s_7, s_7, s_7\}$	$\{s_2, s_2, s_2\}$
	A2	$\{s_4, s_4, s_4\}$	$\{s_7, s_{7.5}, s_8\}$	$\{s_6, s_6, s_6\}$	$\{s_4, s_4, s_4\}$	$\{s_4, s_4, s_4\}$	$\{s_3, s_3, s_3\}$
	A3	$\{s_5, s_{5.5}, s_6\}$	$\{s_6, s_6, s_6\}$	$\{s_5, s_{5.5}, s_6\}$	$\{s_6, s_6, s_6\}$	$\{s_5, s_5, s_5\}$	$\{s_4, s_4, s_4\}$
	A4	$\{s_7, s_7, s_7\}$	$\{s_7, s_7, s_7\}$	$\{s_5, s_5, s_5\}$	$\{s_6, s_6, s_6\}$	$\{s_4, s_4, s_4\}$	$\{s_5, s_5, s_5\}$
	A5	$\{s_1, s_1, s_1\}$	$\{s_4, s_4, s_4\}$	$\{s_4, s_4, s_4\}$	$\{s_2, s_3, s_4\}$	$\{s_7, s_7, s_7\}$	$\{s_1, s_1, s_1\}$
	A6	$\{s_1, s_{1.5}, s_2\}$	$\{s_6, s_6, s_6\}$	$\{s_6, s_6, s_6\}$	$\{s_5, s_5, s_5\}$	$\{s_6, s_6, s_6\}$	$\{s_2, s_{2.5}, s_3\}$
Expert 2 (E2)	**A0**	$\{s_7, s_7, s_7\}$	$\{s_8, s_8, s_8\}$	$\{s_8, s_8, s_8\}$	$\{s_8, s_8, s_8\}$	$\{s_8, s_8, s_8\}$	$\{s_5, s_5, s_5\}$
	A1	$\{s_1, s_1, s_1\}$	$\{s_6, s_6, s_6\}$	$\{s_5, s_5, s_5\}$	$\{s_4, s_4, s_4\}$	$\{s_7, s_7, s_7\}$	$\{s_2, s_2, s_2\}$
	A2	$\{s_3, s_{3.5}, s_4\}$	$\{s_8, s_8, s_8\}$	$\{s_6, s_6, s_6\}$	$\{s_6, s_6, s_6\}$	$\{s_6, s_6, s_6\}$	$\{s_3, s_3, s_3\}$
	A3	$\{s_6, s_6, s_6\}$	$\{s_5, s_{5.5}, s_6\}$	$\{s_6, s_6, s_6\}$	$\{s_8, s_8, s_8\}$	$\{s_4, s_4, s_4\}$	$\{s_3, s_{3.5}, s_4\}$
	A4	$\{s_7, s_7, s_7\}$	$\{s_8, s_8, s_8\}$	$\{s_5, s_5, s_5\}$	$\{s_6, s_6, s_6\}$	$\{s_4, s_4, s_4\}$	$\{s_5, s_5, s_5\}$
	A5	$\{s_1, s_{1.5}, s_2\}$	$\{s_4, s_4, s_4\}$	$\{s_4, s_{4.5}, s_5\}$	$\{s_2, s_2, s_2\}$	$\{s_8, s_8, s_8\}$	$\{s_1, s_1, s_1\}$
	A6	$\{s_2, s_2, s_2\}$	$\{s_6, s_6, s_6\}$	$\{s_8, s_8, s_8\}$	$\{s_3, s_3, s_3\}$	$\{s_6, s_6, s_6\}$	$\{s_4, s_4, s_4\}$
Expert 3 (E3)	**A0**	$\{s_7, s_7, s_7\}$	$\{s_7, s_{7.5}, s_8\}$	$\{s_7, s_7, s_7\}$	$\{s_8, s_8, s_8\}$	$\{s_8, s_8, s_8\}$	$\{s_6, s_6, s_6\}$
	A1	$\{s_0, s_{0.5}, s_1\}$	$\{s_5, s_{5.5}, s_6\}$	$\{s_6, s_6, s_6\}$	$\{s_4, s_{4.5}, s_5\}$	$\{s_7, s_{7.5}, s_8\}$	$\{s_2, s_2, s_2\}$
	A2	$\{s_4, s_4, s_4\}$	$\{s_7, s_7, s_7\}$	$\{s_6, s_6, s_6\}$	$\{s_6, s_{6.5}, s_7\}$	$\{s_4, s_4, s_4\}$	$\{s_4, s_4, s_4\}$
	A3	$\{s_7, s_7, s_7\}$	$\{s_7, s_{7.5}, s_8\}$	$\{s_6, s_6, s_6\}$	$\{s_8, s_8, s_8\}$	$\{s_4, s_{4.5}, s_5\}$	$\{s_6, s_6, s_6\}$
	A4	$\{s_5, s_{5.5}, s_6\}$	$\{s_7, s_{7.5}, s_8\}$	$\{s_6, s_{6.5}, s_7\}$	$\{s_7, s_7, s_7\}$	$\{s_4, s_4, s_4\}$	$\{s_6, s_6, s_6\}$
	A5	$\{s_2, s_2, s_2\}$	$\{s_5, s_5, s_5\}$	$\{s_4, s_4, s_4\}$	$\{s_2, s_2, s_2\}$	$\{s_8, s_8, s_8\}$	$\{s_0, s_{0.5}, s_1\}$
	A6	$\{s_2, s_2, s_2\}$	$\{s_7, s_7, s_7\}$	$\{s_7, s_7, s_7\}$	$\{s_4, s_{4.5}, s_5\}$	$\{s_6, s_{6.5}, s_7\}$	$\{s_1, s_1, s_1\}$

Step 3: *Normalizing HFL decision matrices.*
Each decision maker's HFL decision matrix is normalized using Equation (3.6), considering each criterion as benefit based.

$$\bar{H}_s^{E_1}:\left\{\begin{array}{cccccc}
\{s_{0.259},s_{0.250},s_{0.241}\} & \{s_{0.170},s_{0.178},s_{0.186}\} & \{s_{0.157},s_{0.155},s_{0.153}\} & \{s_{0.187},s_{0.181},s_{0.176}\} & \{s_{0.175},s_{0.175},s_{0.175}\} & \{s_{0.227},s_{0.222},s_{0.217}\} \\
\{s_{0.074},s_{0.071},s_{0.069}\} & \{s_{0.097},s_{0.095},s_{0.093}\} & \{s_{0.157},s_{0.155},s_{0.153}\} & \{s_{0.093},s_{0.090},s_{0.088}\} & \{s_{0.175},s_{0.175},s_{0.175}\} & \{s_{0.090},s_{0.088},s_{0.086}\} \\
\{s_{0.148},s_{0.142},s_{0.137}\} & \{s_{0.170},s_{0.178},s_{0.186}\} & \{s_{0.157},s_{0.155},s_{0.153}\} & \{s_{0.125},s_{0.121},s_{0.117}\} & \{s_{0.100},s_{0.100},s_{0.100}\} & \{s_{0.136},s_{0.133},s_{0.130}\} \\
\{s_{0.185},s_{0.196},s_{0.206}\} & \{s_{0.146},s_{0.142},s_{0.139}\} & \{s_{0.131},s_{0.142},s_{0.153}\} & \{s_{0.187},s_{0.181},s_{0.176}\} & \{s_{0.125},s_{0.125},s_{0.125}\} & \{s_{0.181},s_{0.177},s_{0.173}\} \\
\{s_{0.259},s_{0.250},s_{0.241}\} & \{s_{0.170},s_{0.166},s_{0.162}\} & \{s_{0.131},s_{0.129},s_{0.128}\} & \{s_{0.187},s_{0.181},s_{0.176}\} & \{s_{0.100},s_{0.100},s_{0.100}\} & \{s_{0.227},s_{0.222},s_{0.217}\} \\
\{s_{0.037},s_{0.035},s_{0.034}\} & \{s_{0.097},s_{0.095},s_{0.093}\} & \{s_{0.105},s_{0.103},s_{0.102}\} & \{s_{0.062},s_{0.090},s_{0.117}\} & \{s_{0.175},s_{0.175},s_{0.175}\} & \{s_{0.045},s_{0.044},s_{0.043}\} \\
\{s_{0.037},s_{0.053},s_{0.068}\} & \{s_{0.146},s_{0.142},s_{0.139}\} & \{s_{0.157},s_{0.155},s_{0.153}\} & \{s_{0.156},s_{0.151},s_{0.147}\} & \{s_{0.150},s_{0.150},s_{0.150}\} & \{s_{0.090},s_{0.111},s_{0.130}\}
\end{array}\right\};$$

$$\bar{H}_S^{E_2}:\left\{\begin{array}{cccccc}
\{s_{0.259},s_{0.250},s_{0.241}\} & \{s_{0.195},s_{0.190},s_{0.186}\} & \{s_{0.210},s_{0.207},s_{0.205}\} & \{s_{0.250},s_{0.242},s_{0.235}\} & \{s_{0.200},s_{0.200},s_{0.200}\} & \{s_{0.227},s_{0.222},s_{0.217}\} \\
\{s_{0.037},s_{0.035},s_{0.034}\} & \{s_{0.146},s_{0.142},s_{0.139}\} & \{s_{0.131},s_{0.129},s_{0.128}\} & \{s_{0.125},s_{0.121},s_{0.117}\} & \{s_{0.175},s_{0.175},s_{0.175}\} & \{s_{0.090},s_{0.088},s_{0.086}\} \\
\{s_{0.111},s_{0.125},s_{0.137}\} & \{s_{0.195},s_{0.190},s_{0.186}\} & \{s_{0.157},s_{0.155},s_{0.153}\} & \{s_{0.187},s_{0.181},s_{0.176}\} & \{s_{0.150},s_{0.150},s_{0.150}\} & \{s_{0.136},s_{0.133},s_{0.130}\} \\
\{s_{0.222},s_{0.214},s_{0.206}\} & \{s_{0.121},s_{0.130},s_{0.139}\} & \{s_{0.157},s_{0.155},s_{0.153}\} & \{s_{0.250},s_{0.242},s_{0.235}\} & \{s_{0.100},s_{0.100},s_{0.100}\} & \{s_{0.136},s_{0.155},s_{0.173}\} \\
\{s_{0.259},s_{0.250},s_{0.241}\} & \{s_{0.195},s_{0.190},s_{0.186}\} & \{s_{0.131},s_{0.129},s_{0.128}\} & \{s_{0.187},s_{0.181},s_{0.176}\} & \{s_{0.100},s_{0.100},s_{0.100}\} & \{s_{0.227},s_{0.222},s_{0.217}\} \\
\{s_{0.037},s_{0.053},s_{0.068}\} & \{s_{0.097},s_{0.095},s_{0.093}\} & \{s_{0.105},s_{0.116},s_{0.128}\} & \{s_{0.062},s_{0.060},s_{0.058}\} & \{s_{0.200},s_{0.200},s_{0.200}\} & \{s_{0.045},s_{0.044},s_{0.043}\} \\
\{s_{0.074},s_{0.071},s_{0.068}\} & \{s_{0.146},s_{0.142},s_{0.139}\} & \{s_{0.210},s_{0.207},s_{0.205}\} & \{s_{0.093},s_{0.090},s_{0.088}\} & \{s_{0.150},s_{0.150},s_{0.150}\} & \{s_{0.181},s_{0.177},s_{0.173}\}
\end{array}\right\};$$

$$\bar{H}_S^{E_3}:\left\{\begin{array}{llll}
\{s_{0.259},s_{0.250},s_{0.241}\} & \{s_{0.170},s_{0.178},s_{0.186}\} & \{s_{0.184},s_{0.181},s_{0.179}\} & \{s_{0.250},s_{0.242},s_{0.235}\}\\
\{s_{0.000},s_{0.017},s_{0.034}\} & \{s_{0.121},s_{0.130},s_{0.139}\} & \{s_{0.157},s_{0.155},s_{0.153}\} & \{s_{0.125},s_{0.136},s_{0.147}\}\\
\{s_{0.148},s_{0.142},s_{0.137}\} & \{s_{0.170},s_{0.166},s_{0.162}\} & \{s_{0.157},s_{0.155},s_{0.153}\} & \{s_{0.187},s_{0.196},s_{0.205}\}\\
\{s_{0.259},s_{0.250},s_{0.241}\} & \{s_{0.170},s_{0.178},s_{0.186}\} & \{s_{0.157},s_{0.155},s_{0.153}\} & \{s_{0.250},s_{0.242},s_{0.235}\}\\
\{s_{0.185},s_{0.196},s_{0.206}\} & \{s_{0.170},s_{0.178},s_{0.186}\} & \{s_{0.157},s_{0.168},s_{0.179}\} & \{s_{0.218},s_{0.212},s_{0.205}\}\\
\{s_{0.074},s_{0.071},s_{0.068}\} & \{s_{0.121},s_{0.119},s_{0.116}\} & \{s_{0.105},s_{0.103},s_{0.102}\} & \{s_{0.062},s_{0.060},s_{0.058}\}\\
\{s_{0.074},s_{0.071},s_{0.068}\} & \{s_{0.170},s_{0.166},s_{0.162}\} & \{s_{0.184},s_{0.181},s_{0.179}\} & \{s_{0.125},s_{0.136},s_{0.147}\}
\end{array}\right.$$

$$\left.\begin{array}{ll}
\{s_{0.200},s_{0.200},s_{0.200}\} & \{s_{0.272},s_{0.266},s_{0.260}\}\\
\{s_{0.175},s_{0.187},s_{0.200}\} & \{s_{0.090},s_{0.088},s_{0.086}\}\\
\{s_{0.100},s_{0.100},s_{0.100}\} & \{s_{0.181},s_{0.177},s_{0.173}\}\\
\{s_{0.100},s_{0.112},s_{0.125}\} & \{s_{0.272},s_{0.266},s_{0.260}\}\\
\{s_{0.100},s_{0.100},s_{0.100}\} & \{s_{0.272},s_{0.266},s_{0.260}\}\\
\{s_{0.200},s_{0.200},s_{0.200}\} & \{s_{0.000},s_{0.022},s_{0.043}\}\\
\{s_{0.150},s_{0.162},s_{0.175}\} & \{s_{0.045},s_{0.044},s_{0.043}\}
\end{array}\right\}$$

Step 4: *Aggregating normalized HFL decision matrices.*

Each expert's normalized HFL decision matrices are aggregated into a single HFL decision matrix using the generalized hesitant fuzzy linguistic weighted averaging (GHFLWA) operator, as expressed in Equation (3.12). Here, $\in[0,1]$ is defined, and in practice, the value of λ is taken as 0.4.

In the aggregation process, the first decision maker is regarded as the team leader due to her extensive expertise in sustainable urban mobility planning (SUMP). With over 20 years of professional experience as a consultant in urban mobility and transportation systems, she has been actively involved in policy development, large-scale urban mobility projects, and strategic planning initiatives. Consequently, her opinion is assigned a greater weight in the decision-making process compared to the other decision makers, who are academics specializing in urban mobility and transportation planning. The remaining two experts, while highly qualified, primarily contribute from an academic research perspective rather than direct implementation experience. Therefore, the expert weights are assigned as follows: $\mu_{U_1}=0.50;\mu_{U_2}=0.25;\mu_{U_3}=0.25$. The normalized HFL decision matrices are then aggregated into a single matrix as shown below.

$$\hat{H}_S : \left\{ \begin{array}{llll} \{s_{0.259}, s_{0.250}, s_{0.241}\} & \{s_{0.176}, s_{0.181}, s_{0.186}\} & \{s_{0.176}, s_{0.174}, s_{0.172}\} & \{s_{0.217}, s_{0.210}, s_{0.204}\} \\ \{s_{0.029}, s_{0.045}, s_{0.049}\} & \{s_{0.114}, s_{0.114}, s_{0.114}\} & \{s_{0.151}, s_{0.149}, s_{0.147}\} & \{s_{0.108}, s_{0.108}, s_{0.108}\} \\ \{s_{0.138}, s_{0.138}, s_{0.137}\} & \{s_{0.176}, s_{0.178}, s_{0.180}\} & \{s_{0.157}, s_{0.155}, s_{0.153}\} & \{s_{0.154}, s_{0.152}, s_{0.151}\} \\ \{s_{0.211}, s_{0.213}, s_{0.215}\} & \{s_{0.145}, s_{0.148}, s_{0.150}\} & \{s_{0.144}, s_{0.149}, s_{0.153}\} & \{s_{0.217}, s_{0.210}, s_{0.204}\} \\ \{s_{0.239}, s_{0.235}, s_{0.232}\} & \{s_{0.176}, s_{0.175}, s_{0.174}\} & \{s_{0.137}, s_{0.139}, s_{0.140}\} & \{s_{0.195}, s_{0.189}, s_{0.183}\} \\ \{s_{0.044}, s_{0.047}, s_{0.049}\} & \{s_{0.103}, s_{0.100}, s_{0.098}\} & \{s_{0.105}, s_{0.107}, s_{0.108}\} & \{s_{0.062}, s_{0.074}, s_{0.085}\} \\ \{s_{0.053}, s_{0.062}, s_{0.068}\} & \{s_{0.152}, s_{0.148}, s_{0.145}\} & \{s_{0.176}, s_{0.174}, s_{0.172}\} & \{s_{0.131}, s_{0.131}, s_{0.130}\} \end{array} \right.$$

$$\left. \begin{array}{ll} \{s_{0.187}, s_{0.187}, s_{0.187}\} & \{s_{0.238}, s_{0.232}, s_{0.227}\} \\ \{s_{0.175}, s_{0.178}, s_{0.181}\} & \{s_{0.090}, s_{0.088}, s_{0.086}\} \\ \{s_{0.111}, s_{0.111}, s_{0.111}\} & \{s_{0.147}, s_{0.143}, s_{0.140}\} \\ \{s_{0.112}, s_{0.115}, s_{0.118}\} & \{s_{0.189}, s_{0.191}, s_{0.193}\} \\ \{s_{0.100}, s_{0.100}, s_{0.100}\} & \{s_{0.238}, s_{0.232}, s_{0.227}\} \\ \{s_{0.187}, s_{0.187}, s_{0.187}\} & \{s_{0.022}, s_{0.038}, s_{0.043}\} \\ \{s_{0.150}, s_{0.153}, s_{0.156}\} & \{s_{0.095}, s_{0.104}, s_{0.112}\} \end{array} \right\}$$

Step 5: *Weighting the aggregated HFL decision matrix.*

Taking into account the criteria weights determined in Step 1 and the aggregated HFL decision matrix obtained in Step 4, the weighted normalized HFL decision matrix is derived as expressed in Equation (3.14), as shown below.

$$\hat{H}_S : \left\{ \begin{array}{llll} \{s_{0.053}, s_{0.051}, s_{0.049}\} & \{s_{0.047}, s_{0.048}, s_{0.050}\} & \{s_{0.042}, s_{0.041}, s_{0.041}\} & \{s_{0.024}, s_{0.023}, s_{0.022}\} \\ \{s_{0.006}, s_{0.009}, s_{0.010}\} & \{s_{0.030}, s_{0.031}, s_{0.031}\} & \{s_{0.035}, s_{0.035}, s_{0.035}\} & \{s_{0.012}, s_{0.012}, s_{0.012}\} \\ \{s_{0.028}, s_{0.028}, s_{0.028}\} & \{s_{0.047}, s_{0.048}, s_{0.048}\} & \{s_{0.037}, s_{0.037}, s_{0.036}\} & \{s_{0.017}, s_{0.017}, s_{0.016}\} \\ \{s_{0.043}, s_{0.044}, s_{0.044}\} & \{s_{0.039}, s_{0.039}, s_{0.040}\} & \{s_{0.034}, s_{0.035}, s_{0.036}\} & \{s_{0.024}, s_{0.023}, s_{0.022}\} \\ \{s_{0.049}, s_{0.048}, s_{0.047}\} & \{s_{0.047}, s_{0.047}, s_{0.047}\} & \{s_{0.032}, s_{0.033}, s_{0.033}\} & \{s_{0.021}, s_{0.021}, s_{0.020}\} \\ \{s_{0.009}, s_{0.009}, s_{0.010}\} & \{s_{0.027}, s_{0.027}, s_{0.026}\} & \{s_{0.025}, s_{0.025}, s_{0.025}\} & \{s_{0.006}, s_{0.008}, s_{0.009}\} \\ \{s_{0.011}, s_{0.012}, s_{0.014}\} & \{s_{0.041}, s_{0.040}, s_{0.039}\} & \{s_{0.042}, s_{0.041}, s_{0.041}\} & \{s_{0.014}, s_{0.014}, s_{0.014}\} \end{array} \right.$$

$$\left. \begin{array}{ll} \{s_{0.023}, s_{0.023}, s_{0.023}\} & \{s_{0.011}, s_{0.011}, s_{0.010}\} \\ \{s_{0.022}, s_{0.022}, s_{0.022}\} & \{s_{0.004}, s_{0.004}, s_{0.004}\} \\ \{s_{0.014}, s_{0.014}, s_{0.014}\} & \{s_{0.007}, s_{0.006}, s_{0.006}\} \\ \{s_{0.014}, s_{0.014}, s_{0.015}\} & \{s_{0.009}, s_{0.009}, s_{0.009}\} \\ \{s_{0.012}, s_{0.012}, s_{0.012}\} & \{s_{0.011}, s_{0.011}, s_{0.010}\} \\ \{s_{0.023}, s_{0.023}, s_{0.023}\} & \{s_{0.001}, s_{0.001}, s_{0.002}\} \\ \{s_{0.019}, s_{0.019}, s_{0.019}\} & \{s_{0.004}, s_{0.004}, s_{0.005}\} \end{array} \right\}$$

Step 6: *Calculating the total performance value for each alternative*

In this step, the performance value of each alternative $\left(\tilde{P}_i\right)$ is determined as shown below using Equation (3.15).

$$\tilde{P}_i : \left\{ \begin{array}{l} \left\{ s_{0.202}, s_{0.200}, s_{0.198} \right\} \\ \left\{ s_{0.111}, s_{0.114}, s_{0.115} \right\} \\ \left\{ s_{0.152}, s_{0.151}, s_{0.151} \right\} \\ \left\{ s_{0.164}, s_{0.166}, s_{0.168} \right\} \\ \left\{ s_{0.175}, s_{0.173}, s_{0.172} \right\} \\ \left\{ s_{0.093}, s_{0.096}, s_{0.098} \right\} \\ \left\{ s_{0.132}, s_{0.133}, s_{0.134} \right\} \end{array} \right\}$$

Step 7: *Ranking the alternatives*

Subsequently, Equation (3.16) is used to obtain the proximity value of each alternative relative to the performance value of the ideal reference alternative, and the results are presented in Table 3.8.

The alternatives are ranked in descending order based on their benefit degrees $\left(Q_i\right)$ in Table 3.8. The urban mobility alternative with the highest benefit degree $\left(T_2\right)$ is considered the top priority in terms of importance.

3.5 RESULTS AND DISCUSSION

The evaluation of urban mobility alternatives within smart cities is crucial in addressing the pressing challenges of urbanization, sustainability, and efficiency in transportation systems. As cities expand and populations grow, the need for innovative, sustainable, and user-friendly mobility solutions becomes increasingly essential. In this study, the HFL-ARAS method is used to assess six urban mobility

TABLE 3.8
Ranking of the alternatives

Alternatives	P_i	Q_i	*Ranking*
A_0	0.200	1.000	—
A_1	0.114	0.569	5
A_2	0.152	0.757	3
A_3	0.167	0.832	2
A_4	0.174	0.868	1
A_5	0.096	0.480	6
A_6	0.133	0.665	4

alternatives—self-driving vehicles, electric public transport, car-sharing systems, smart bike-sharing systems, autonomous drones, and intelligent traffic management systems—across a set of six criteria: low-cost implementation, sustainability, user satisfaction, ease of implementation, time efficiency, and ease of maintenance and operation.

Based on the rankings obtained through the HFL-ARAS method, smart bike-sharing systems (A4) emerged as the most favorable alternative, achieving the highest benefit degree (Q_i= 0.868). This outcome indicates that, among the evaluated alternatives, smart bike-sharing systems offer a balanced performance across the criteria, particularly excelling in sustainability and user satisfaction, which are critical for sustainable urban mobility. Following smart bike-sharing systems, car-sharing systems (A3) and electric public transport (A2) ranked second and third, respectively, demonstrating significant potential in terms of sustainability and ease of integration within existing urban infrastructures.

The results emphasize that sustainability, cost-effectiveness, and user satisfaction play a decisive role in determining the most suitable urban mobility alternatives. These findings indicate that future urban mobility strategies should prioritize environmentally friendly, user-adaptive, and economically feasible solutions to align with smart city development goals. Moreover, the structured approach employed in this study provides a replicable framework for urban planners and policymakers, enabling them to make data-driven mobility decisions that balance technological innovation with environmental and social considerations. By integrating hesitant fuzzy linguistic assessments within the ARAS method, this study offers a more nuanced and flexible evaluation framework, which can be applied to a wide range of urban mobility planning contexts.

Overall, this research contributes to the growing discourse on smart city mobility solutions, demonstrating the effectiveness of an integrated decision-making approach in evaluating alternative transportation strategies. The proposed methodology enhances decision-making precision under linguistic uncertainty, offering valuable insights for future implementations within the evolving landscape of smart cities.

REFERENCES

Ahad, M. A., Paiva, S., Tripathi, G., & Feroz, N. (2020). Enabling technologies and sustainable smart cities. *Sustainable Cities and Society*, *61*, 102301.

Alsattar, H. A., Qahtan, S., Mourad, N., Zaidan, A. A., Deveci, M., Jana, C., & Ding, W. (2023). Three-way decision-based conditional probabilities by opinion scores and Bayesian rules in circular-Pythagorean fuzzy sets for developing sustainable smart living framework. *Information Sciences*, *649*, 119681.

Andreou, A., Mavromoustakis, C. X., Markakis, E., & Song, H. (2023). On the integration of user preferences by using a hybrid methodology for multi-criteria decision making. *IEEE Access*, *11*, 139157–139170.

Çakır, E., & Demircioğlu, E. (2023, August). Circular intuitionistic Fuzzy PROMETHEE methodology: A case of smart cities evaluation. In *International Conference on Intelligent and Fuzzy Systems* (pp. 353–361). Cham: Springer Nature Switzerland.

Ceder, A. (2021). Urban mobility and public transport: Future perspectives and review. *International Journal of Urban Sciences*, *25*(4), 455–479.

Da Silva, R. R., Santos, G. D., & Setti, D. (2022). A multi-criteria approach for urban mobility project selection in medium-sized cities. *Sustainable Cities and Society*, *86*, 104096.

Das, P., & Ahmed, M. A. (2025). Factors influencing the car-sharing system through structural equation model in Guwahati city, Assam. *Transportation Planning and Technology*, *48*(2), 342–365.

Demircan, B. G., & Yetilmezsoy, K. (2023). A hybrid fuzzy AHP-TOPSIS approach for implementation of smart sustainable waste management strategies. *Sustainability*, *15*(8), 6526.

Eren, E., & Katanalp, B. Y. (2022). Fuzzy-based GIS approach with new MCDM method for bike-sharing station site selection according to land-use types. *Sustainable Cities and Society*, *76*, 103434.

Gallo, M., & Marinelli, M. (2020). Sustainable mobility: A review of possible actions and policies. *Sustainability*, *12*(18), 7499.

Glotz-Richter, M., & Koch, H. (2016). Electrification of public transport in cities (Horizon 2020 ELIPTIC Project). *Transportation Research Procedia*, *14*, 2614–2619.

Gündoğdu, F. K. (2022). Analyzing critical barriers of smart energy city in Turkey based on two-dimensional uncertainty by hesitant z-fuzzy linguistic terms. *Engineering Applications of Artificial Intelligence*, *113*, 104935.

Hanine, M., Boutkhoum, O., El Barakaz, F., Lachgar, M., Assad, N., Rustam, F., & Ashraf, I. (2021). An intuitionistic fuzzy approach for smart city development evaluation for developing countries: Moroccan context. *Mathematics*, *9*(21), 2668.

Ibrahim, H. A., Qahtan, S., Zaidan, A. A., Deveci, M., Hajiaghaei-Keshteli, M., Mohammed, R. T., & Alamoodi, A. H. (2024). Sustainability in mobility for autonomous vehicles over smart city evaluation; using interval-valued Fermatean fuzzy rough set-based decision-making model. *Engineering Applications of Artificial Intelligence*, *129*, 107609.

Khanna, A., Goyal, R., Verma, M., & Joshi, D. (2019). Intelligent traffic management system for smart cities. In *Futuristic Trends in Network and Communication Technologies: First International Conference, FTNCT 2018, Solan, India, February 9–10, 2018, Revised Selected Papers 1* (pp. 152–164). Springer Singapore.

Krishankumar, R., Pamucar, D., Deveci, M., & Ravichandran, K. S. (2021). Prioritization of zero-carbon measures for sustainable urban mobility using integrated double hierarchy decision framework and EDAS approach. *Science of The Total Environment*, *797*, 149068.

Li, Z., Al Hassan, R., Shahidehpour, M., Bahramirad, S., & Khodaei, A. (2017). A hierarchical framework for intelligent traffic management in smart cities. *IEEE Transactions on Smart Grid*, *10*(1), 691–701.

Liao, H. C., Wen, Z., & Liu, L. (2019). Integrating BWM and ARAS under hesitant linguistic environment for digital supply chain finance supplier section. *Technological and Economic Development of Economy*, *25*(6), 1188–1212.

Liao, H. C., Xu, Z. S., & Zeng, X. J. (2015b). Hesitant fuzzy linguistic VIKOR method and its application in qualitative multiple criteria decision making. *IEEE Transactions on Fuzzy Systems*, *23*(5), 1343–1355.

Liao, H. C., Xu, Z. S., Zeng, X. J., & Merigó, J. M. (2015a). Qualitative decision making with correlation coefficients of hesitant fuzzy linguistic term sets. *Knowledge-Based Systems*, *76*, 127–138.

Liu, H., Cai, J., & Jiang, L. (2014). On improving the additive consistency of the fuzzy preference relations based on comparative linguistic expressions. *International Journal of Intelligent Systems*, *29*(6), 544–559.

Lyu, H. M., Yin, Z. Y., Zhou, A., & Shen, S. L. (2023). MCDM-based flood risk assessment of metro systems in smart city development: A review. *Environmental Impact Assessment Review*, *101*, 107154.

Maldonado Silveira Alonso Munhoz, P. A., da Costa Dias, F., Kowal Chinelli, C., Azevedo Guedes, A. L., Neves dos Santos, J. A., da Silveira e Silva, W., & Pereira Soares, C. A. (2020). Smart mobility: The main drivers for increasing the intelligence of urban mobility. *Sustainability*, *12*(24), 10675.

Manirathinam, T., Narayanamoorthy, S., Geetha, S., Ahmadian, A., Ferrara, M., & Kang, D. (2024). Assessing performance and satisfaction of micro-mobility in smart cities for sustainable clean energy transportation using novel APPRESAL method. *Journal of Cleaner Production*, *436*, 140372.

Manju, P., Pooja, D., & Dutt, V. (2021). Drones in smart cities. In A. K. Dubey, A. Kumar, S. R. Kumar, N. Gayathri, & P. Das (Eds.), *AI and IoT-Based Intelligent Automation in Robotics*, (pp. 205–228). https://doi.org/10.1002/9781119711230.ch12

Midgley, P. (2009). The role of smart bike-sharing systems in urban mobility. *Journeys*, *2*(1), 23–31.

Mishra, A. R., Rani, P., Saha, A., Hezam, I. M., Pamucar, D., Marinović, M., & Pandey, K. (2022). Assessing the adaptation of Internet of Things (IoT) barriers for smart cities' waste management using Fermatean fuzzy combined compromise solution approach. *IEEE Access*, *10*, 37109–37130.

Miskolczi, M., Földes, D., Munkacsy, A., & Jaszberenyi, M. (2021). Urban mobility scenarios until the 2030s. *Sustainable Cities and Society*, *72*, 103029.

Morfoulaki, M., & Papathanasiou, J. (2021). Use of PROMETHEE MCDA method for ranking alternative measures of sustainable urban mobility planning. *Mathematics*, *9*(6), 602.

Nguyen, D. D., Rohacs, J., & Rohacs, D. (2021). Autonomous flight trajectory control system for drones in smart city traffic management. *ISPRS International Journal of Geo-Information*, *10*(5), 338.

Otay, İ., Onar, S. Ç., Öztayşi, B., & Kahraman, C. (2024). Evaluation of sustainable energy systems in smart cities using a Multi-Expert Pythagorean fuzzy BWM & TOPSIS methodology. *Expert Systems with Applications*, *250*, 123874.

Ozaki, H. (2018). Technical standardization of ITS and Asian initiatives for intelligent mobility. *IATSS Research*, *42*(2), 72–75.

Pandya, S., Srivastava, G., Jhaveri, R., Babu, M. R., Bhattacharya, S., Maddikunta, P. K. R., ... & Gadekallu, T. R. (2023). Federated learning for smart cities: A comprehensive survey. *Sustainable Energy Technologies and Assessments*, *55*, 102987.

Poszler, F., Geisslinger, M., Betz, J., & Lütge, C. (2023). Applying ethical theories to the decision-making of self-driving vehicles: A systematic review and integration of the literature. *Technology in Society*, *75*, 102350.

Rani, R., & Potika, K. (2024). Smart city wildfire risk analysis with fuzzy multi-criteria decision making. *International Journal of Semantic Computing*, *18*(03), 347–381.

Ribeiro, P. J., Dias, G., & Mendes, J. F. (2024). Public transport decarbonization: An exploratory approach to bus electrification. *World Electric Vehicle Journal*, *15*(3), 81.

Richter, M. A., Hagenmaier, M., Bandte, O., Parida, V., & Wincent, J. (2022). Smart cities, urban mobility and autonomous vehicles: How different cities needs different sustainable investment strategies. *Technological Forecasting and Social Change*, *184*, 121857.

Rodríguez, R. M., Martínez, L., & Herrera, F. (2012). Hesitant fuzzy linguistic term sets for decision making. *IEEE Transactions on Fuzzy Systems*, *20*(1), 109–119.

Rodríguez, R. M., Martínez, L., & Herrera, F. (2013). A group decision making model dealing with comparative linguistic expressions based on hesitant fuzzy linguistic term sets. *Information Sciences*, *241*, 28–42.

Shao, Q. G., Jiang, C. C., Lo, H. W., & Liou, J. J. (2023). Establishing a sustainable development assessment framework for a smart city using a hybrid Z-fuzzy-based decision-making approach. *Clean Technologies and Environmental Policy*, *25*(9), 3027–3044.

Shir, B., Prakash Verma, J., & Bhattacharya, P. (2023). Mobility prediction for uneven distribution of bikes in bike sharing systems. *Concurrency and Computation: Practice and Experience*, *35*(2), e7465.

Standing, C., Standing, S., & Biermann, S. (2019). The implications of the sharing economy for transport. *Transport Reviews*, *39*(2), 226–242.

Torra, V. (2010). Hesitant fuzzy sets. *International Journal of Intelligent Systems*, *25*, 529–539.

UNESCAP. (2019). Data collection guideline of SUTI. https://www.unescap.org/sites/default/files/SUTI%20Data%20Collection%20Guideline%20_%202019%20update.pdf (Accessed 20 February 2023).

URL-1. https://www.sae.org/standards/content/j3016_202104/

Wang, N., Tian, H., Guo, J., Wu, H., & Lyu, J. (2024). Efficiency-oriented vehicle relocation of shared autonomous electric fleet in station-based car-sharing system. *Journal of Traffic and Transportation Engineering (English Edition)*, *11*(4), 681–699.

Yaqoob, I., Khan, L. U., Kazmi, S. A., Imran, M., Guizani, N., & Hong, C. S. (2019). Autonomous driving cars in smart cities: Recent advances, requirements, and challenges. *IEEE Network*, *34*(1), 174–181.

Yörükoğlu, M., & Aydın, S. (2021, August). Assessment of smart waste management systems with spherical AHP method. In *IFIP International Workshop on Artificial Intelligence for Knowledge Management* (pp. 32–43). Cham: Springer International Publishing.

Yucesan, M., Özkan, B., Mete, S., Gul, M., & Özceylan, E. (2024). Evaluating sustainability of urban mobility of Asian cities: An integrated approach of interval type-2 fuzzy best-worst method and MULTIMOORA. *Engineering Applications of Artificial Intelligence*, *127*, 107266.

Zavadskas, E. K., & Turskis, Z. (2010). A new Additive Ratio Assessment (ARAS) method in multicriteria decision-making. *Technological and Economic Development of Economy*, *16*(2), 159–172.

Zhang, Z., & Wu, C. (2014). Hesitant fuzzy linguistic aggregation operators and their applications to multiple attribute group decision making. *Journal of Intelligent & Fuzzy Systems*, *26*, 2185–2202.

Zhu, B., & Xu, Z. S. (2014). Consistency measures for hesitant fuzzy linguistic preference relations. *IEEE Transactions on Fuzzy Systems*, *22*, 35–45.

4 Choquet Integral-Based q-ROF Entropy and Its Application to Information Technologies

Mahmut Can Bozyiğit and Mehmet Ünver

4.1 INTRODUCTION

Fuzzy sets (FSs), introduced by Zadeh (1965), extend the concept of crisp sets to enable the representation of uncertainty and vagueness in data. Formally, a fuzzy set A in a universe of discourse X is defined by a membership function $\mu_A \to [0,1]$, where $\mu_A(x)$ represents the degree of membership of an element x in A. This framework is particularly effective in dealing with uncertainty or ambiguity in data. Intuitionistic fuzzy sets (IFSs), introduced by Atanassov (1999), extend FSs to address their limitations in representing uncertainty and vagueness. While FSs only accommodate partial membership, IFSs define each element with two components: a membership degree and a non-membership degree. Unlike FSs, where the sum of membership and non-membership degrees for each element equals 1, IFSs allow this sum to be less than or equal to 1, thereby offering greater flexibility in modeling uncertainty. Pythagorean fuzzy sets (PFSs) further build upon the concept of IFSs (Yager, 2013a, b). In this framework, each element is characterized by a membership function and a non-membership function, with the constraint that the sum of their squares does not exceed 1. This additional flexibility makes PFSs suitable for more complex uncertainty modeling. q-rung orthopair fuzzy sets (q-ROFSs), proposed by Yager (2016), generalize both IFSs and PFSs. In a q-ROFS, each element is characterized by two values: a membership degree and a non-membership degree, which must satisfy the condition that the q-th power sum of these degrees does not exceed 1, where $q \geq 1$ is a fixed real number. By accommodating higher levels of uncertainty, q-ROFSs provide a more versatile framework with applications in diverse fields such as decision-making, pattern recognition, medical diagnosis, and clustering analysis (Bozyiğit et al., 2023, 2024; Li et al., 2020; Peng and Liu, 2019; Wei et al., 2018; Yager, 2016).

Multi-criteria decision-making (MCDM) is a branch of operations research that addresses complex problems involving multiple, often conflicting criteria. When uncertainty and vagueness are inherent in the criteria and alternatives, MCDM

DOI: 10.1201/9781003529088-4

based on q-ROFSs offers a powerful approach to better represent such imprecisions. The use of q-ROFSs enables decision makers to handle uncertainty more effectively, providing a more realistic representation of their preferences and judgments to support informed decision-making. Liu and Wang (2018) developed several q-rung orthopair fuzzy (q-ROF) aggregation operators and demonstrated their applicability in solving MCDM problems. Liu et al. (2018) further advanced this field by introducing the q-ROF extended Bonferroni mean operator, its weighted variant, and a knowledge-based entropy measure for q-ROFSs, which they applied to an MCDM scenario. Wang and Li (2018) proposed the q-ROF power-weighted average operator and used it in a green supplier selection problem. Furthermore, Wei et al. (2018) introduced the generalized Heronian mean and geometric Heronian mean operators for q-ROFSs, using them to evaluate MCDM challenges, specifically focusing on enterprise resource planning system selection. To improve this framework, the Choquet integral, a mathematical operator introduced by Choquet (1954), plays a critical role. The Choquet integral generalizes the arithmetic mean and is particularly advantageous in scenarios where decision maker preferences involve interaction and dependence among criteria. Unlike traditional linear aggregation methods, it allows for non-linear combination of criteria, accounting for their interdependencies and varying importance. Integrating the Choquet integral into MCDM based on q-ROFSs provides a more flexible and comprehensive method for aggregating preferences, improving the decision-making process. This combined approach has found applications in diverse fields such as decision theory, artificial intelligence, and economics (Bhowal et al., 2021; Grabisch and Roubens, 2000; Heilpern, 2002), demonstrating its effectiveness in tackling complex decision-making scenarios.

Mobile edge caching (MEC) is a technology and strategy designed to improve the performance and efficiency of content delivery in mobile networks (Zhang, 2022). It involves storing frequently accessed content, such as videos, images, applications, and web pages, at servers or data storage points closer to the edge of the mobile network, in proximity to the mobile devices (Yao et al., 2019). The primary objective of MEC is to reduce latency and improve the user experience by bringing the content closer to the end users. Instead of fetching content from remote data centers or cloud servers, which can result in delays and increased network traffic, mobile edge caching allows users to access the content directly from nearby edge servers (Reiss-Mirzaei et al., 2023). This approach significantly reduces the content retrieval time and decreases the load on the core network and backhaul links. MEC is particularly beneficial in scenarios with high data traffic and user density, where large-sized data need to be distributed efficiently to a large number of mobile devices (Thar et al., 2018). By utilizing edge servers with caching capabilities, MEC optimizes content delivery and reduces the burden on the mobile core network, leading to a more efficient and reliable mobile network infrastructure (Peng et al., 2021; Wang et al., 2017).

Mobile edge caching scheme (MECS) refers to a strategy or approach used to implement MEC in a mobile network (Peng et al., 2021; Yao et al., 2019). It involves the management and optimization of content caching at the edge of the network, closer to the end users and their devices (Zhu et al., 2018). The primary goal of the scheme is to reduce latency and develop the efficiency of content delivery to mobile users. In a MECS, frequently accessed content, such as videos, images, applications, and web pages, is strategically stored and cached at servers or data storage points located at the edge of the mobile network (Yao et al., 2019; Zhang, 2022). This positioning allows mobile devices to access the content more quickly, as it is located in close proximity to the users, reducing the need to fetch data from remote data centers or cloud servers (Thar et al., 2018). To advance the performance of MEC, six fundamental criteria must be considered: achieving a high cache hit probability, improving spectrum efficiency, improving energy efficiency, optimizing network throughput, minimizing content retrieval delay, and maximizing traffic offloading (Zhang et al., 2018). Recently, seven main styles of MECSs have been extensively researched and applied, based on six common criteria and their corresponding scenarios. Figure 4.1 illustrates these styles (Peng et al., 2021; Wu et al., 2020). A MCDM problem related to MECS is analyzed in the q-ROF setting after some information measures are proposed by Peng et al. (2021). Iqbal and Asaduzzaman (2024) examined the selection of the optimal caching option as a MCDM problem using the AHP-TOPSIS method.

Entropy serves as a valuable tool for quantifying uncertain information. Shannon entropy is a concept introduced by Shannon (1948). It is a measure of the average amount of information or uncertainty contained in a random variable or a probability distribution. On the other hand, Kullback and Leibler (1951) put forth the idea of cross-entropy (KL divergence) as a tool for quantifying the distinction between two probability distributions. In the realm of fuzzy set theory, the notion of entropy

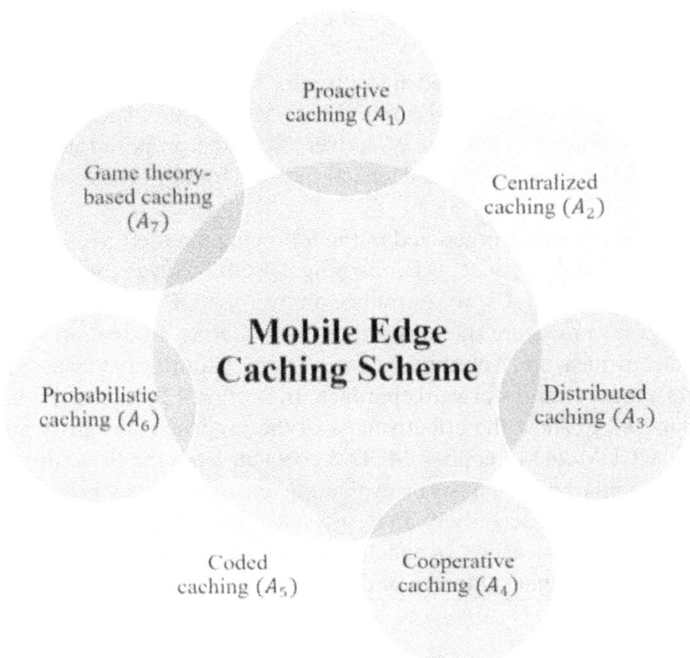

FIGURE 4.1 Seven main styles of MECSs

was initially introduced by Zadeh (1968) to quantify the fuzziness stemming from the elements' degree of membership in FS. De Lucca and Termini (1972) further refined this concept by proposing axioms that fuzzy entropy should adhere to, building on the foundations laid by the Shannon entropy measure. Fuzzy entropy has found applications across diverse fields, including decision-making, face recognition systems, pattern recognition, data science, classification, and medical diagnosis (Bozyiğit et al., 2024a; Garg et al., 2023, 2025; Hung and Yang, 2008; Jaganathan and Kuppuchamy, 2013; Joshi and Kumar, 2014; Pujol et al., 2017; Ünver and Aydoğan, 2024). As a generalization of De Luca and Termini's definition of fuzzy entropy, the axioms of intuitionistic fuzzy entropy were given by Szmidt and Kacprzyk (2001). Intuitionistic fuzzy entropy measure is a concept that quantifies the uncertainty of IFSs. Peng et al. (2017) established a formal axiomatic definition for entropy measures of PFSs. Then, Peng and Liu (2019) extended the concept of entropy from PFSs to q-ROFSs by defining axioms that must be satisfied.

This study's primary contributions can be summarized as follows:

- This study presents a novel q-ROF entropy measure based on Choquet integral defined by a fuzzy measure that allows to take into account the interaction between criteria. Thus, it is aimed to obtain more precise results in the decision-making process compared to the existing entropy measures in the literature.
- The method of obtaining a similarity measure from entropy is introduced.
- A new similarity measure is defined thanks to the proposed Choquet integral-based entropy measure.
- We propose an entropy-based algorithm for MCDM in q-rung fuzzy orthopair fuzzy environment and apply it to a MCDM problem about MECSs from the literature. Thus, the effectiveness of the proposed novel entropy and similarity measure in the decision-making framework is demonstrated.

The remaining sections are organized in the following manner: Section 4.2 presents a concise overview of key concepts concerning q-ROFSs, fuzzy measure theory, and fuzzy entropy. In Section 4.3, we introduce a novel q-ROF entropy that utilizes the Choquet integral to measure the uncertainty of q-ROFSs. In Section 4.4, using the proposed entropy measure, a method for establishing a similarity measure is devised, which offers a more straightforward approach. In Section 4.5, an entropy-based algorithm is proposed to show the effectiveness of the proposed similarity and entropy measures in MCDM and to apply a MCDM problem from the literature. Moreover, we conduct a comparative analysis by evaluating existing entropy measures from the literature and perform a sensitivity analysis concerning the parameter λ within the context of the λ-fuzzy measure. In conclusion, Section 4.6 serves as the final section of the chapter, providing a summary of the principal findings and contributions.

4.2 PRELIMINARIES

This section provides a brief overview of essential concepts from fuzzy set theory, fuzzy measure theory, and information measures that are employed throughout the chapter. Unless otherwise stated, we assume that $X=\{x_1,\ldots,x_n\}$ is a finite set and 2^X denotes its power set.

Definition 1. (Yager, 2016) Let $q\geq 1$ be a real number. A q-ROFS is given by:

$$A=\left\{\left\langle x_i,\mu_A(x_i),\nu_A(x_i)\right\rangle : i=1,..,n\right\},$$

where $\mu_A : X\rightarrow[0,1]$ indicates membership function and $\nu_A : X\rightarrow[0,1]$ indicates nonmembership function satisfying $0\leq\mu_A^q(x_i)+\nu_A^q(x_i)\leq 1$ for each $i=1,\ldots,n$. The degree of hesitancy is calculated by utilizing $\pi_A(x_i)=\left(1-\mu_A^q(x_i)-\nu_A^q(x_i)\right)^{\frac{1}{q}}$. q-ROFSs(X) represents the family of all q-ROFSs on X. A q-ROF value (q-ROFV) is a pair $A=\langle\mu_A,\nu_A\rangle$, where μ_A and ν_A are both within the range of $[0,1]$ and satisfy the condition $\mu_A^q+\nu_A^q\leq 1$.

The cornerstone of fuzzy measure theory is the concept of a fuzzy measure, initially introduced by Choquet (1954) and independently defined by Sugeno (1974) within the framework of fuzzy integrals.

Definition 2. (Sugeno, 1974) A fuzzy measure is described as a set function $\zeta : 2^X\rightarrow[0,1]$ that adheres to the following conditions:

(1) $\zeta(\varnothing)=0$ and $\zeta(X)=1$,

(2) For any subsets A and B of X, if $A\subset B$, then $\zeta(A)\leq\zeta(B)$.

Sugeno (1974) introduced the concept of λ-fuzzy measure, which is constructed with interaction index and criterion weighting. A λ-fuzzy measure helps to capture the synergy between criteria and provides more precise solutions in complex decision-making scenarios.

Definition 3. (Sugeno, 1974) Let $\lambda>-1$. A set function ζ is termed a λ-fuzzy measure on X if, for any disjoint subsets A and B of X, we have $\zeta(A\cup B)=\zeta(A)+\zeta(B)+\lambda\zeta(A)\zeta(B)$.

To construct such a measure, Takahagi (2000, 2005) introduced several algorithms. These algorithms serve as effective tools for establishing a λ-fuzzy measure based on the given interaction index and the weights assigned to elements.

The following definition introduces the Choquet integral, a fundamental aggregation operator with respect to a fuzzy measure.

Definition 4. (Choquet, 1954) Let $\zeta : 2^X\rightarrow[0,1]$ be a fuzzy measure. The Choquet integral of a function $g : X\rightarrow[0,1]$ with respect to ζ is defined as follows:

$$C_\zeta(g)=(C)\int_X g\,d\zeta := \sum_{i=1}^{n}\left[g\left(x_{(i)}\right)-g\left(x_{(i-1)}\right)\right]\zeta\left(D_{(i)}\right),$$

where the sequence $\{x_{(i)}\}_{i=0}^{n}$ indicates the indices permuted such that $0 := g(x_{(0)}) \le g(x_{(1)}) \le \ldots \le g(x_{(n)})$ and $D_{(i)} := \{x_{(i)}, x_{(i+1)}, \ldots, x_{(n)}\}$.

Theorem 1. (Sugeno, 1974) Consider a finite set X, a fuzzy measure $\zeta : 2^X \to [0,1]$, and two functions $g, h : X \to [0,1]$. The following properties hold:

(a) If $g = 0$, then the integral of g with respect to ζ is zero. For the other side to be true, ζ must satisfy $\zeta(E) > 0$ for $\varnothing \ne E \subset X$.

(b) If $\mathbf{1}_E$ is the characteristic function of a subset $E \subset X$, then $(C)\int_X 1_X d\zeta = 1$.

For the proposition to be true in both directions, ζ must satisfy $\zeta(E) < 1$ for $E \subsetneq X$.

(c) If $g \le h$, then $(C)\int_X g d\zeta \le (C)\int_X h d\zeta$ (monotonicity).

To measure the degree of similarity between two q-ROFSs, the following definition was proposed by Peng and Liu (2019).

Definition 5. (Peng and Liu, 2019) Let A, B, C be three q-ROFSs. A mapping S : q-ROFSs(X) × q-ROFSs(X) $\to [0,1]$ is called a similarity measure if it holds the following conditions:

(S1) $S(A,B) = S(B,A)$,
(S2) $S(A, A^c) = 0$ iff A is a crisp set,
(S3) $S(A,B) = 1$ iff $A = B$,
(S4) If $A \subset B \subset C$, then $S(A,C) \le S(A,B)$ and $S(A,C) \le S(B,C)$.

The axioms that the q-ROF entropy must satisfy were defined by Peng and Liu (2019) as follows:

Definition 6. (Peng and Liu, 2019) Let A and B be two q-ROFSs. A mapping E : q-ROFSs(X) $\to [0,1]$ is called a q-ROF entropy if it satisfies the following conditions:

(E1) $E(A) = 0$ iff A is a crisp set, i.e., $\mu_A(x_i) = 0, \nu_A(x_i) = 1$ or $\mu_A(x_i) = 1, \nu_A(x_i) = 0$ for all $i = 1, \ldots, n$,
(E2) $E(A) = 1$ iff $\mu_A(x_i) = \nu_A(x_i)$ for all $i = 1, \ldots, n$,
(E3) $E(A) = E(A^c)$,
(E4) $E(A) \le E(B)$ if $\mu_A(x_i) \le \mu_B(x_i) \le \nu_B(x_i) \le \nu_A(x_i)$ or $\nu_A(x_i) \le \nu_B(x_i) \le \mu_B(x_i) \le \mu_A(x_i)$ for any $i = 1, \ldots, n$.

We review the entropy measures previously introduced for q-ROFS with their applications, which are presented as follows:

- Khan et al. (2021) introduced a q-ROF entropy measure defined by

$$E_K(A)=1-\frac{1}{n}\sum_{i=1}^{n}\left[1-\frac{1}{2}(1-\left(\mu_A^{2q}(x_i)+\nu_A^{2q}(x_i)\right)+\frac{4}{\pi}\tan^{-1}\{(1-\mu_A^q(x_i)-\nu_A^q(x_i))^2\})\right]$$

to analyze a MCDM problem.

- Kumar and Sharma (2022) gave a q-ROF entropy measure:

$$E_{KS}(A)=\frac{1}{3n}\sum_{i=1}^{n}\left(\sqrt{\mu_A^q(x_i)\nu_A^q(x_i)}+2\pi_A^q(x_i)+\sqrt{\left(1-\mu_A^q(x_i)\right)\left(1-\nu_A^q(x_i)\right)}\right)$$

and applied it to a MCDM problem about investment.

- Liu et al. (2018) established a q-ROF entropy measure defined by

$$E_L(A)=1-\frac{1}{n\sqrt{2}}\sum_{i=1}^{n}\sqrt{\left(\mu_A^q(x_i)\right)^2+\left(\nu_A^q(x_i)\right)^2+\left(\pi_A^q(x_i)-1\right)^2}$$

and proposed a method based on this novel entropy for MCDM.

- Peng and Liu (2019) proposed some q-ROF entropy measures defined by:

$$E_{PL}^1(A)=1-\frac{1}{n}\sum_{i=1}^{n}|\mu_A^q(x_i)-\nu_A^q(x_i)|,$$

$$E_{PL}^2(A)=\frac{1}{n}\sum_{i=1}^{n}\frac{\min\left(\mu_A^q(x_i),\nu_A^q(x_i)\right)}{\max\left(\mu_A^q(x_i),\nu_A^q(x_i)\right)},$$

$$E_{PL}^3(A)=\frac{1}{n}\sum_{i=1}^{n}\frac{1-|\mu_A^q(x_i)-\nu_A^q(x_i)|}{1+|\mu_A^q(x_i)-\nu_A^q(x_i)|}|$$

and

$$E_{PL}^4(A)=\frac{1}{n}\sum_{i=1}^{n}\tan\left\{\frac{\pi}{4}-\frac{|\mu_A^q(x_i)-\nu_A^q(x_i)|}{4\left(1+\pi_A^q(x_i)\right)}\pi\right\},$$

to investigate problems about clustering analysis, pattern recognition, and medical diagnosis.

- Liu et al. (2020) proposed a q-ROF entropy by:

$$\hat{E}(A)=\frac{1}{n}\sum_{i=1}^{n}\left[1-\left(\mu_A^q(x_i)+\nu_A^q(x_i)\right)|\mu_A^q(x_i)-\nu_A^q(x_i)|\right].$$

4.3 A NOVEL ENTROPY BASED ON CHOQUET INTEGRAL

In this section, we propose a Choquet-based entropy for q-ROFSs motivated by Liu et al. (2020).

Definition 7. Let ζ be a fuzzy measure satisfying $0<\zeta(Y)<1$ for every $\varnothing \neq Y \subsetneq X$, and let A be a q-ROFS. A q-ROF entropy based on Choquet integral is defined by:

$$\mathbb{E}_{C,\zeta}(A)=(C)\int_X E_A d\zeta,$$

where

$$E_A(x_i)=1-\left(\mu_A^q(x_i)+\nu_A^q(x_i)\right)\left|\mu_A^q(x_i)-\nu_A^q(x_i)\right|$$

for each $i = 1,\ldots,n$.

Remark 1. Let B be a q-ROFS given by $B=\{\langle x_1,\mu_B,\nu_B\rangle,\langle x_2,0.5,0.5\rangle\}$, and let τ be the fuzzy measure defined by $\tau(\varnothing)=0$, $\tau(\{x_1\})=0.3$, $\tau(\{x_2\})=0.9$, and, $\tau(X)=1$. From Definition 7 we get:

$$\mathbb{E}_{C,\tau}(B)=1-0.1\left(\mu_B^q+\nu_B^q\right)|\mu_B^q-\nu_B^q|.$$

where $E_B(x_1)=1-\left(\mu_B^q+\nu_B^q\right)|\mu_B^q-\nu_B^q|$ and $E_B(x_2)=1$. It can be concluded from Figure 4.2 that $\mathbb{E}_{C,\tau}(B)$ provides the properties of the definition of q-ROF entropy.

Theorem 2. Let ζ be a fuzzy measure satisfying $0<\zeta(Y)<1$ for every $\varnothing \neq Y \subsetneq X$, and let A be a q-ROFSs. $\mathbb{E}_{C,\zeta}$ is a q-ROF entropy.

Proof. As $0\le \mu_A^q(x_i)+\nu_A^q(x_i)\le 1$ and $0\le|\mu_A^q(x_i)-\nu_A^q(x_i)|\le 1$ for each $x_i\in X$, it is seen that $0\le E_A(x_i)\le 1$. From Theorem 1, we derive $0\le \mathbb{E}_{C,\zeta}(A)\le 1$.

(E1) If A is crips set, which indicates then $E(x_i)=0$ for any $x_i\in X$. We obtain $\mathbb{E}_{C,\zeta}(A)=0$. On the other hand, if $\mathbb{E}_{C,\zeta}(A)=0$ and τ is a strictly positive fuzzy measure, then $E_A(x_i)=1-\left(\mu_A^q(x_i)+\nu_A^q(x_i)\right)|\mu_A^q(x_i)-\nu_A^q(x_i)|=0$ for each $i\in\{1,\ldots,n\}$. Since $\mu_A^q(x_i)+\nu_A^q(x_i)=1$, $|\mu_A^q(x_i)-\nu_A^q(x_i)|=1$, and $\mu_A(x_i)$, $\nu_A(x_i)\in[0,1]$ for $i=1,\ldots,n$, we can derive $\mu_A(x_i)=1,\nu_A(x_i)=0$ or $\mu_A(x_i)=0,\nu_A(x_i)=1$, which means A is a crisp set.

(E2) If $\mu_A(x_i)=\nu_A(x_i)$ for each $x_i\in X$, we get $E(x_i)=1-\left(\mu_A^q(x_i)+\nu_A^q(x_i)\right)$ $|\mu_A^q(x_i)-\nu_A^q(x_i)|=1$. From Theorem 1, it is concluded that $\mathbb{E}_{C,\zeta}(A)=1$. On the other part, if it is assumed that $\mathbb{E}_{C,\tau}(A)=1$, we attain that $E_A(x_i)=1-\left(\mu_A^q(x_i)+\nu_A^q(x_i)\right)|$ $\mu_A^q(x_i)-\nu_A^q(x_i)|=1$ for each $x_i\in X$, which indicates $\mu_A^q(x_i)+\nu_A^q(x_i)=0$ or $|\mu_A^q(x_i)-\nu_A^q(x_i)|=0$. Therefore, we get $\mu_A(x_i)=\nu_A(x_i)$ for all $x_i\in X$.

(E3) It is clear from the definition.

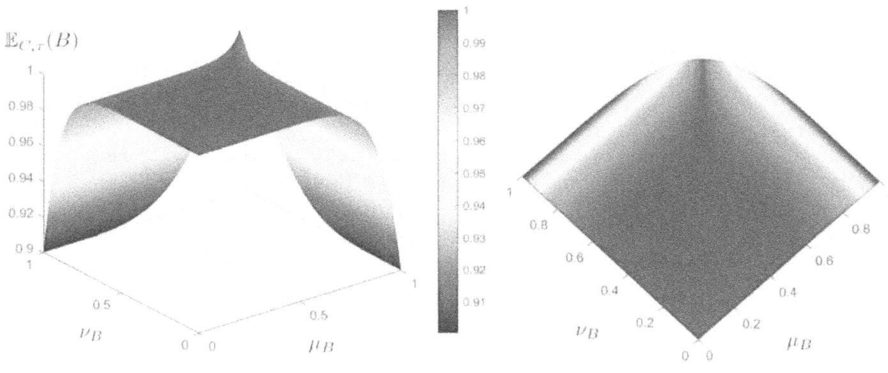

FIGURE 4.2 The graph and the projection of $\mathbb{E}_{C,\tau}(B)$ for $q = 5$ in Remark 1

(E4) If $\mu_A(x_i) \le \mu_B(x_i) \le \nu_B(x_i) \le \nu_A(x_i)$ for any $i = 1,\dots,n$, then we get

$$\begin{aligned}
E_B(x_i) - E_A(x_i) &= 1 - \left(\mu_B^q(x_i) + \nu_B^q(x_i)\right) | \mu_B^q(x_i) - \nu_B^q(x_i) | \\
&\quad - \left(1 - \left(\mu_A^q(x_i) + \nu_A^q(x_i)\right) | \mu_A^q(x_i) - \nu_A^q(x_i) |\right) \\
&= \left(\mu_A^q(x_i) + \nu_A^q(x_i)\right) | \mu_A^q(x_i) + \nu_A^q(x_i) | - \left(\mu_B^q(x_i) + \nu_B^q(x_i)\right) | \mu_B^q(x_i) - \nu_B^q(x_i) | \\
&= \left(\mu_A^q(x_i) + \nu_A^q(x_i)\right)\left(\nu_A^q(x_i) - \mu_A^q(x_i)\right) - \left(\mu_B^q(x_i) + \nu_B^q(x_i)\right)\left(\nu_B^q(x_i) - \mu_B^q(x_i)\right) \\
&= \nu_A^{2q}(x_i) - \mu_A^{2q}(x_i) - \nu_B^{2q}(x_i) - \mu_B^{2q}(x_i) \\
&= \left(\nu_A^q(x_i) + \nu_B^q(x_i)\right)\left(\nu_A^q(x_i) - \nu_B^q(x_i)\right) + \left(\mu_A^q(x_i) + \mu_B^q(x_i)\right)\left(\mu_B^q(x_i) - \mu_A^q(x_i)\right)
\end{aligned}$$

Since $\nu_A^q(x_i) - \nu_B^q(x_i) \ge 0$ and $\mu_B^q(x_i) - \mu_A^q(x_i) \ge 0$, it is seen that $E_A(x_i) \le E_B(x_i)$ for any $x_i \in X$. From Theorem 1, we derive $\mathbb{E}_{C,\zeta}(A) \le \mathbb{E}_{C,\zeta}(B)$.

4.4 TRANSFORMATION RELATIONSHIPS AMONG ENTROPY AND SIMILARITY MEASURES

Peng and Liu (2019) proposed a method to construct a similarity measure via entropy in q-ROF environment. Let A and B be two q-ROFSs. Then:

$$F(A,B) = \left\{\left\langle x_i, \mu_{F(A,B)}(x_i), \nu_{F(A,B)}(x_i)\right\rangle : i = 1,\dots,n\right\}$$

defined by

$$\mu_{F(A,B)}(x_i)=\sqrt{\frac{1+\min\{|\mu_A^q(x_i)-\mu_B^q(x_i)|,|\nu_A^q(x_i)-\nu_B^q(x_i)|\}}{2}},$$

and

$$\nu_{F(A,B)}(x_i)=\sqrt{\frac{1-\max\{|\mu_A^q(x_i)-\mu_B^q(x_i)|,|\nu_A^q(x_i)-\nu_B^q(x_i)|\}}{2}},$$

is a q-ROFSs. Peng and Liu (2019) showed that $E(F(A,B))$ is a q-rung orthopair fuzzy similarity measure when E is a q-rung orthopair fuzzy entropy.

For convenience, $G(A,B)$ can be defined by

$$\mu_{G(A,B)}(x_i)=\left(\frac{1+\min\{|\mu_B^q(x_i)-\mu_B^q(x_i)|,|\nu_A^q(x_i)-\nu_B^q(x_i)|\}}{2}\right)^{\frac{1}{q}}$$

and

$$\nu_{G(A,B)}(x_i)=\left(\frac{1-\max\{|\mu_B^q(x_i)-\mu_B^q(x_i)|,|\nu_A^q(x_i)-\nu_B^q(x_i)|\}}{2}\right)^{\frac{1}{q}},$$

which is a q-ROFSs, as $\mu_{G(A,B)}^q(x_i)+\nu_{G(A,B)}^q(x_i)\le 1$ for every $i=1,\ldots,n$.

Theorem 3. Let E be a q-ROF entropy, and let A and B be two q-ROFSs. Then $S(A,B)=E(G(A,B))$ is a similarity measure.

Proof. (S1) and (S2) are clear.

(S3) Using the entropy definition of q-ROFSs, we obtain for any $i=1,\ldots,n$ that

$$\begin{aligned} S(A,B)=E(G(A,B))=1 \quad &\Leftrightarrow \quad \mu_{G(A,B)}(x_i)=\nu_{G(A,B)}(x_i)\\ &\Leftrightarrow \quad |\mu_A^q(x_i)-\mu_B^q(x_i)|=0 \text{ and } |\nu_A^q(x_i)-\nu_B^q(x_i)|=0\\ &\Leftrightarrow \quad \mu_A(x_i)=\mu_B(x_i) \text{ and } \nu_A(x_i)=\nu_B(x_i)\\ &\Leftrightarrow \quad A=B. \end{aligned}$$

(S4) Since $A\subset B\subset C$, it is easily seen that $\mu_A(x_i)\le\mu_B(x_i)\le\mu_C(x_i)$ and $\nu_C(x_i)\le\nu_B(x_i)\le\nu_A(x_i)$ for any $i=1,\ldots,n$. Thus we derive:

$$|\mu_A^q(x_i)-\mu_C^q(x_i)|\ge|\mu_A^q(x_i)-\mu_B^q(x_i)| \text{ and } |\nu_A^q(x_i)-\nu_C^q(x_i)|\ge|\nu_A^q(x_i)-\nu_B^q(x_i)|.$$

From here we obtain that

$$\min\{|\mu_A^q(x_i)-\mu_C^q(x_i)|,|\nu_A^q(x_i)-\nu_C^q(x_i)|\}\ge\min\{|\mu_A^q(x_i)-\mu_B^q(x_i)|,|\nu_A^q(x_i)-\nu_B^q(x_i)|\},$$

$$\max\{|\mu_A^q(x_i)-\mu_C^q(x_i)|,|\nu_A^q(x_i)-\nu_C^q(x_i)|\}\ge\max\{|\mu_A^q(x_i)-\mu_B^q(x_i)|,|\nu_A^q(x_i)-\nu_B^q(x_i)|\}.$$

Therefore, we have $\mu_{G(A,B)}(x_i) \leq \mu_{G(A,C)}(x_i)$, $\nu_{G(A,B)}(x_i) \geq \nu_{G(A,C)}(x_i)$, and $\nu_{G(A,B)}(x_i) \leq \mu_{G(A,B)}(x_i)$, in other words, $\nu_{G(A,C)}(x_i) \leq \nu_{G(A,B)}(x_i) \leq \mu_{G(A,B)}(x_i) \leq \mu_{G(A,C)}(x_i)$. From the definition of q-ROF entropy, we attain $E(G(A,C)) \leq E(G(A,B))$, that is, $S(A,C) \leq S(A,B)$. In a similar way, one can easily prove $S(A,C) \leq S(B,C)$.

Definition 8. A Choquet similarity measure based on $\mathbb{E}_{C,\zeta}$ is defined by:

$$\mathbb{S}_{C,\zeta}(A,B) = \mathbb{E}_{C,\zeta}(G(A,B)) = (C)\int_X g_{A,B}d\zeta = \sum_{i=1}^{n}\left(g_{A,B}(x_{(i)}) - g_{A,B}(x_{(i-1)})\right)\zeta(D_{(i)}),$$

where

$$\begin{aligned} g_{A,B}(x_i) \;=\; 1-\frac{1}{4}(\; & 2+\min\{|\mu_A^q(x_i)-\mu_B^q(x_i)|,|\nu_A^q(x_i)-\nu_B^q(x_i)|\} \\ & -\max\{|\mu_A^q(x_i)-\mu_B^q(x_i)|,|\nu_A^q(x_i)-\nu_B^q(x_i)|\}) \\ & \times(|\mu_A^q(x_i)-\mu_B^q(x_i)|+|\nu_A^q(x_i)-\nu_B^q(x_i)|) \end{aligned}$$

for $i = 1,\dots,n$.

Remark 2. Let B be a q-ROFS given by $B = \{\langle x_1,\mu_B,\nu_B\rangle,\langle x_2,0.5,0.5\rangle\}$, and let τ be a fuzzy measure defined by $\tau(\varnothing)=0$, $\tau(\{x_1\})=0.4$, $\tau(\{x_2\})=0.2$, and $\tau(X)=1$. From Definition 8 we obtain that

$$\mathbb{S}_{C,\tau}(B,B^c) = 1-0.8|\mu_B^q-\nu_B^q|,$$

where $g_{B,B^c}(x_1) = 1-|\mu_B^q-\nu_B^q|$ and $g_{B,B^c}(x_2)=1$. $\mathbb{S}_{C,\tau}(B,B^c)$ is visualized in Figure 4.3 for $q=\frac{5}{2}$. From Figure 4.3 it is seen that $\mathbb{S}_{C,\tau}(B,B^c)$ satisfies properties of similarity measure.

4.5 AN APPLICATION

MCDM is a structured approach used to make decisions when multiple criteria or factors need to be considered simultaneously. MCDM techniques provide a systematic and transparent framework for decision-making, allowing decision makers to consider multiple factors and their interrelationships. In this section, an algorithm is proposed in the decision-making framework and then applied to a problem from the literature on MECS.

In Algorithm 1, a method based on entropy is proposed as an approach to identify the optimal alternative within the MCDM process. The algorithm utilizes a similarity measure based on entropy to evaluate and compare the alternatives, aiming to determine the alternative that best aligns with the decision criteria. By incorporating

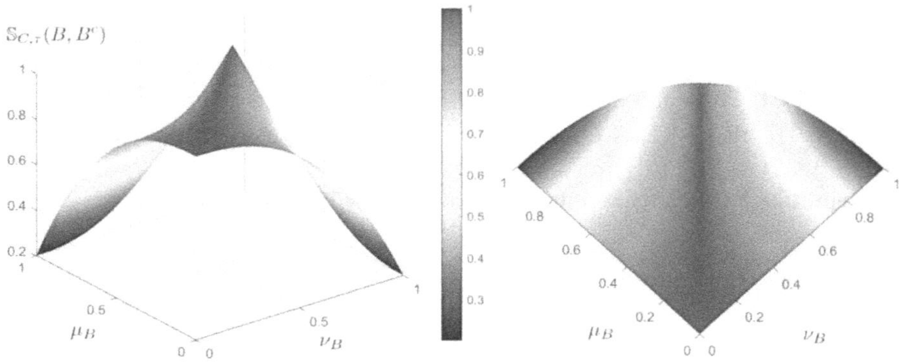

FIGURE 4.3 The graph and the projection of $\mathbb{S}_{C,\tau}\left(B,B^c\right)$ for $q=\frac{5}{2}$ in Remark 2

entropy-based calculations, the algorithm provides a systematic and objective framework for decision makers to assess the alternatives and make informed choices within the MCDM context.

ALGORITHM 1 AN ENTROPY-BASED ALGORITHM FOR MCDM ACCORDING TO PARTIAL CRITERION WEIGHT INFORMATION

1: **Input:**

2: Set of alternatives $\mathcal{A}$: $\{\mathcal{A}_1,\ldots,\mathcal{A}_n\}$

3: Set of criteria $\mathcal{C}$: $\{\mathcal{C}_1,\ldots,\mathcal{C}_m\}$

4: **Output:**

5: Best alternative based on Choquet similarity measure

6: **begin**

7: Step 1: Create a MCDM problem with a set of alternatives $\mathcal{A}$ and a set of criteria $\mathcal{C}$.

8: Step 2: Form the q-ROF assessment matrix $\mathcal{E}$ based on the expert's input as:

$$\mathcal{E}=\begin{bmatrix} e_{11} & \cdots & e_{1m} \\ \vdots & \ddots & \vdots \\ e_{n1} & \cdots & e_{nm} \end{bmatrix},$$

where $e_{ij}=\left\langle \mu_{e_{ij}},\nu_{e_{ij}}\right\rangle$ is a q-ROFV for each $i=1,\ldots,n$ and $j=1,\ldots,m$.

9: Step 3: Convert the q-ROF assessment matrix $\mathcal{E}$ into the standardized matrix $\mathcal{F}$ as:

$$\mathcal{F}=\begin{bmatrix} f_{11} & \cdots & f_{1m} \\ \vdots & \ddots & \vdots \\ f_{n1} & \cdots & f_{nm} \end{bmatrix},$$

where

$$f_{ij} = \left\langle \mu_{f_{ij}}, \nu_{f_{ij}} \right\rangle = \begin{cases} \left\langle \mu_{e_{ij}}, \nu_{e_{ij}} \right\rangle, & \text{if } C_j \in \mathcal{B} \\ \left\langle \nu_{e_{ij}}, \mu_{e_{ij}} \right\rangle, & \text{if } C_j \in \mathcal{C} \end{cases}$$

such that $\mathcal{B}$ represents a set of benefit criteria and $\mathcal{C}$ represents a set of cost criteria.

10: Step 4: Calculate the score matrix $\mathcal{M}$ using one of the existing score functions. The score matrix is represented as follows:

$$\mathcal{M}_{nm} = \begin{bmatrix} \mathcal{S}(f_{11}) & \cdots & \mathcal{S}(f_{1m}) \\ \vdots & \ddots & \vdots \\ \mathcal{S}(f_{n1}) & \cdots & \mathcal{S}(f_{nm}) \end{bmatrix}.$$

11: Step 5: By utilizing the score matrix $\mathcal{M}_{nm}$, an optimization model is formulated to determine the weights of the criteria. The model can be represented as follows:

$$\max g = \sum_{j=1}^{m} \sum_{i=1}^{n} \omega_j \mathcal{M}_{ij} \text{ such that } \omega_j \in P \text{ and} \sum_{j=1}^{m} \omega_j = 1.$$

Here, P represents "partial information about the given criteria."

12: Step 6: Utilizing the standardized decision matrix from Step 3, ascertain the positive ideal solution defined by:

$$\mathcal{A}^{+} = \left(\left\langle \max_{i=1,\ldots,n} \mu_{f_{i,1}}, \min_{i=1,\ldots,n} \nu_{f_{i,1}} \right\rangle, \ldots, \left\langle \max_{i=1,\ldots,n} \mu_{f_{i,m}}, \min_{i=1,\ldots,n} \nu_{f_{i,m}} \right\rangle \right).$$

13: Step 7: Construct a λ-fuzzy measure to compute Choquet similarity measure based on entropy.

14: Step 8: Determine the values of $g_{\mathcal{A}_i,\mathcal{A}^+}(C_j)$ for each alternative $\mathcal{A}_i$ $(i = 1,\ldots,n)$ given in Definition 8.

15: Step 9: Calculate Choquet similarity measure $\mathbb{S}_{C,\zeta}$ between $\mathcal{A}_i$ $(i = 1,\ldots,n)$ and $\mathcal{A}^+$.

16: Step 10: The alternative with the highest similarity measure value to $\mathcal{A}^+$ is considered the best.

17: **end**

4.5.1 Evaluation of the Problem of MECS Preferences

MECSs utilize their own storage to determine the optimal location, method, and content to cache on user equipment. However, when evaluating the performance of

MECS, there are frequently inherent uncertainties that need to be considered. Mobile networks are experiencing a rapid surge in traffic due to the proliferation of mobile devices and the advent of innovative applications. Unfortunately, traditional centralized network architectures struggle to meet the demands of mobile users, leading to heavy backhaul link loads and substantial delays. To address these challenges, MEC has emerged, revolutionizing network capabilities by shifting them from the core network to the edge network. This section aims to construct an evaluation system that is both scientifically rigorous and useful in assessing MECS performance. To ensure the accuracy and practicality of the evaluation results, it is crucial to design a dependable and reasonable evaluation system. In this regard, the evaluation criteria for MECS, denoted as $\mathcal{C}_j\left(j=1,\ldots,6\right)$, are established and summarized as follows:

- **Cache hit probability (CHP)** ($\mathcal{C}_1$) refers to the proportion of user-requested files that are already cached in the system, compared to the total number of files present in the cache. A higher CHP indicates that a larger percentage of user requests can be fulfilled by the cached content. By increasing the cache size, the CHP can be raised, resulting in a decrease in the required backhaul capacity (Peng et al., 2021).
- **Spectrum efficiency** ($\mathcal{C}_2$) relates to the data rate achieved within a specified frequency bandwidth. Network densification, achieved through the deployment of additional small base stations in the macrocell, has been employed to evolve spectrum efficiency in the region (Peng et al., 2021).
- **Energy efficiency** ($\mathcal{C}_3$) is a performance metric that measures the supported data rate achieved with a given energy consumption. By implementing caching at mobile edges and eliminating duplicate transmission, the energy consumption can be reduced (Peng et al., 2021).
- **Network throughput** ($\mathcal{C}_4$) is the maximum data rate that the network can provide and has a significant impact on overall network performance. Higher network throughput can effectively decrease content download latency, which is closely related to the caching storage capacities (Peng et al., 2021).
- **Traffic offloading** ($\mathcal{C}_5$) is the term used to describe the reduction in traffic on the backhaul link when comparing an uncached system to a cached system. Mobile edge caching enables the offloading of traffic from the backhaul link. Maximizing traffic offloading leads to improved caching performance (Peng et al., 2021).
- **Content retrieval delay** ($\mathcal{C}_6$) refers to the time it takes for a user to retrieve content, usually measured as the round-trip time. This delay has a direct impact on the quality of service experienced by the user (Peng et al., 2021).

Based on six common criteria and their corresponding scenarios, researchers have recently investigated and widely utilized seven main styles of MECSs. These seven main styles constitute the alternatives in the decision-making process, denoted as $\mathcal{A}_i\ (i = 1,\ldots,7)$ and summarized as follows (Peng et al., 2021):

- **Proactive caching ($\mathcal{A}_1$)** involves preemptively storing content in the cache based on predictions of user needs, before it is actually requested. This strategy utilizes an analysis of request patterns to optimize cache performance and meet quality of service requirements. By proactively downloading popular content during off-peak hours and delivering it during predictable peak hours, proactive caching ensures efficient resource utilization and improves the overall user experience (Han et al., 2019; Peng et al., 2021).
- **Centralized caching ($\mathcal{A}_2$)** utilizes a central controller that has a comprehensive view of the entire network's status to make decisions regarding caching schemes. This approach enables optimal caching performance by using the central controller's ability to make optimal caching decisions. However, in the context of 5G wireless networks with a large number of mobile users, centralized caching faces significant challenges (Peng et al., 2021; Yang et al., 2019).
- **Distributed caching ($\mathcal{A}_3$)**, also referred to as decentralized caching, operates without a central controller and involves cache nodes making decisions based on local information and information obtained from neighboring nodes. In this approach, the cache nodes are jointly optimized to build up the cache hit probability across neighboring base stations (Li et al., 2015; Peng et al., 2021).
- **Cooperative caching ($\mathcal{A}_4$)** involves the sharing of cached content among base stations to improve cache utilization and reduce inefficiencies. In this approach, base stations collaborate and exchange information about their cache status. However, information exchange can lead to significant signaling overhead. To address this challenge, a solution is proposed to enable cache state sharing with minimal overhead, ultimately aiming to minimize average download delay (Chen et al., 2020; Peng et al., 2021).
- **Coded caching ($\mathcal{A}_5$)** involves classifying files in the file library as coding packets, where any linear combination of these packets can reconstruct the entire original object. This approach helps minimize unnecessary network traffic during the coding and decoding processes. A hierarchical coded caching scheme has been developed, which utilizes a two-layer hierarchical cache and applies it to each layer (Karamchandani et al., 2016; Peng et al., 2021).
- **Probabilistic caching ($\mathcal{A}_6$)** addresses the challenges of uncertainty and user location mobility by employing random distributions, such as spatial randomness, to optimize cache hit probability. By using these random distributions, the issue of uncertain user movements and locations is effectively managed. To improve the accuracy of predicting average content popularity within a given time window, a dynamic probabilistic caching approach is proposed (Gao et al., 2020; Peng et al., 2021).

- **Game theory-based caching ($\mathcal{A}_7$)** examines the interaction among multiple coexisting parties, such as service providers (SPs). Each party selfishly optimizes their own interests, which can lead to conflicts between them. An illustrative example is an auction game, where SPs compete for limited cache space to improve their caching performance (Peng et al., 2021; Xu et al., 2020).

All alternatives and criteria for MECSs are shown in Figure 4.4. The aforementioned mobile caching schemes have their own advantages and disadvantages and are primarily evaluated based on individual criteria such as cache hit probability, spectrum efficiency, and content retrieval delay. However, it becomes evident that a comprehensive evaluation of given MECSs in real-world environments is necessary. In other words, the practical evaluation process of MECS should be considered as a MCDM problem. Furthermore, the increasingly complex decision-making environment poses challenges in expressing uncertain information when solving the evaluation of mobile caching schemes. To evaluate MECSs, we have employed entropy-based algorithm.

4.5.2 A MECS Problem in Decision-Making Framework

Within this subsection, we will conduct an analysis on a MCDM problem related to MECS sourced from existing literature (Peng et al., 2021). The objective is to observe and assess the impact of the proposed algorithm within the context of the MEC problem. Through this examination, we aim to gain insights into how the algorithm influences the decision-making process and its effectiveness in addressing MEC challenges.

- Step 1: The MECS evaluation involves the assessment of seven distinct styles of MECSs: proactive caching ($\mathcal{A}_1$), centralized caching ($\mathcal{A}_2$), distributed caching ($\mathcal{A}_3$), cooperative caching ($\mathcal{A}_4$), coded caching ($\mathcal{A}_5$), probabilistic caching ($\mathcal{A}_6$), and game theory-based caching ($\mathcal{A}_7$). These seven styles, which constitute the alternatives, are evaluated based on six criteria, which include cache hit probability ($\mathcal{C}_1$), spectrum efficiency ($\mathcal{C}_2$), energy efficiency ($\mathcal{C}_3$), network throughput ($\mathcal{C}_4$), traffic offloading ($\mathcal{C}_5$), and content retrieval delay ($\mathcal{C}_6$), mentioned in Subsection 5.1.
- Step 2: The seven main styles of MECSs $\mathcal{A}_i$ are evaluated by an expert as q-ROFVs according to six criteria $\mathcal{C}_j$ as q-ROF assessment matrix $\mathcal{E}$ shown in Table 4.1.
- Step 3: Since $\mathcal{C}_6$ is the cost criterion among the criteria, the evaluation matrix $\mathcal{E}$ is converted into the standard matrix $\mathcal{F}$ shown in Table 4.2 using the complement operation.
- Step 4: By utilizing the score function $\mathcal{S}(\langle \mu, \nu \rangle) = \frac{1}{2}(1 + \mu^q - \nu^q)$ for $q = 3$ (Wei et al., 2018), score matrix $\mathcal{M}$ of standardized assessment matrix $\mathcal{F}$ is obtained and listed in Table 4.3.

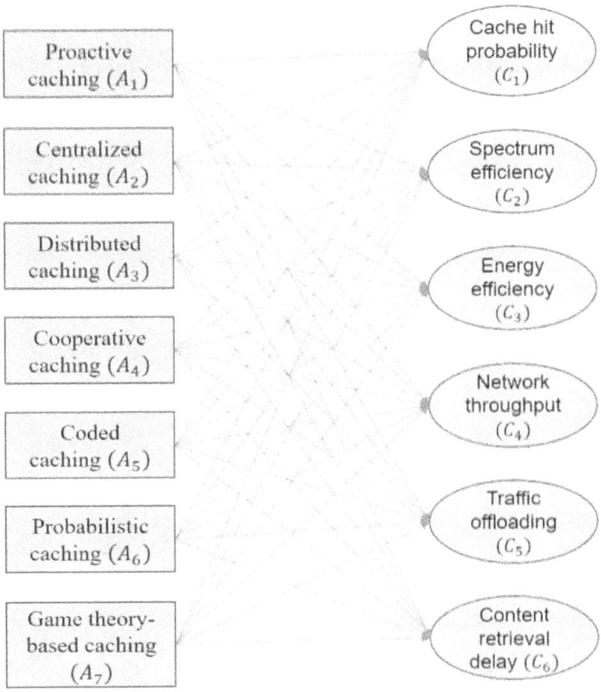

FIGURE 4.4 Hierarchical organization of MECSs

- Step 5: Utilizing the score matrix $\mathcal{M}$ provided in Table 4.3 and considering the available partial information about the criteria, we formulate the following optimization model:

$$\begin{aligned} \max \quad g \quad & = (3.42)\omega_1 + (3.10)\omega_2 + (2.71)\omega_3 + (2.94)\omega_4 + (2.87)\omega_5 + (4.3)\omega_6 \\ \text{such that} \quad & 0.25 \le \omega_1 \le 0.35 \quad ; \quad 0.10 \le \omega_2 \le 0.16 \quad ; \quad 0.15 \le \omega_3 \le 0.18 \\ & 0.05 \le \omega_4 \le 0.15 \quad ; \quad 0.14 \le \omega_5 \le 0.18 \quad ; \quad 0.12 \le \omega_6 \le 0.15 \\ & \sum_{j=1}^{6} \omega_j = 1. \end{aligned}$$

- After solving the optimization problem, we obtain the weight vector $\omega = (\omega_1, \omega_2, \omega_3, \omega_4, \omega_5, \omega_6) = (0.35, 0.16, 0.15, 0.05, 0.14, 0.15)$.
- Step 6: If we utilize the standardized decision matrix $\mathcal{F}$ in Table 4.2, the positive ideal solution $\mathcal{A}^+ = (\langle 0.6, 0.4 \rangle, \langle 0.5, 0.5 \rangle, \langle 0.6, 0.4 \rangle, \langle 0.5, 0.5 \rangle, \langle 0.5, 0.5 \rangle, \langle 0.8, 0.2 \rangle)$ is obtained.
- Step 7: A λ-fuzzy measure can be constructed using the algorithm proposed by Takahagi. For specific construction, we set $\lambda = 0.6$ and

TABLE 4.1
***q*-ROF assessment matrix $\mathcal{E}$**

$\mathcal{E}$	$\mathcal{C}_1$	$\mathcal{C}_2$	$\mathcal{C}_3$	$\mathcal{C}_4$	$\mathcal{C}_5$	$\mathcal{C}_6$
$\mathcal{A}_1$	$\langle 0.3,0.7\rangle$	$\langle 0.5,0.5\rangle$	$\langle 0.3,0.7\rangle$	$\langle 0.4,0.6\rangle$	$\langle 0.3,0.7\rangle$	$\langle 0.5,0.5\rangle$
$\mathcal{A}_2$	$\langle 0.5,0.5\rangle$	$\langle 0.2,0.8\rangle$	$\langle 0.3,0.7\rangle$	$\langle 0.2,0.8\rangle$	$\langle 0.3,0.7\rangle$	$\langle 0.4,0.6\rangle$
$\mathcal{A}_3$	$\langle 0.6,0.4\rangle$	$\langle 0.5,0.5\rangle$	$\langle 0.3,0.7\rangle$	$\langle 0.5,0.5\rangle$	$\langle 0.4,0.6\rangle$	$\langle 0.3,0.7\rangle$
$\mathcal{A}_4$	$\langle 0.6,0.4\rangle$	$\langle 0.5,0.5\rangle$	$\langle 0.4,0.6\rangle$	$\langle 0.4,0.6\rangle$	$\langle 0.3,0.7\rangle$	$\langle 0.3,0.7\rangle$
$\mathcal{A}_5$	$\langle 0.6,0.4\rangle$	$\langle 0.5,0.5\rangle$	$\langle 0.6,0.4\rangle$	$\langle 0.5,0.5\rangle$	$\langle 0.5,0.5\rangle$	$\langle 0.2,0.8\rangle$
$\mathcal{A}_6$	$\langle 0.4,0.6\rangle$	$\langle 0.4,0.6\rangle$	$\langle 0.3,0.7\rangle$	$\langle 0.3,0.7\rangle$	$\langle 0.4,0.6\rangle$	$\langle 0.4,0.6\rangle$
$\mathcal{A}_7$	$\langle 0.4,0.6\rangle$	$\langle 0.4,0.6\rangle$	$\langle 0.3,0.7\rangle$	$\langle 0.5,0.5\rangle$	$\langle 0.5,0.5\rangle$	$\langle 0.4,0.6\rangle$

TABLE 4.2
Standardized *q*-ROF assessment matrix $\mathcal{F}$

$\mathcal{F}$	$\mathcal{C}_1$	$\mathcal{C}_2$	$\mathcal{C}_3$	$\mathcal{C}_4$	$\mathcal{C}_5$	$\mathcal{C}_6$
$\mathcal{A}_1$	$\langle 0.3,0.7\rangle$	$\langle 0.5,0.5\rangle$	$\langle 0.3,0.7\rangle$	$\langle 0.4,0.6\rangle$	$\langle 0.3,0.7\rangle$	$\langle 0.5,0.5\rangle$
$\mathcal{A}_2$	$\langle 0.5,0.5\rangle$	$\langle 0.2,0.8\rangle$	$\langle 0.3,0.7\rangle$	$\langle 0.2,0.8\rangle$	$\langle 0.3,0.7\rangle$	$\langle 0.6,0.4\rangle$
$\mathcal{A}_3$	$\langle 0.6,0.4\rangle$	$\langle 0.5,0.5\rangle$	$\langle 0.3,0.7\rangle$	$\langle 0.5,0.5\rangle$	$\langle 0.4,0.6\rangle$	$\langle 0.7,0.3\rangle$
$\mathcal{A}_4$	$\langle 0.6,0.4\rangle$	$\langle 0.5,0.5\rangle$	$\langle 0.4,0.6\rangle$	$\langle 0.4,0.6\rangle$	$\langle 0.3,0.7\rangle$	$\langle 0.7,0.3\rangle$
$\mathcal{A}_5$	$\langle 0.6,0.4\rangle$	$\langle 0.5,0.5\rangle$	$\langle 0.6,0.4\rangle$	$\langle 0.5,0.5\rangle$	$\langle 0.5,0.5\rangle$	$\langle 0.8,0.2\rangle$
$\mathcal{A}_6$	$\langle 0.4,0.6\rangle$	$\langle 0.4,0.6\rangle$	$\langle 0.3,0.7\rangle$	$\langle 0.3,0.7\rangle$	$\langle 0.4,0.6\rangle$	$\langle 0.6,0.4\rangle$
$\mathcal{A}_7$	$\langle 0.4,0.6\rangle$	$\langle 0.4,0.6\rangle$	$\langle 0.3,0.7\rangle$	$\langle 0.5,0.5\rangle$	$\langle 0.5,0.5\rangle$	$\langle 0.6,0.4\rangle$

$w=(0.35,0.16,0.15,0.05,0.14,0.15)$, resulting in the following fuzzy measure ζ, as shown in Table 4.4.

- Step 8: To compute the Choquet similarity measure based on entropy, we begin by determining the values of $g_{\mathcal{A}_i,\mathcal{A}^+}(\mathcal{C}_j)$ for each alternative $\mathcal{A}_i$ ($i=1,2,3,4,5,6,7$) given in Definition 8. Once these values are obtained, we can then proceed to fill in Table 4.5 with the calculated results.

TABLE 4.3
Score matrix $\mathcal{M}$ of standardized assessment matrix $\mathcal{F}$

$\mathcal{M}$	$\mathcal{C}_1$	$\mathcal{C}_2$	$\mathcal{C}_3$	$\mathcal{C}_4$	$\mathcal{C}_5$	$\mathcal{C}_6$
$\mathcal{A}_1$	0.342	0.500	0.342	0.424	0.342	0.500
$\mathcal{A}_2$	0.500	0.248	0.342	0.248	0.342	0.576
$\mathcal{A}_3$	0.576	0.500	0.342	0.500	0.424	0.658
$\mathcal{A}_4$	0.576	0.500	0.424	0.424	0.342	0.658
$\mathcal{A}_5$	0.576	0.500	0.576	0.500	0.500	0.752
$\mathcal{A}_6$	0.424	0.424	0.342	0.342	0.424	0.576
$\mathcal{A}_7$	0.424	0.424	0.342	0.500	0.500	0.576

- Step 9: Using the information from Table 4.5, we proceed to compute Choquet similarity measure-based entropy $\mathbb{S}_{C,\zeta}$ between $\mathcal{A}_i$ $(i = 1,2,3,4,5,6,7)$ and $\mathcal{A}^+$. The results are listed in Table 4.6.
- Step 10: Since the alternative with the highest similarity measure with A_i is the best, we have the ranking $\mathcal{A}_1 \preceq \mathcal{A}_2 \preceq \mathcal{A}_6 \preceq \mathcal{A}_7 \preceq \mathcal{A}_4 \preceq \mathcal{A}_3 \preceq \mathcal{A}_5$. Therefore, the best selection with MECS is coded caching $(\mathcal{A}_5)$.

4.5.3 The Sensitivity Analysis of the Parameter λ

In this subsection, we evaluate the influence stemming from the choice of the parameter λ within the λ-fuzzy measure. We juxtapose and contrast the outcomes against each other. To perform a comprehensive sensitivity analysis, we systematically adjust the parameter λ within the range of –0.6 to 3.0, with increments of 0.4. The findings of this analysis are presented in Table 4.7. Under each λ value, A_5 emerges as the optimal choice, while A_1 falls on the other end as the least favorable option. The shifting of λ parameter results in alterations to the rankings of alternatives A_2, A_3, A_4, and A_6. Some shifts in the order occur due to the utilization of the Choquet integral method in this study, enabling a more intricate evaluation.

4.5.4 An Evaluative Comparison with Several Established MCDM Approaches

In this subsection, a comparative analysis of the MECS problem with the existing entropy measures mentioned in Section 1 is given. By utilizing entropy measures, the weights of the criteria are calculated. Then we compute weighted cosine similarity measure $WC_{q\text{-}ROF}$,which was proposed by Liu et al. (2019), between $\mathcal{A}_i$ $(i = 1,2,3,4,5,6,7)$ and $\mathcal{A}^+$. Table 4.8 presents the results of the weighted cosine similarity metric $WC_{q\text{-}ROF}$ for the entropy measures $E_{KS}, E_{PL}^1, E_{PL}^2, E_{PL}^3$, and E_{PL}^4. As

TABLE 4.4
Fuzzy measure

$\varsigma(\varnothing)=0$	$\varsigma(\{C_1\})=0.298008$	$\varsigma(\{C_2\})=0.130167$
$\varsigma(\{C_1,C_2\})=0.45145$	$\varsigma(\{C_3\})=0.121742$	$\varsigma(\{C_1,C_3\})=0.441518$
$\varsigma(\{C_2,C_3\})=0.261417$	$\varsigma(\{C_1,C_2,C_3\})=0.606168$	$\varsigma(\{C_4\})=0.0396307$
$\varsigma(\{C_1,C_4\})=0.344725$	$\varsigma(\{C_2,C_4\})=0.172893$	$\varsigma(\{C_1,C_2,C_4\})=0.501815$
$\varsigma(\{C_3,C_4\})=0.164267$	$\varsigma(\{C_1,C_3,C_4\})=0.491647$	$\varsigma(\{C_2,C_3,C_4\})=0.307264$
$\varsigma(\{C_1,C_2,C_3,C_4\})=0.660213$	$\varsigma(\{C_5\})=0.113356$	$\varsigma(\{C_1,C_5\})=0.431633$
$\varsigma(\{C_2,C_5\})=0.252376$	$\varsigma(\{C_1,C_2,C_5\})=0.595511$	$\varsigma(\{C_3,C_5\})=0.243378$
$\varsigma(\{C_1,C_3,C_5\})=0.584903$	$\varsigma(\{C_2,C_3,C_5\})=0.392553$	$\varsigma(\{C_1,C_2,C_3,C_5\})=0.760752$
$\varsigma(\{C_4,C_5\})=0.155682$	$\varsigma(\{C_1,C_4,C_5\})=0.481527$	$\varsigma(\{C_2,C_4,C_5\})=0.298008$
$\varsigma(\{C_1,C_2,C_4,C_5\})=0.649302$	$\varsigma(\{C_3,C_4,C_5\})=0.288796$	$\varsigma(\{C_1,C_3,C_4,C_5\})=0.638442$
$\varsigma(\{C_2,C_3,C_4,C_5\})=0.441518$	$\varsigma(\{C_1,C_2,C_3,C_4,C_5\})=0.818472$	$\varsigma(\{C_6\})=0.121742$
$\varsigma(\{C_1,C_6\})=0.441518$	$\varsigma(\{C_2,C_6\})=0.261417$	$\varsigma(\{C_1,C_2,C_6\})=0.606168$
$\varsigma(\{C_3,C_6\})=0.252376$	$\varsigma(\{C_1,C_3,C_6\})=0.595511$	$\varsigma(\{C_2,C_3,C_6\})=0.402254$
$\varsigma(\{C_1,C_2,C_3,C_6\})=0.772188$	$\varsigma(\{C_4,C_6\})=0.164267$	$\varsigma(\{C_1,C_4,C_6\})=0.491647$
$\varsigma(\{C_2,C_4,C_6\})=0.307264$	$\varsigma(\{C_1,C_2,C_4,C_6\})=0.660213$	$\varsigma(\{C_3,C_4,C_6\})=0.298008$
$\varsigma(\{C_1,C_3,C_4,C_6\})=0.649302$	$\varsigma(\{C_2,C_3,C_4,C_6\})=0.45145$	$\varsigma(\{C_1,C_2,C_3,C_4,C_6\})=0.83018$
$\varsigma(\{C_5,C_6\})=0.243378$	$\varsigma(\{C_1,C_5,C_6\})=0.584903$	$\varsigma(\{C_2,C_5,C_6\})=0.392553$
$\varsigma(\{C_1,C_2,C_5,C_6\})=0.760752$	$\varsigma(\{C_3,C_5,C_6\})=0.382897$	$\varsigma(\{C_1,C_3,C_5,C_6\})=0.74937$
$\varsigma(\{C_2,C_3,C_5,C_6\})=0.542969$	$\varsigma(\{C_1,C_2,C_3,C_5,C_6\})=0.938063$	$\varsigma(\{C_4,C_5,C_6\})=0.288796$
$\varsigma(\{C_1,C_4,C_5,C_6\})=0.638442$	$\varsigma(\{C_2,C_4,C_5,C_6\})=0.441518$	$\varsigma(\{C_1,C_2,C_4,C_5,C_6\})=0.818472$
$\varsigma(\{C_3,C_4,C_5,C_6\})=0.431633$	$\varsigma(\{C_1,C_3,C_4,C_5,C_6\})=0.80682$	$\varsigma(\{C_2,C_3,C_4,C_5,C_6\})=0.595511$
$\varsigma(\{C_1,C_2,C_3,C_4,C_5,C_6\})=1$		

shown in Table 4.9, the most favorable alternative based on these entropy measures is $\mathcal{A}_5$.

4.6 CONCLUSION

Although there are weighted entropy measures in different fuzzy environments, entropy measures that take into account the interaction between criteria are limited.

TABLE 4.5
The values of $g_{\mathcal{A}_i,\mathcal{A}^+}(\mathcal{C}_j)$ for $q = 3$

$g_{\mathcal{A}_i,\mathcal{A}^+}(\mathcal{C}_j)$	$\mathcal{C}_1$	$\mathcal{C}_2$	$\mathcal{C}_3$	$\mathcal{C}_4$	$\mathcal{C}_5$	$\mathcal{C}_6$
$\mathcal{A}_1$	0.77653	1	0.77653	0.92514	0.85148	0.78202
$\mathcal{A}_2$	0.92514	0.78202	0.77653	0.78202	0.85148	0.84512
$\mathcal{A}_3$	1	1	0.77653	1	0.92514	0.91305
$\mathcal{A}_4$	1	1	0.84800	0.92514	0.85148	0.91305
$\mathcal{A}_5$	1	1	1	1	1	1
$\mathcal{A}_6$	0.84800	0.92514	0.77653	0.85148	0.92514	0.84512
$\mathcal{A}_7$	0.84800	0.92514	0.77653	1	1	0.84512

TABLE 4.6
Choquet similarity measure $\mathbb{S}_{C,\zeta}$ between $\mathcal{A}_i$ and $\mathcal{A}^+$

	$\mathcal{A}_1$	$\mathcal{A}_2$	$\mathcal{A}_3$	$\mathcal{A}_4$	$\mathcal{A}_5$	$\mathcal{A}_6$	$\mathcal{A}_7$
$\mathbb{S}_{C,\zeta}(\mathcal{A}_i,\mathcal{A}^+)$	0.822133	0.842627	0.933684	0.93136	**1**	0.854166	0.869182

TABLE 4.7
Ranking results for different λ values from −0.6 to 3

λ parameters	Rankings	Best alternative
$\lambda = -0.6$	$\mathcal{A}_1 \preceq \mathcal{A}_6 \preceq \mathcal{A}_2 \preceq \mathcal{A}_7 \preceq \mathcal{A}_4 \preceq \mathcal{A}_3 \preceq \mathcal{A}_5$	$\mathcal{A}_5$
$\lambda = -0.2$	$\mathcal{A}_1 \preceq \mathcal{A}_6 \preceq \mathcal{A}_2 \preceq \mathcal{A}_7 \preceq \mathcal{A}_4 \preceq \mathcal{A}_3 \preceq \mathcal{A}_5$	$\mathcal{A}_5$
$\lambda = 0.2$	$\mathcal{A}_1 \preceq \mathcal{A}_6 \preceq \mathcal{A}_2 \preceq \mathcal{A}_7 \preceq \mathcal{A}_4 \preceq \mathcal{A}_3 \preceq \mathcal{A}_5$	$\mathcal{A}_5$
$\lambda = 0.6$	$\mathcal{A}_1 \preceq \mathcal{A}_2 \preceq \mathcal{A}_6 \preceq \mathcal{A}_7 \preceq \mathcal{A}_4 \preceq \mathcal{A}_3 \preceq \mathcal{A}_5$	$\mathcal{A}_5$
$\lambda = 1.0$	$\mathcal{A}_1 \preceq \mathcal{A}_2 \preceq \mathcal{A}_6 \preceq \mathcal{A}_7 \preceq \mathcal{A}_4 \preceq \mathcal{A}_3 \preceq \mathcal{A}_5$	$\mathcal{A}_5$
$\lambda = 1.4$	$\mathcal{A}_1 \preceq \mathcal{A}_2 \preceq \mathcal{A}_6 \preceq \mathcal{A}_7 \preceq \mathcal{A}_4 \preceq \mathcal{A}_3 \preceq \mathcal{A}_5$	$\mathcal{A}_5$
$\lambda = 1.8$	$\mathcal{A}_1 \preceq \mathcal{A}_2 \preceq \mathcal{A}_6 \preceq \mathcal{A}_7 \preceq \mathcal{A}_3 = \mathcal{A}_4 \preceq \mathcal{A}_5$	$\mathcal{A}_5$
$\lambda = 2.2$	$\mathcal{A}_1 \preceq \mathcal{A}_2 \preceq \mathcal{A}_6 \preceq \mathcal{A}_7 \preceq \mathcal{A}_3 \preceq \mathcal{A}_4 \preceq \mathcal{A}_5$	$\mathcal{A}_5$
$\lambda = 2.6$	$\mathcal{A}_1 \preceq \mathcal{A}_2 \preceq \mathcal{A}_6 \preceq \mathcal{A}_7 \preceq \mathcal{A}_3 \preceq \mathcal{A}_4 \preceq \mathcal{A}_5$	$\mathcal{A}_5$
$\lambda = 3.0$	$\mathcal{A}_1 \preceq \mathcal{A}_2 \preceq \mathcal{A}_6 \preceq \mathcal{A}_7 \preceq \mathcal{A}_3 \preceq \mathcal{A}_4 \preceq \mathcal{A}_5$	$\mathcal{A}_5$

The Choquet integral also generalizes the weighted arithmetic mean. Considering these, in this study, we propose a q-ROF Choquet integral-based entropy measure that considers the interaction between criteria since it is defined by a fuzzy measure. Thus, more sensitive results can be obtained in the decision-making process compared to weighted entropy. This study introduces a simple method for obtaining similarity measures, which are frequently used in decision-making, pattern recognition, medical diagnosis, and data processing, by using entropy measure. With this method, a Choquet integral-based similarity measure is constructed. Furthermore, we give an algorithm based on entropy and similarity measure for MCDM and utilize it to solve a problem about MECS. Finally, a comparison analysis of the proposed entropy and similarity measures with other existing measures and a sensitivity analysis by varying parameter λ are analyzed. In future studies, the Choquet integral-based entropy and similarity measures proposed here can be also used to investigate problems in pattern recognition, medical diagnosis, clustering analysis, data processing, and many other fields. In addition, these measures can be an

TABLE 4.8

Results of weighted cosine similarity measure $WC_{q\text{-}ROF}$ for E_{KS} (Kumar and Sharma, 2022), $E_{PL}^1, E_{PL}^2, E_{PL}^3, E_{PL}^4$ (Peng and Liu, 2019)

$q=3$	E_{KS}	E_{PL}^1	E_{PL}^2	E_{PL}^3	E_{PL}^4
$WC_{q\text{-}ROF}(\mathcal{A}_1, \mathcal{A}^+)$	0.678	0.641	0.622	0.636	0.640
$WC_{q\text{-}ROF}(\mathcal{A}_2, \mathcal{A}^+)$	0.737	0.722	0.723	0.722	0.721
$WC_{q\text{-}ROF}(\mathcal{A}_3, \mathcal{A}^+)$	0.864	0.827	0.826	0.826	0.825
$WC_{q\text{-}ROF}(\mathcal{A}_4, \mathcal{A}^+)$	0.859	0.835	0.835	0.835	0.833
$WC_{q\text{-}ROF}(\mathcal{A}_5, \mathcal{A}^+)$	**1**	**1**	**1**	**1**	**1**
$WC_{q\text{-}ROF}(\mathcal{A}_6, \mathcal{A}^+)$	0.733	0.711	0.697	0.706	0.709
$WC_{q\text{-}ROF}(\mathcal{A}_7, \mathcal{A}^+)$	0.793	0.764	0.748	0.759	0.762

TABLE 4.9

The results of ranking for $E_{KS}, E_{PL}^1, E_{PL}^2, E_{PL}^3, E_{PL}^4$

Entropy measures	Rankings	Best alternative
E_{KS} (Kumar and Sharma, 2022)	$\mathcal{A}_1 \preceq \mathcal{A}_6 \preceq \mathcal{A}_2 \preceq \mathcal{A}_7 \preceq \mathcal{A}_4 \preceq \mathcal{A}_3 \preceq \mathcal{A}_5$	$\mathcal{A}_5$
E_{PL}^1 (Peng and Liu, 2019)	$\mathcal{A}_1 \preceq \mathcal{A}_6 \preceq \mathcal{A}_2 \preceq \mathcal{A}_7 \preceq \mathcal{A}_3 \preceq \mathcal{A}_4 \preceq \mathcal{A}_5$	$\mathcal{A}_5$
E_{PL}^2(Peng and Liu, 2019)	$\mathcal{A}_1 \preceq \mathcal{A}_6 \preceq \mathcal{A}_2 \preceq \mathcal{A}_7 \preceq \mathcal{A}_3 \preceq \mathcal{A}_4 \preceq \mathcal{A}_5$	$\mathcal{A}_5$
E_{PL}^3 (Peng and Liu, 2019)	$\mathcal{A}_1 \preceq \mathcal{A}_6 \preceq \mathcal{A}_2 \preceq \mathcal{A}_7 \preceq \mathcal{A}_3 \preceq \mathcal{A}_4 \preceq \mathcal{A}_5$	$\mathcal{A}_5$
E_{PL}^4 (Peng and Liu, 2019)	$\mathcal{A}_1 \preceq \mathcal{A}_6 \preceq \mathcal{A}_2 \preceq \mathcal{A}_7 \preceq \mathcal{A}_3 \preceq \mathcal{A}_4 \preceq \mathcal{A}_5$	$\mathcal{A}_5$

effective tool in decision-making methods. Thus, the outputs of this study have the potential to be utilized in many applications.

REFERENCES

K. T. Atanassov. *Intuitionistic Fuzzy Sets*. Springer-Verlag Berlin Heidelberg, 1999.

P. Bhowal, S. Sen, J. H. Yoon, Z. W. Geem, and Ram Sarkar. Choquet integral and coalition game-based ensemble of deep learning models for COVID-19 screening from chest x-ray images. *IEEE Journal of Biomedical and Health Informatics*, 25(12):4328–4339, 2021.

M. C. Bozyiğit, M. Olgun, and M. Ünver. Circular Pythagorean fuzzy sets and applications to multi-criteria decision making. *Informatica*, 34(4):713–742, 2023.

M. C. Bozyiğit, M. Olgun, M. Ünver, and D. Söylemez. Parametric picture fuzzy cross-entropy measures based on d-Choquet integral for building material recognition. *Applied Soft Computing*, 112167, 2024a.

M. C. Bozyiğit, M. Ünver, and M. Olgun. Circular Pythagorean fuzzy Choquet integral operators and applications to multi-criteria decision making. In *International Conference on Intelligent and Fuzzy Systems*, pages 635–643. Springer, 2024b.

J. Chen, H. Wu, P. Yang, F. Lyu, and X. Shen. Cooperative edge caching with location-based and popular contents for vehicular networks. *IEEE Transactions on Vehicular Technology*, 69(9):10291–10305, 2020.

G. Choquet. Theory of capacities. *Annales de l'institut Fourier*, 5:131–295, 1954.

A. De Lucca and S. Termini. A definition of a nonprobabilistic entropy in the setting of fuzzy sets theory. *Information and Control*, 20(4):301–312, 1972.

J. Gao, S. Zhang, L. Zhao, and X. Shen. The design of dynamic probabilistic caching with time-varying content popularity. *IEEE Transactions on Mobile Computing*, 20(4):1672–1684, 2020.

H. Garg, M. Ünver, B. Aydoğan, and M. Olgun. An extended TOPSIS and entropy measure based on Sugeno integral in Pythagorean fuzzy set setting. *Journal of Intelligent & Fuzzy Systems*, 45(2):2537–2549, 2023.

H. Garg, M. C. Bozyiğit, and M. Ünver, (2025). An integrated group decision-making approach with a cross-entropy measure under circular q-rung orthopair fuzzy set information. *Information Sciences*, 122841.

M. Grabisch and M. Roubens. Application of the Choquet integral in multicriteria decision making. In Michel Grabisch, Toshiaki Murofushi, Michio Sugeno (Eds.), *Fuzzy Measures and Integrals-Theory and Applications*, pages 348–374, Physica-Verlag HD, 2000.

S. Han, H. Su, C. Yang, and A. F. Molisch. Proactive edge caching for video on demand with quality adaptation. *IEEE Transactions on Wireless Communications*, 19(1):218–234, 2019.

S. Heilpern. Using Choquet integral in economics. *Statistical Papers*, 43(1):53–73, 2002.

W. L. Hung and M. S. Yang. On the J-divergence of intuitionistic fuzzy sets with its application to pattern recognition. *Information Sciences*, 178(6):1641–1650, 2008.

S. M. A. Iqbal and Asaduzzaman. Cache-MCDM: A hybrid caching scheme in mobile named data networks based on multi-criteria decision making. *Future Generation Computer Systems*, 154:344–358, 2024.

P. Jaganathan and R. Kuppuchamy. A threshold fuzzy entropy based feature selection for medical database classification. *Computers in Biology and Medicine*, 43(12):2222–2229, 2013.

D. Joshi and S. Kumar. Intuitionistic fuzzy entropy and distance measure based TOPSIS method for multi-criteria decision making. *Egyptian Informatics Journal*, 15(2):97–104, 2014.

N. Karamchandani, U. Niesen, M. A. Maddah-Ali, and S. N. Diggavi. Hierarchical coded caching. *IEEE Transactions on Information Theory*, 62(6):3212–3229, 2016.

M. J. Khan, P. Kumam, and M. Shutaywi. Knowledge measure for the q-rung orthopair fuzzy sets. *International Journal of Intelligent Systems*, 36(2):628–655, 2021.

S. Kullback and R. A. Leibler. On information and sufficiency. *The Annals of Mathematical Statistics*, 22(1):79–86, 1951.

K. Kumar and A. Sharma. q-rung orthopair fuzzy entropy measure and its application in multi-attribute decision-making. *Meta-Heuristic Optimization Techniques: Applications in Engineering*, 10:117, 2022.

H. Li, Y. Yang, and S. Yin. Two λ-correlation coefficients of q-rung orthopair fuzzy sets and their application to clustering analysis. *Journal of Intelligent & Fuzzy Systems*, 39(1):581–591, 2020.

J. Li, Y. Chen, Z. Lin, W. Chen, B. Vucetic, and L. Hanzo. Distributed caching for data dissemination in the downlink of heterogeneous networks. *IEEE Transactions on communications*, 63(10):3553–3568, 2015.

D. Liu, X. Chen, and D. Peng. Some cosine similarity measures and distance measures between q-rung orthopair fuzzy sets. *International Journal of Intelligent Systems*, 34(7):1572–1587, 2019.

L. Liu, J. Wu, G. Wei, C. Wei, J. Wang, and Y. Wei. Entropy-based GLDS method for social capital selection of a PPP project with q-rung orthopair fuzzy information. *Entropy*, 22(4):414, 2020.

P. Liu and P. Wang. Some q-rung orthopair fuzzy aggregation operators and their applications to multiple-attribute decision making. *International Journal of Intelligent Systems*, 33(2):259–280, 2018.

Z. Liu, P. Liu, and X. Liang. Multiple attribute decision-making method for dealing with heterogeneous relationship among attributes and unknown attribute weight information under q-rung orthopair fuzzy environment. *International Journal of Intelligent Systems*, 33(9):1900–1928, 2018.

X. Peng and L. Liu. Information measures for q-rung orthopair fuzzy sets. *International Journal of Intelligent Systems*, 34(8):1795–1834, 2019.

X. Peng, H. Yuan, and Y. Yang. Pythagorean fuzzy information measures and their applications. *International Journal of Intelligent Systems*, 32(10):991–1029, 2017.

X. Peng, H. Huang, and Z. Luo. q-rung orthopair fuzzy decision-making framework for integrating mobile edge caching scheme preferences. *International Journal of Intelligent Systems*, 36(5):2229–2266, 2021.

F. A. Pujol, M. Pujol, A. Jimeno-Morenilla, and M. J. Pujol. Face detection based on skin color segmentation using fuzzy entropy. *Entropy*, 19(1):26, 2017.

M. Reiss-Mirzaei, M. Ghobaei-Arani, and L. Esmaeili. A review on the edge caching mechanisms in the mobile edge computing: A social-aware perspective. *Internet of Things*, 22:100690, 2023.

C. E. Shannon. A mathematical theory of communication. *The Bell System Technical Journal*, 27(3):379–423, 1948.

M. Sugeno. Theory of fuzzy integrals and its applications. *Doctoral Thesis, Tokyo Institute of Technology*, 1974.

E. Szmidt and J. Kacprzyk. Entropy for intuitionistic fuzzy sets. *Fuzzy Sets and Systems*, 118(3):467–477, 2001.

E. Takahagi. On identification methods of λ-fuzzy measures using weights and λ. *Japanese Journal of Fuzzy Sets and Systems*, 12(5):665–676, 2000.

E. Takahagi. λ fuzzy measure identification methods using λ and weights. Senshu University, Japan, 2005. Accessed on Nov. 2016. [Online]. http://www.isc.senshu-u.ac.jp/~thc0456/Efuzzyweb/mant2/fz_sou.pdf.

K. Thar, N. H. Tran, T. Z. Oo, and C. S. Hong. DeepMEC: Mobile edge caching using deep learning. *IEEE Access*, 6:78260–78275, 2018.

M. Ünver and B. Aydoğan. Information measures for continuous function-valued q-rung orthopair fuzzy sets and an extended TOPSIS. *Journal of Uncertain Systems*, 17(1):2350012, 2024.

R. Wang and Y. Li. A novel approach for green supplier selection under a q-rung orthopair fuzzy environment. *Symmetry*, 10(12):687, 2018.

S. Wang, X. Zhang, Y. Zhang, L. Wang, W. Yang, and J. Wang. A survey on mobile edge networks: Convergence of computing, caching and communications. *IEEE Access*, 5:6757–6779, 2017.

G. Wei, H. Gao, and Y. Wei. Some q-rung orthopair fuzzy Heronian mean operators in multiple attribute decision making. *International Journal of Intelligent Systems*, 33(7):1426–1458, 2018.

H. Wu, J. Zhang, Z. Cai, F. Liu, Y. Li, and A. Liu. Toward energy-aware caching for intelligent connected vehicles. *IEEE Internet of Things Journal*, 7(9):8157–8166, 2020.

Q. Xu, Z. Su, and R. Lu. Game theory and reinforcement learning based secure edge caching in mobile social networks. *IEEE Transactions on Information Forensics and Security*, 15:3415–3429, 2020.

R. R. Yager. Pythagorean membership grades in multicriteria decision making. *IEEE Transactions on Fuzzy Systems*, 22(4):958–965, 2013a.

R. R. Yager. Pythagorean fuzzy subsets. In *2013 Joint IFSA World Congress and NAFIPS Annual Meeting (IFSA/NAFIPS)*, pages 57–61, 2013b.

R. R. Yager. Generalized orthopair fuzzy sets. *IEEE Transactions on Fuzzy Systems*, 25(5):1222–1230, 2016.

Q. Yang, P. Hassanzadeh, D. Gündüz, and E. Erkip. Centralized caching and delivery of correlated contents over gaussian broadcast channels. *IEEE Transactions on Communications*, 68(1):122–136, 2019.

J. Yao, T. Han, and N. Ansari. On mobile edge caching. *IEEE Communications Surveys & Tutorials*, 21(3):2525–2553, 2019.

L. A. Zadeh. Fuzzy sets. *Information and Control*, 8(3):338–353, 1965.

L. A. Zadeh. Probability measures of fuzzy events. *Journal of Mathematical Analysis and Applications*, 23(2):421–427, 1968.

K. Zhang, K. Zhang, S. Leng, Y. He, S. Maharjan, and Y. Zhang. Cooperative content caching in 5G networks with mobile edge computing. *IEEE Wireless Communications*, 25(3):80–87, 2018.

Y. Zhang. *Mobile Edge Computing. Vol. 9.* Springer International Publishing, 2022.

H. Zhu, Y. Cao, T. Wang, W. Jiang, and S. Jin. Deep reinforcement learning for mobile edge caching: Review, new features, and open issues. *IEEE Network*, 32(6):50–57, 2018.

5 Hybrid Approach Using Interval Type-2 Fuzzy TOPSIS and Unsupervised Machine Learning for Water Security and Water Source Area Challenges

Nurnadiah Zamri, Wan Nur Amira Wan Azman, Mohamad Ammar Pairan, Sukono, and Lazim Abdullah

5.1 INTRODUCTION

Globally, water security is facing severe challenges due to a combination of factors, including rapid population growth, changing hydrological conditions, higher per capita water usage, migration from rural to urban areas, pollution of water resources, excessive groundwater extraction, and climate change and variability (GWP, 2000b; Jones et al., 2009; Vörösmarty et al., 2010; Evengard et al., 2011). Over 1.1 billion people lack access to a secure water supply, and 2.6 billion need improved sanitation (WWC, 2000; UNESCO, 2009; Onda et al., 2012). Efforts to improve water security have been taken by a number of organizations and entities, including the World Water Forum, the Group of Eight (G8), the United Nations, and a group of researchers (Jones et al., 2009). These researchers have explored water security from various angles and perspectives. For example, Chapagain et al. (2022) conducted a thorough analysis of the outcomes of water security indicators and connected the strategies, plans, and policies of five cities with the indicators of a combined framework to identify policy gaps and propose measures to enhance urban water management. Demerdash et al. (2022) proposed a new theory for water security, the Irrigation Water Security Quality-based index, which evaluates irrigation water security based on both quality

DOI: 10.1201/9781003529088-5

and quantity. They also estimated the impact of water quality on soil degradation, agricultural productivity, and public health. Veettil et al. (2022) presented an overview of water security assessment by focusing on various water security indicators and the water footprint theory. They also examined the use of physically based hydrological models, such as the Variable Infiltration Capacity (VIC) and the Soil and Water Assessment Tool (SWAT), in water security assessment at both regional and continental levels. Castro-Pardo et al. (2022) calculated a composite index to assess water security through the application of Data Envelopment Analysis (DEA). The model takes into consideration nine indicators across the four dimensions of water security: human health and well-being, water environment state, society's functions and responsibilities, and the sustainability of livelihoods and stability.

In addition to that, some studies in the field of water also utilized Fuzzy Multi-Criteria Decision-Making (FMCDM) for further analysis. For example, Torynyeviadzi et al. (2021) applied Fuzzy Analytic Hierarchy Process-Technique for Order Preference by Similarity to Ideal Solution (Fuzzy AHP-TOPSIS) framework to holistically evaluate the vulnerability of water distribution networks (WDNs) by incorporating topological, hydraulic, and water quality parameters. The results can be used as a tool by water engineers to analyze critical nodes, schedule maintenance strategies, and assess the impact of failures on the distribution networks. Elleuch et al. (2019) combined FMCDM and Mathematical Optimization Programming (MOP) to solve the water allocation problem in the Sfax irrigation district in southern Tunisia. The results showed that this combination of methods effectively linked irrigation expert opinions, economic returns, and water resource objectives. Noori et al. (2021) expanded upon the MCDM approach with the fuzzy Elimination Et Choice Translating Reality (FELECTRE) III for water supply choice optimization. Then, they applied the method to the water supply system in the Gamasiab Basin in Iran and found that it is a suitable technique for similar complex water supply engineering projects. Tu et al. (2021) developed a comprehensive evaluation methodology for assessing the level of Regional Water Resources Coordination (RWRC) using Decision-Making Trial and Evaluation Laboratory (DEMATEL) and VlseKriterijumska Optimizacija I Kompromisno Resenje (VIKOR) with Hesitant Fuzzy Linguistic Term Sets (HFLTSs). Their final ranking obtained the six relevant regions in North China Plain for water resources coordination evaluation. Fernández-López (2021) used TOPSIS-based FMCDM for the assessment of wastewater treatment plants (WWTPs) to remove organic matter, phosphorus, and nitrogen but no other pollutants such as pharmaceutical active compounds (PhACs). Sahoo et al. (2024) proposed an FMCDM under fuzzy TOPSIS's min and max operations for optimal college location selection. Their study presented a straightforward case study identifying a suitable location for establishing a college. Gao and Zhou (2025) combined Fuzzy DEMATEL with Failure Mode and Effect Analysis (FMEA) to assess the risk in the converter steelmaking process. Their result provided a refined and reliable risk assessment framework, offering valuable insights for industrial practitioners to implement targeted preventive measures and ultimately improve the safety and reliability of converter steelmaking.

Interval type-2 fuzzy TOPSIS (IT2FTOPSIS)-based FMCDM has been the subject of discussion among various authors. Chen and Lee (2010) were first to develop

IT2FTOPSIS that can handle FMCDM problems using IT2FSs, making it a more flexible and intelligent way of handling FMCDM. Zamri and Abdullah (2014) introduced a new positive and negative linguistic variable for IT2FTOPSIS, and their results showed that it was highly beneficial in terms of applicability and provided a new dimension to problem-solving techniques for type-2 fuzzy group decision-making. Wu et al. (2018) proposed a new IT2FTOPSIS model to solve large-scale group decision-making (LGDM) problems in complex and uncertain environments. Celik and Akyuz (2018) combined AHP and TOPSIS under IT2FSs to address the uncertainty of DM's judgments in decision-making and conducted sensitivity analysis to investigate the impact of key performance indicators. Chen et al. (2019) developed a proportional interval type-2 hesitant fuzzy TOPSIS based on Hamacher aggregation operators and andness optimization models. Their proposed method represented a novel paradigm for linguistic group decision-making for solving computing with words problems. Meniz (2021) proposed an advanced TOPSIS method with new fuzzy metric based on interval type-2 fuzzy sets (IT2FSs). Then, he applied it to a video chat program selection problem. Yilmaz et al. (2023) integrated IT2FTOPSIS with goal programming for solving celiac disease multipurpose diet plan. Their results retrieved the best diet model in line with the calories and cost value based on the proposed method. IT2FTOPSIS has not yet been applied to any studies related to water security, and IT2FTOPSIS is known as a robust decision-making technique. However, it might struggle to address highly complex and dynamic water security, where adaptive, data-driven methods are required for identifying patterns, trends, and relationships in large water security datasets. Therefore, there is a need for an advanced intelligent method that can be integrated with IT2FTOPSIS. Machine learning (ML) is one of the suitable methods that can be integrated with IT2FTOPSIS to achieve this advanced intelligent method.

ML has been extensively studied in various FMCDM methods. For instance, Zamri et al. (2023) combined IT2FVIKOR with ML to determine the best water security strategies for the Terengganu River areas. This study offered an intelligent model for addressing water security issues and locating the most contaminated areas. Chowdhury et al. (2022) presented an ensemble-based MCDM approach for selecting the best ML techniques for COVID-19 cough classification. Using four cough datasets, the study showed promising results with high accuracy. Saha and Mondal (2022) integrated MCDM and ML procedures to evaluate the viability of agricultural land in the Gangarampur subdivision, taking into account soil fertility and site suitability. Yazici et al. (2022) prioritized individual tacit knowledge criteria using ML methods and FAHP, with support vector machine (SVM) and FAHP defining communication, time efficiency, and innovative capability as the most important factors. The results indicated that SVM and FAHP identified communication between employees and supervisors, employee time efficiency, and employee innovation capability as the crucial tacit knowledge criteria. Alazemi et al. (2022) focused on the role of internal factors in small- and medium-scale supply chains, proposing a three-phase fuzzy-based framework for addressing product completion time. The three phases consist of using supervised ML to categorize production options, conducting quantitative research, and implementing a fuzzy-based heuristic to generate production. Mohsin et al. (2022) introduced a novel method for landfill site selection and sustainable SWM in India using integrated FAHP-SVM and FAHP-RF, providing guidance for decision makers and planners in other developing countries. Our study improved IT2FTOPSIS with ML to address water security issues, due to the compatibility of ML with MCDM.

Due to the unstructured and high-dimensional data to uncover hidden structures and relationships on water security issues, this chapter focuses on integrating IT2FTOPSIS with unsupervised ML to enhance the decision-making process in water security. IT2FTOPSIS with unsupervised ML is capable of analyzing complex water security datasets to group regions based on similar related challenges (such as pollution areas, pollution levels, and many more). The proposed integration combines IT2FTOPSIS with positive and negative IT2TrFNs (Zamri et al., 2015), IT2 fuzzy entropy, hybrid averaging of linear order, and IT2FELECTRE. Then, unsupervised ML includes hierarchical clustering and self-organizing map (SOM) for clustering, silhouette analysis to find the most optimal number of clusters, and lastly Adjusted Rand Index (ARI) to find the best comparison within two clusters. The methodology is divided into two phases. Phase 1 uses the IT2FTOPSIS method to evaluate and select the best water quality strategies from six alternatives based on five criteria and five decision makers. Phase 2 employs hierarchical clustering and SOM to cluster the polluted areas in the Selangor River. Silhouette analysis is used to determine the optimal number of clusters, while ARI is used to compare the results of the two clustering algorithms. The best clusters and their corresponding polluted areas are highlighted. This integrated IT2FTOPSIS with ML approach provides a systematic and rational solution for the water security problem in the Selangor River area.

5.2 METHODOLOGY FRAMEWORK

This section focuses on using MCDM methods including IT2FTOPSIS and unsupervised machine learning algorithms, hierarchical clustering, and SOM. The proposed method is depicted in Figure 5.1 as a schematic diagram. In Phase 1, IT2FTOPSIS is used to identify the best water security strategies based on linguistic variables data collected from decision makers, considering all the human and nonhuman factors related to water security. In Phase 2, data from the Department of Environment Malaysia are used to apply two different clustering algorithms and perform silhouette analysis to obtain a list of clusters and the optimal number of clusters. The ARI is used to compare the agreement between the cluster assignments produced by each algorithm, with a score ranging from 1 to –1 indicating the extent of agreement between the two sets of groupings. Finally, the list of polluted areas in each cluster is obtained.

5.3 THE INTERVAL TYPE-2 FUZZY TOPSIS PROCEDURES

The IT2FTOPSIS with positive and negative IT2TrFNs, IT2 fuzzy entropy, hybrid averaging of linear order, and IT2FELECTRE is developed to achieve a higher rational, systematic decision-making process. IT2TrFNs help to cope with the competence

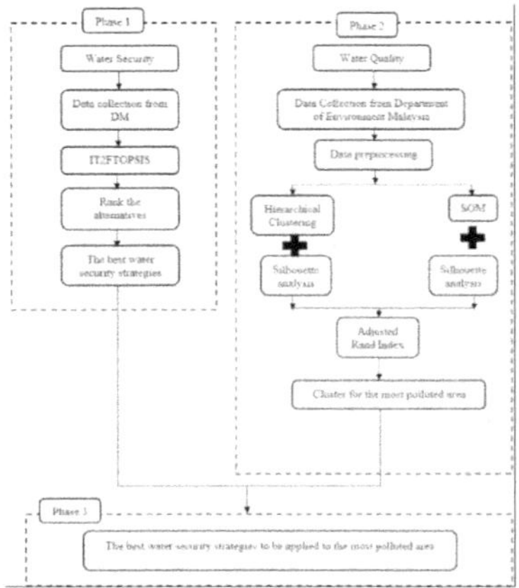

FIGURE 5.1 A process of integrated IT2FTOPSIS with machine learning to the real application

of the DM's decisions. IT2 fuzzy entropy helps reducing the decision-making burden in the subjective weighting process and achieving a rather consistent weighting outcome, counting the uncertainty and ambiguity that can affect the evaluation of the alternatives and attributes performance and at the same time verifying the existence of a concordance of attribute in favor of the assertion that an alternative is as good as another one. The IT2FTOPSIS procedure is explained step-by-step as follows:

Assume that there is a finite set X of alternatives, where $X = \{x_1, x_2, \ldots, x_n\}$, and assume that there is a finite set F of attributes, where $F = \{f_1, f_2, \ldots, f_m\}$. Assume that there are k DMs $D_1, D_2, \ldots,$ and D_k. The proposed method is now presented as follows:

Step 1: Construct a hierarchical diagram of MADM problem.

Construct the decision matrix Y_p of the pth DM and construct the average decision matrix $\tilde{Y}$, respectively, shown as follows:

$$Y_p = \left[\tilde{\tilde{f}}_{ij}^{\,p}\right]_{m\times n} = \begin{array}{c} \\ f_1 \\ f_2 \\ \vdots \\ f_m \end{array} \begin{array}{c} \begin{array}{cccc} x_1 & x_2 & \cdots & x_n \end{array} \\ \begin{bmatrix} \tilde{\tilde{f}}_{11}^{\,p} & \tilde{\tilde{f}}_{12}^{\,p} & \cdots & \tilde{\tilde{f}}_{1n}^{\,p} \\ \tilde{\tilde{f}}_{21}^{\,p} & \tilde{\tilde{f}}_{22}^{\,p} & \cdots & \tilde{\tilde{f}}_{2n}^{\,p} \\ \vdots & \vdots & \vdots & \vdots \\ \tilde{\tilde{f}}_{m1}^{\,p} & \tilde{\tilde{f}}_{m2}^{\,p} & \cdots & \tilde{\tilde{f}}_{mn}^{\,p} \end{bmatrix} \end{array} \tag{5.1}$$

$$\tilde{Y} = \left[\tilde{\tilde{f}}_{ij}\right]_{m\times n}, \tag{5.2}$$

where $\tilde{\tilde{f}}_{ij} = \left(\frac{\tilde{\tilde{f}}_{ij}^{1} \oplus \tilde{\tilde{f}}_{ij}^{2} \oplus \ldots \oplus \tilde{\tilde{f}}_{ij}^{k}}{k} \right)$, $\tilde{\tilde{f}}_{ij}$ is a positive and negative IT2TrFNs (Table 5.1), $1 \le i \le m, 1 \le j \le n, 1 \le p \le k$, and k denotes the number of DMs.

Step 2: Construct the weighted DMs' matrix.

Therefore, the proposed entropy weight method based on IT2FSs is defined as follows:

Let assign $\tilde{\tilde{f}}_{ij} = \left(\tilde{\tilde{f}}_{ij}^{L}, \tilde{\tilde{f}}_{ij}^{U} \right)$,
thus,

$$E\left(\tilde{\tilde{A}}_{ij}^{L} \right) = \frac{\sqrt{\sum_{i=1}^{n} \left[\tilde{\tilde{f}}_{ij}^{near} - \tilde{\tilde{f}}_{ij}^{L} \right]}}{\sqrt{\sum_{i=1}^{n} \left[\tilde{\tilde{f}}_{ij}^{far} - \tilde{\tilde{f}}_{ij}^{L} \right]}} \tag{5.3a}$$

$$E\left(\tilde{\tilde{A}}_{ij}^{U} \right) = \frac{\sqrt{\sum_{i=1}^{n} \left[\tilde{\tilde{f}}_{ij}^{near} - \tilde{\tilde{f}}_{ij}^{U} \right]}}{\sqrt{\sum_{i=1}^{n} \left[\tilde{\tilde{f}}_{ij}^{far} - \tilde{\tilde{f}}_{ij}^{U} \right]}} \tag{5.3b}$$

where

$$E\left(\tilde{\tilde{A}}_{ij} \right) = \left[E\left(\tilde{\tilde{A}}_{ij}^{L} \right), E\left(\tilde{\tilde{A}}_{ij}^{U} \right) \right] \tag{5.3c}$$

Step 3: Divide using maximal entropy value.

Then, all the entropy values are divided by using the maximal entropy value and value of h_{ij} is used to represent the outcomes of maximal entropy value. Therefore, it can be defined as follows:

TABLE 5.1
Linguistic scale of rating for the extended IT2FTOPSIS

Linguistic terms	Linguistic scale
Very poor (VP)	((−1.0, −0.9, −0.8, −0.7; 0.8, 0.8), (−1.0, −1.0, −0.8, −0.6; 1, 1))
Poor (P)	((−0.8, −0.7, −0.5, −0.4; 0.8, 0.8), (−0.9, −0.7, −0.5, −0.3; 1, 1))
Medium poor (MP)	((−0.5, −0.4, −0.2, −0.1; 0.8, 0.8), (−0.6, −0.4, −0.2, 0; 1, 1))
Medium (M)	((−0.2, −0.1, 0.1, 0.2; 0.8, 0.8), (−0.3, −0.2, 0.2, 0.3; 1, 1))
Medium good (MG)	((0.1, 0.2, 0.4, 0.5; 0.8, 0.8), (0, 0.2, 0.4, 0.6; 1, 1))
Good (G)	((0.4, 0.5, 0.7, 0.8; 0.8, 0.8), (0.3, 0.5, 0.7, 0.9; 1, 1))
Very good (VG)	((0.7, 0.8, 1.0, 1.0; 0.8, 0.8), (0.6, 0.8, 1.0, 1.0; 1, 1))

$$\tilde{\tilde{h}}_{ij} = 1 / FOU\left(\tilde{\tilde{h}}_{ij}\right) = \left[\tilde{\tilde{h}}_{ij}^{\,L}, \tilde{\tilde{h}}_{ij}^{\,U}\right] \tag{5.4a}$$

$$\tilde{\tilde{h}}_{i1} = \left[\left(\frac{E\left(\tilde{\tilde{A}}_{i1}^{L}\right)}{\max\left(E\left(\tilde{\tilde{A}}_{i1}^{L}\right)\right)}\right), \left(\frac{E\left(\tilde{\tilde{A}}_{i1}^{U}\right)}{\max\left(E\left(\tilde{\tilde{A}}_{i1}^{U}\right)\right)}\right)\right], \tilde{\tilde{h}}_{i2} = \left[\left(\frac{E\left(\tilde{\tilde{A}}_{i2}^{L}\right)}{\max\left(E\left(\tilde{\tilde{A}}_{i2}^{L}\right)\right)}\right), \left(\frac{E\left(\tilde{\tilde{A}}_{i2}^{U}\right)}{\max\left(E\left(\tilde{\tilde{A}}_{i2}^{U}\right)\right)}\right)\right]$$

$$\tilde{\tilde{h}}_{in} = \left[\left(\frac{E\left(\tilde{\tilde{A}}_{in}^{L}\right)}{\max\left(E\left(\tilde{\tilde{A}}_{in}^{L}\right)\right)}\right), \left(\frac{E\left(\tilde{\tilde{A}}_{in}^{U}\right)}{\max\left(E\left(\tilde{\tilde{A}}_{in}^{U}\right)\right)}\right)\right] \tag{5.4b}$$

Then, the decision matrix can be expressed as follows:

$$D = \begin{bmatrix} \tilde{\tilde{h}}_{11} & \tilde{\tilde{h}}_{12} & \cdots & \tilde{\tilde{h}}_{1n} \\ \tilde{\tilde{h}}_{21} & \tilde{\tilde{h}}_{22} & \cdots & \tilde{\tilde{h}}_{2n} \\ \vdots & \vdots & \ddots & \vdots \\ \tilde{\tilde{h}}_{m1} & \tilde{\tilde{h}}_{m2} & \cdots & \tilde{\tilde{h}}_{mn} \end{bmatrix} \tag{5.4c}$$

Step 4: Weight of attributes.

Finally, calculate the weight of attributes by using the IT2 fuzzy entropy weight formula. The w_j is used to present the outcome of weight value of attribute *j*. Thus, it can be defined as:

$$\left(\tilde{\tilde{w}}_j\right) = 1 / F\left(\tilde{\tilde{w}}_j\right) = \left[\tilde{\tilde{w}}_j^{L}, \tilde{\tilde{w}}_j^{U}\right] \tag{5.5a}$$

$$\left(\tilde{\tilde{w}}_j^{L}, \tilde{\tilde{w}}_j^{U}\right) = \left[\left(\frac{1 - \tilde{\tilde{a}}_j^{L}}{\tilde{\tilde{T}}^{L}}\right), \left(\frac{1 - \tilde{\tilde{a}}_j^{U}}{\tilde{\tilde{T}}^{U}}\right)\right] \tag{5.5b}$$

where $\tilde{\tilde{a}}_j^{L} = \sum_{j=1}^{m} \tilde{\tilde{h}}_{ij}^{L} / n$, $\tilde{\tilde{a}}_j^{U} = \sum_{j=1}^{m} \tilde{\tilde{h}}_{ij}^{U} / n$, $\tilde{\tilde{T}}^{L} = \sum_{j=1}^{m} \tilde{\tilde{a}}_j^{L}$, $\tilde{\tilde{T}}^{U} = \sum_{j=1}^{m} \tilde{\tilde{a}}_j^{U}$, and

F is a FOU. The $\tilde{\tilde{a}}_j^{L}$ and $\tilde{\tilde{a}}_j^{U}$ represents the summation of the normalized entropy values, which correspond to the attribute *j*. The $\tilde{\tilde{T}}_j^{L}$ and $\tilde{\tilde{T}}_j^{U}$ is the summation of $\tilde{\tilde{a}}_j^{L}$ and $\tilde{\tilde{a}}_j^{U}$, and n is the number of attributes. The sum of attribute weight is $\Sigma\left(\tilde{\tilde{w}}_j^{L}, \tilde{\tilde{w}}_j^{U}\right) = (1, 1), \left(\tilde{\tilde{w}}_j^{L}, \tilde{\tilde{w}}_j^{U}\right) \in [0,1]$.

Step 5: Construct the weighted of decision matrices.

Construct the weighted decision matrix $\overline{Y}_w$,

$$\overline{Y}_W = \left[\tilde{\tilde{v}}_{ij}\right]_{m\times n} = \begin{array}{c} \\ f_1 \\ f_2 \\ \vdots \\ f_m \end{array}\begin{array}{c} \begin{array}{cccc} x_1 & x_2 & \cdots & x_n \end{array} \\ \begin{bmatrix} \tilde{\tilde{v}}_{11} & \tilde{\tilde{v}}_{12} & \cdots & \tilde{\tilde{v}}_{1n} \\ \tilde{\tilde{v}}_{21} & \tilde{\tilde{v}}_{22} & \cdots & \tilde{\tilde{v}}_{2n} \\ \vdots & \vdots & \ddots & \vdots \\ \tilde{\tilde{v}}_{m1} & \tilde{\tilde{v}}_{m2} & \cdots & \tilde{\tilde{v}}_{mn} \end{bmatrix} \end{array}, \tag{5.6}$$

where $\tilde{\tilde{v}}_{ij} = \tilde{\tilde{w}}_i \otimes \tilde{\tilde{f}}_{ij}$, $1 \le i \le m$, and $1 \le j \le n$.

Step 6: Construct the hybrid averaging approach of linear order method.

Employ the linear order hybrid averaging operation to determine the collective evaluation value of weighted decision matrix $\tilde{\tilde{v}}_{ij}$ and the collective importance weight $\tilde{\tilde{w}}_j$.

$$\tilde{\tilde{v}}_{ij}^{h'} = \frac{1}{2}\left[\left(\left(\frac{-6l+6u-x-y}{12}\right)^L /2\right)\left(\frac{H_1\left(\tilde{A}_1^L\right)}{H_2\left(\tilde{A}_1^L\right)}\right)+\left(\left(\frac{-6l+6u-x-y}{12}\right)^U /2\right)\left(\frac{H_1\left(\tilde{A}_1^U\right)}{H_2\left(\tilde{A}_1^U\right)}\right)\right] \tag{5.7}$$

where

$\left(\frac{-6l+6u-x-y}{12}\right)^L$ represented as $\left(\underline{y}_l(x)+\overline{y}_l(x)\right)$,

$\left(\frac{-6l+6u-x-y}{12}\right)^U$ represented as $\left(\underline{y}_r(x)+\overline{y}_r(x)\right)$,

and h' represented as hybrid.

Step 7: Calculate the concordance matrix.

The concordance set $\tilde{\tilde{c}}_{gf}$ of the pair of alternatives A_g and A_j is the index set corresponding to the attribute for which A_g is preferred to A_j by comparing their linear order $\tilde{\tilde{v}}_{ij}^{h'}(A_g)$ and $\tilde{\tilde{v}}_{ij}^{h'}(A_j)$. Specifically,

$$P_c = \left\{ j \,\middle|\, \tilde{\tilde{v}}_{ij}^{h'}(A_g) \ge \tilde{\tilde{v}}_{ij}^{h'}(A_j)\right\} \tag{5.8}$$

where P_c is the index of all attributes belonging to the concordance coalition with the outranking relation:

$$\tilde{\tilde{v}}_{ij}^{h'}(A_g)\, S\, \tilde{\tilde{v}}_{ij}^{h'}(A_j). \tag{5.9}$$

The relative evaluative value for the concordance set is measured using the concordance index. The concordance index is proportional to the sum of positive and negative IT2TrFN weights $\tilde{\tilde{w}}_j^{h'}$ associated with the attributes whose indices are contained in the concordance set. This is denoted as:

$$\tilde{\tilde{w}}_{j=1}^{n\ h'} = \tilde{\tilde{w}}_1^{h'} \oplus \tilde{\tilde{w}}_2^{h'} \oplus \ldots \oplus \tilde{\tilde{w}}_n^{h'}. \tag{5.10}$$

Thus, the concordance matrix is defined as follows:

$$\tilde{\tilde{C}} = \begin{bmatrix} - & \cdots & \tilde{\tilde{c}}_{1f} & \cdots & \tilde{\tilde{c}}_{1(m-1)} & \tilde{\tilde{c}}_{1m} \\ \vdots & \ddots & \vdots & \ddots & \vdots & \vdots \\ \tilde{\tilde{c}}_{g1} & \cdots & \tilde{\tilde{c}}_{gf} & \cdots & \tilde{\tilde{c}}_{g(m-1)} & \tilde{\tilde{c}}_{gm} \\ \vdots & \ddots & \vdots & \ddots & \vdots & \vdots \\ \tilde{\tilde{c}}_{m1} & \cdots & \tilde{\tilde{c}}_{mf} & \cdots & \tilde{\tilde{c}}_{m(m-1)} & - \end{bmatrix} \tag{5.11}$$

where

$$\tilde{\tilde{c}}_{g1} = \sum_{j \in P_c} \tilde{\tilde{w}}_n^{h'} \tag{5.12}$$

or can be says as, the concordance level, $\tilde{\tilde{C}}$, can be defined as the average of the elements in the concordance matrix, represented by $\tilde{\tilde{c}}_{gf} = \sum_{g=1}^{m} \sum_{f=1}^{m} c_{gf} / m(m-1)$.

Step 8: Calculate the discordance matrix.

By contrast, the complementary part known as the discordance set P_D can be defined as $P_D = \left\{ j \middle| \tilde{\tilde{v}}_{ij}^{h'}(A_g) \le \tilde{\tilde{v}}_{ij}^{h'}(A_j) \right\}$, where P_D is the index of all attributes belonging to the discordance coalition and it is against the assertion "A_g is at least as good as A_f."

Thus, the discordance matrix is defined as:

$$\tilde{\tilde{D}} = \begin{bmatrix} - & \cdots & \tilde{\tilde{d}}_{1f} & \cdots & \tilde{\tilde{d}}_{1(m-1)} & \tilde{\tilde{d}}_{1m} \\ \vdots & \ddots & \vdots & \ddots & \vdots & \vdots \\ \tilde{\tilde{d}}_{g1} & \cdots & \tilde{\tilde{d}}_{gf} & \cdots & \tilde{\tilde{d}}_{g(m-1)} & \tilde{\tilde{d}}_{gm} \\ \vdots & \ddots & \vdots & \ddots & \vdots & \vdots \\ \tilde{\tilde{d}}_{m1} & \cdots & \tilde{\tilde{d}}_{mf} & \cdots & \tilde{\tilde{d}}_{m(m-1)} & - \end{bmatrix} \tag{5.13}$$

where

$$\tilde{\tilde{d}}_{gf} = \frac{\max\limits_{j \in P_D} \left| \tilde{\tilde{v}}_{ij}^{h'}(A_g) - \tilde{\tilde{v}}_{ij}^{h'}(A_j) \right|}{\max\limits_{j} \left| \tilde{\tilde{v}}_{ij}^{h'}(A_g) - \tilde{\tilde{v}}_{ij}^{h'}(A_j) \right|} = \frac{\max\limits_{j \in P_D} \left| d\left(\tilde{\tilde{v}}_{ij}^{h'}(A_g), \tilde{\tilde{v}}_{ij}^{h'}(A_j) \right), \tilde{\tilde{v}}_{ij}^{h'}(A_j) \right|}{\max\limits_{j} \left| d\left(\tilde{\tilde{v}}_{ij}^{h'}(A_g), \tilde{\tilde{v}}_{ij}^{h'}(A_j) \right), \tilde{\tilde{v}}_{ij}^{h'}(A_j) \right|} \tag{5.14}$$

Note that there are prominent differences between the elements of $\tilde{\tilde{\boldsymbol{C}}}$ and $\tilde{\tilde{\boldsymbol{D}}}$. The concordance matrix $\tilde{\tilde{\boldsymbol{C}}}$reflects weights of the concordance attribute, and the asymmetric discordance matrix $\tilde{\tilde{\boldsymbol{D}}}$ reflects most relative differences according to the discordance attribute. Both concordance and discordance indices have to be calculated for every pair of actions $(\boldsymbol{g}, \boldsymbol{f})$, where $\boldsymbol{g} \neq \boldsymbol{f}$.

Step 9: Construct the global matrix.

Next, the global matrix $\tilde{\tilde{Z}}$ is calculated by peer-to-peer subtraction of the elements of the matrices $\tilde{\tilde{C}}$ and $\tilde{\tilde{D}}$ as follows:

$$\tilde{\tilde{Z}} = \tilde{\tilde{C}} \otimes \tilde{\tilde{D}} \tag{5.15}$$

The step of this procedure consists of exploitation of the above outranking relation (matrix $\tilde{\tilde{Z}}$) in order to identify as small as possible a subset of actions, from which the best compromise action could be selected.

Step 10: Determine the PIS and NIS.

Determine the positive ideal solution (PIS) and negative ideal solution (NIS), respectively:

$$f^{+} = \left\{\tilde{\tilde{Z}}_1^{+}, \ldots, \tilde{\tilde{Z}}_n^{+}\right\} = \left\{\left(\max_j \tilde{\tilde{Z}}_{ij} \mid i \in I'\right), \left(\min_j \tilde{\tilde{Z}}_{ij} \mid i \in I''\right)\right\}, \tag{5.16a}$$

$$f^{-} = \left\{\tilde{\tilde{Z}}_1^{-}, \ldots, \tilde{\tilde{Z}}_n^{-}\right\} = \left\{\left(\min_j \tilde{\tilde{Z}}_{ij} \mid i \in I'\right), \left(\max_j \tilde{\tilde{Z}}_{ij} \mid i \in I''\right)\right\}, \tag{5.16b}$$

where I' is associated with the positive attribute, and I'' is associated with the negative attribute.

Step 11: Calculate the distance of PIS and NIS.

Calculate the separation measures using the n-dimensional Euclidean distance. The separation of each alternative from the PIS is given as:

$$D_j^{+} = \sqrt{\sum_{i=1}^{n} (\tilde{\tilde{Z}}_{ij} - \tilde{\tilde{Z}}_i^{+})^2}, j = 1, 2, \ldots, J. \tag{5.17a}$$

Similarly, the separation from the NIS is given as:

$$D_j^{-} = \sqrt{\sum_{i=1}^{n} (\tilde{\tilde{Z}}_{ij} - \tilde{\tilde{Z}}_i^{-})^2}, j = 1, 2, \ldots, J. \tag{5.17b}$$

Step 12: Calculate the relative closeness.

Calculate the relative closeness to the ideal solution. The relative closeness of the alternative x_i with respect to f^* is defined as:

$$C_j^{*} = \frac{D_j^{+} + D_j^{-}}{2}, j = 1, 2, \ldots, J. \tag{5.18}$$

Step 13: Normalize the relative closeness.

Lastly, relative closeness is normalized as follows:

$$N_j = \frac{C_j^*}{\sum_{j=1}^{n} C_j^*}, j = 1, 2, \ldots, n. \tag{5.19}$$

Step 14: Sort the values.

Rank the preference order. A large value of closeness coefficient N_j indicates a good performance of the alternative $\tilde{\tilde{f}}_i$. The best alternative is the one with the greatest relative closeness to the ideal solution.

5.4 THE UNSUPERVISED MACHINE LEARNING PROCEDURES

In unsupervised machine learning, data are analyzed without predefined labels to identify new patterns. These algorithms can identify relationships and similarities in data without human involvement. Clustering is a subset of unsupervised learning algorithms, which also includes anomaly detection and latent variable models. The focus of this study is on clustering algorithms, which will be further explained in the next section.

5.5 HIERARCHICAL CLUSTERING ALGORITHM

Ward (1963) researched hierarchical clustering methods that aim to reduce the loss of information from merging two groups. This approach is commonly used with the loss of information calculated as an increase in the error sum of squares (ESS) criterion. The ESS is defined as follows:

$$o(t) = \sum_{k=1}^{K} \sum_{x_i \in C_k} \sum_{j=1}^{n} \left(x_{ij} - x_{kj}\right)^2 \tag{5.20}$$

where K is the number of clusters, j is the sample index $(1, 2, \cdots, n)$, C_k is a k-th cluster, x_i indicates elements in C_k, x_{ij} denotes the i-th item in the j-th cluster C_j, and μ_{kj} is a mean vector in C_k.

Initially, each cluster consists of a single item. During each stage of the analysis, all possible pairs of clusters are evaluated, and the pair that results in the smallest increase in the ESS criterion (representing a minimal loss of information) is merged. The clustering analysis is completed once all clusters have been combined into a single group of N items. The outcome of the Ward's method can be represented in the form of a dendrogram, with the vertical axis indicating the ESS values at which the mergers occur (Richard and Dean, 2002).

5.6 SELF-ORGANIZING MAP ALGORITHM

The self-organizing map (SOM) is a well-known unsupervised neural network that maps n-dimensional multivariate data onto a 2D representation (Kohonen, 1984). In other words, it transforms a nonlinear projection of data into a low-dimensional display. The SOM network is composed of nodes, typically arranged in a rectangular or hexagonal grid, with each neuron in the array connected to an input vector x_i and assigned a weight vector w_i. The nodes in adjacent cells are continually updated through a neighborhood function. The SOM algorithm has two main phases: the competitive phase and the cooperative phase (Hoffmann, 2005; Liu et al., 2008). The competitive phase aims to identify the best matching neuron, while the cooperative phase adjusts the weight of the winner and its surrounding neurons. During the SOM training process, a sample vector x is randomly selected and the Euclidean distances between the input vector and all neurons are calculated. The best matching unit is then determined based on the minimization of the distance between the sample and the weighted vectors.

$$\|x - m_c\| = \min_i \|x - m_i\| \tag{5.21}$$

where c represent the winning neuron and m_c is the closest vector to x on the map.

The weight vector m_c associated with the best matching unit is adjusted to bring the sample x even closer. Additionally, the nodes near the best matching unit are also updated. The revised version can be expressed as follows:

$$m_i(t+1) = m_i(t) + h_{ci}(t)^* \left(x - m_i(t)\right) \tag{5.22}$$

where the $h_{ci}(t)$ is the neighbourhood kernel centered on the winner unit, and t denotes the time.

$$h_{ci}(t) = a(t)^* \exp\left(-\frac{\|r_c - r_i\|^2}{2o(t)^2}\right) \tag{5.23}$$

where $a(t)$ represents the learning rate, which serves as a measure of the degree of matching and decreases gradually with the progression of regression steps (time). $o(t)$ represents the radius of the neighborhood function's expansion. r_c and r_i are the positions of neurons c and i on the SOM grid, respectively.

5.7 ADJUSTED RAND INDEX ALGORITHMS

The Adjusted Rand Index (ARI) is considered the top method for comparing two partitions and achieves a score of 1 when the partitions match perfectly and 0 when there is only random agreement. Although the ARI generally outperforms other partition comparison methods, it still faces the challenge of determining an arbitrary cutoff to determine if two nodes should be connected or remain separate.

Fortunately, we can leverage a useful proof by Warrens (2008) that shows the ARI is equivalent to Cohen's K (1960) when there are two partitions with known a priori cluster labels. This is applicable to our scenario, as the two partitions are defined by the edge sets of nodes n_i and n_j, and the labels are static across all nodes, either "connected" or "not connected." With these conditions, the ARI (and K) can be calculated as follows:

$$ARI_{ij} = \frac{2(ad - bc)}{(a+b)(b+d)+(a+c)(c+d)} \tag{5.24}$$

In this application, the ARI assigns a score between –1 and 1 to each potential link, with a score greater than 0 indicating a higher probability of the link being present than random chance. A benefit of ARI over other methods is that it uses 0 as the defined cutoff value for determining links, avoiding the need to predetermine the cutoff as required by other methods.

This section covers the IT2FTOPSIS method formulation, including a systematic step-by-step process, followed by a discussion of unsupervised machine learning algorithms. The evaluation of these algorithms using a real application is then discussed in the next section.

5.8 APPLICATION OF WATER SECURITY USING IT2FTOPSIS

The feasibility and procedures of IT2FTOPSIS are aimed at finding the best strategies to improve water supply security in Selangor River. Six alternatives, $(A_1, A_2, A_3, A_4, A_5, A_6)$ are proposed, including strengthening protection of water source areas (A_1), improving infrastructure for urban and rural water security (A_2), developing a water-saving system (A_3), fully implementing the river chief system (A_4), reinforcing groundwater monitoring and protection (A_5), and strengthening water security policies (A_6). These alternatives are evaluated by five decision makers, $(DM_1, DM_2, DM_3, DM_4, \text{and } DM_5)$, based on five criteria: household water security (C_1), economic water security (C_2), urban water security (C_3), environmental water security (C_4), and resilience to water-related disasters (C_5), to determine the most suitable alternative for enhancing water security. The decision makers used linguistic scales of restriction and reliability, as listed in Table 5.1, to evaluate the ratings of each alternative with respect to each criterion in the form of a decision matrix.

The evaluation of the optimal strategies to improve water supply security in Malaysia is presented in a hierarchical structure in Figure 5.2.

The results for the relative closeness coefficients are shown in Table 5.2.

As a conclusion, the best alternative selection is A_6 and the ranking order of the alternative for selecting the best water security is given by $A_2 > A_4 > A_3 > A_5 > A_1 > A_6$. Thus, after taking into account the five attributes, six alternatives, and the opinion from the five experts, a single measurement for water supply security selection is obtained. The results show that the first rank is improving infrastructure to safeguard urban and rural water security at 0.3431, followed by fully implementing

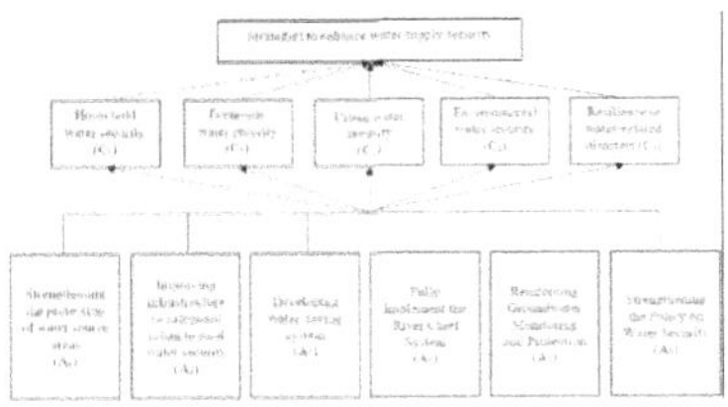

FIGURE 5.2 Water security decision attributes hierarchy

TABLE 5.2
Normalize the relative closeness coefficients

A_1	0.0709
A_2	0.3431
A_3	0.1835
A_4	0.2675
A_5	0.1032
A_6	0.0318

the river chief system, developing water-saving systems, reinforcing groundwater monitoring and protection, and strengthening the protection of water source areas. Strengthening the policy on water security is ranked last.

5.9 APPLICATION OF THE POLLUTION AREA USING MACHINE LEARNING

The Selangor River, which originates from the state of Selangor in Peninsular Malaysia, is selected as the primary focus of this study. With a catchment area of 1,937 km^2 and a length of 110 km, this river provides water to over 1.3 million residents of Selangor state and the capital city, Kuala Lumpur. It also serves as a source of raw water for public water supplies. The Selangor River is surrounded by agricultural activities such as vegetable farms, rubber and oil palm plantations, and aquaculture farms, as well as industrial wastewater from paper manufacturing and oil palm and rubber refineries. This study involves 38 main sampling stations and 19 water quality parameters, including 349 water samples collected from various locations along the Selangor River basin. The 19 water quality parameters include dissolved oxygen (DO), biochemical oxygen demand (BOD), chemical oxygen demand (COD), suspended solids (SSs), pH, ammonia nitrates (NH_3-N), cyanide, temperature (TEMP), nitrates nitrogen (NO_3), phosphate (PO_4), arsenic (As), mercury (Hg), cadmium (Cd), chromium (Cr), lead (Pb), zinc (Zn), oganesson (OG), and *E. coli* (Figure 5.3).

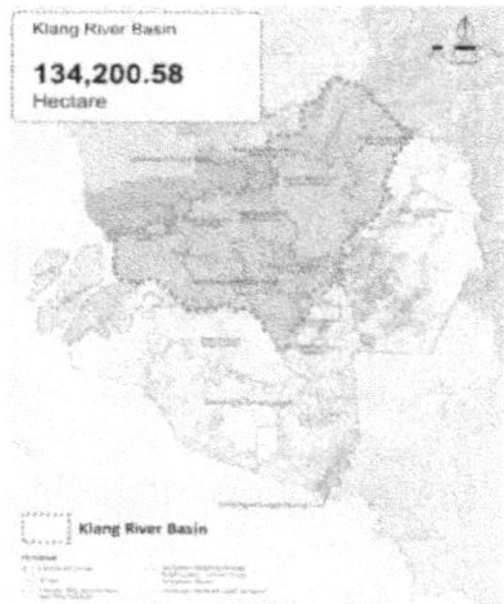

FIGURE 5.3 Selangor River basin (*Source*: ADB, 2007).

5.10 THE POLLUTION AREA USING MACHINE LEARNING

The goal of this study is to identify clusters of polluted areas in the Selangor River data. To achieve this, we employ unsupervised algorithms, including hierarchical clustering and SOMs, followed by K-means with silhouette analysis, to determine the optimal number of clusters. The preprocessed data are used to visualize the intrinsic clusters in the Selangor River data and identify the polluted areas in each cluster. The hierarchical clustering algorithm works by initially considering each data point as its own cluster, which represents the polluted areas, and then merging the most appropriate pair of clusters based on their pollution level. The merging process continues until there is only one cluster. The Ward method is used as the merge criterion in the hierarchical clustering, and the objective function employed is the total within-cluster sum of squared errors (TWSSE).

Initially, each data point is treated as belonging to its own cluster of polluted areas, and the cluster mean is equal to the data point for each cluster. As a result, the TWSSE at this stage is zero. The TWSSE increases as two polluted area clusters are merged, and the actual merging results in the minimum increase in the TWSSE. Figure 5.4 displays the hierarchical maps for 405 water samples, which are divided into three distinct groups, as indicated by their different colors. To determine the optimal number of clusters, we next run K-means with silhouette analysis.

Next, K-means was executed 10 times with a target of 10 clusters. The highest average silhouette score was chosen from each of the 10 cluster solutions. The final clustering solution was determined to be the one with the highest average Silhouette score, based on Table 5.3 and Figure 5.5. The silhouette analysis indicated two clusters of polluted areas with a score of 0.5425 using hierarchical clustering.

The SOMs are a form of unsupervised ANN that works by using a 2D grid of map units, each represented by a prototype vector and connected to its neighboring units through a neighborhood relationship (Vesanto and Alhoniemi, 2000). The map is constructed iteratively, where for each sample of a polluted area, the distance between the input vector (water quality parameters) and the prototype vectors of each map unit is calculated to find the best matching unit (BMU). The BMU and its neighboring units are then adjusted to be closer to the input vector. The final grid of

FIGURE 5.4 Hierarchical maps

TABLE 5.3
Silhouette analysis

No. of cluster	Hierarchical clustering
2	0.5425
3	0.4618
4	0.4369
5	0.4310
6	0.4299
7	0.4491
8	0.4186
9	0.4268
10	0.4353

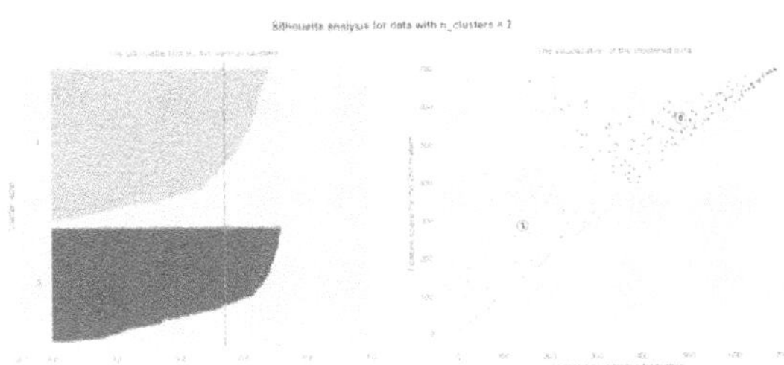

FIGURE 5.5 Silhouette with n_clusters = 2 for hierarchical clustering

map units shows the samples assigned to them, allowing clustering to be performed on the map units themselves. This clustering process results in the optimal grouping of the samples, as identified by four clusters (Figure 5.6) that cover all 349 water samples.

For the purpose of this study, K-means was run 10 times with the value of k set to 10, resulting in 10 clusters. Out of these 10 clusters, the one with the highest average silhouette analysis was chosen as the final clustering solution. As indicated by Table 5.4 and Figure 5.7, the silhouette analysis suggested that four clusters are optimal for

FIGURE 5.6 Visualization of SOM

TABLE 5.4
Silhouette analysis

No. of cluster	SOM
2	0.3996
3	0.4291
4	0.4123
5	0.4062
6	0.3942
7	0.3848
8	0.3988
9	0.4003
10	0.4087

FIGURE 5.7 Silhouette with n_clusters = 3 for SOM

identifying the polluted areas in the Selangor River data using SOM, with a score of 0.3971.

The original Selangor River dataset was divided into three clusters, Clean, Slightly Polluted, and Polluted, based on the water quality index (WQI) analysis. To evaluate the quality of the clustering analysis, an external cluster validity index, the ARI, was used. The ARI, proposed by Hubert and Arabic (1985), calculates the extent of agreement between two sets of groupings and outputs a score between 1 and −1, with 1 being identical and 0 being equivalent to random association. A score of 1 indicates that the cluster membership computed by the clustering algorithm is the same as the true agreement of the clustering result with the true cluster memberships, while negative scores indicate against clustering. In this chapter, even though silhouette analysis showed high scores with two clusters for hierarchical clustering and three clusters for SOM, it was suggested to evaluate the performance using ARI with four different numbers of clusters for both algorithms. These numbers were chosen based on the suggested clusters from the original data and include hierarchical clustering with two clusters, hierarchical clustering with three clusters, SOM with three clusters, and SOM with four clusters.

The ARI scores indicate that the agreement between the original data and hierarchical clustering with two clusters (0.2848) is stronger than the agreement between the original data and hierarchical clustering with three clusters (0.2374). Furthermore, the agreement between the original data and SOM with three clusters (0.1817) is stronger than the agreement between the original data and SOM with four clusters (0.0609). This is shown in Figure 5.8.

The proportion of water samples in each cluster of the final water pollution area identification is shown in Figure 5.9. The original data have 55 samples in Cluster 1, 146 samples in Cluster 2, and 147 samples in Cluster 3. Hierarchical clustering with two clusters has 101 samples in Cluster 1 and 247 samples in Cluster 2. Meanwhile, hierarchical clustering with three clusters has 101 samples in Cluster 1, 34 samples in Cluster 2, and 213 samples in Cluster 3. SOM with three clusters has 115 samples in Cluster 1, 124 samples in Cluster 2, and 109 samples in Cluster 3. Lastly, SOM with

FIGURE 5.8 Adjusted Rand Index for all clusters

FIGURE 5.9 Proportion of each sample in each algorithm

four clusters has 104 samples in Cluster 1, 105 samples in Cluster 2, 71 samples in Cluster 3, and 68 samples in Cluster 4.

As a result, hierarchical clustering and SOM can be considered alternative algorithms that can provide additional solutions to decision-making problems. Previous research in FMCDM primarily focused on analyzing various options in certain scenarios or fields to identify the most preferred alternative (our goal) and present it in a ranked order. This chapter expands the examination of FMCDM problems by incorporating ML techniques.

5.11 CONCLUSION AND DISCUSSION

The aim of this chapter is to create a smart model for identifying optimal strategies for water security by combining the IT2FTOPSIS method with machine learning algorithms. IT2FTOPSIS can accommodate multiple preferences among decision makers and decision problems with multiple strategies, but it only focuses on ranking and selecting alternatives. To address this, unsupervised machine learning was integrated to further analyze and solve decision problems. This chapter presents a two-phase framework that combines fuzzy logic and machine learning. In Phase 1, experiments were conducted to evaluate the water security problems in Malaysia using IT2FTOPSIS, considering six different strategies based on five main criteria. The results showed that improving infrastructure to safeguard urban and rural water security is the best strategy, followed by fully implementing the river chief system, reinforcing groundwater monitoring and protection, and strengthening protection of water source areas. Strengthening the policy on water security was ranked last. The next steps in solving the decision problems were continued in Phase 2 after the best alternatives were selected and ranked. In Phase 2, unsupervised machine learning was applied to the polluted areas in the Selangor River using hierarchical clustering and SOM clustering algorithms. Silhouette analysis was used to determine the optimal number of clusters and found two clusters for hierarchical clustering (with a score of 0.5425) and three clusters for SOM (with a score of 0.4291). The ARI scores were used to compare the original data with the results from both algorithms. The ARI scores indicate that the comparison between the original data and hierarchical clustering with two clusters (0.2848) is better than with three clusters (0.2374). Similarly, the comparison between the original data and SOM with three clusters (0.1817) is better than with four clusters (0.0609). Based on the clustering results, lists of polluted areas were obtained, revealing the number of water samples in each cluster. The original data had 55 water samples in Cluster 1, 146 in Cluster 2, and 147 in Cluster 3. Hierarchical clustering with two clusters resulted in 101 water samples in Cluster 1 and 247 in Cluster 2. Hierarchical clustering with three clusters had 101 water samples in Cluster 1, 34 in Cluster 2, and 213 in Cluster 3. SOM with three clusters resulted in 115 water samples in Cluster 1, 124 in Cluster 2, and 109 in Cluster 3. Finally, SOM with four clusters had 104 water samples in Cluster 1, 105 in Cluster 2, 71 in Cluster 3, and 68 in Cluster 4. Therefore, it can be concluded that the results from SOM with three clusters are more consistent with the original clusters and provide the expected results. In terms of three different categories (Clean,

Slightly Polluted, and Polluted), the results are similar in terms of the number of samples in each cluster. These two-phase fuzzy-ML-based approaches can be considered the best water security strategies for the most polluted areas. Additionally, integrating IT2FTOPSIS with unsupervised ML offers a new solution for decision-making problems. Overall, the results help multiple stakeholders identify the most effective strategies to strengthen water security in Selangor, making complex issues easier to address and prevent. Moreover, they allow stakeholders to track water pollution and quality levels over time and prepare them to handle evolving water security challenges. For future work, the goal is to integrate IT2FTOPSIS and ML to increase the uncertainty in the decision-making system and to gather opinions from a larger group of decision makers.

ACKNOWLEDGMENT

We would like to extend gratitude to the Department of Environment Malaysia for the permission to conduct this study. Enormous appreciation and special thanks to the experts from the Department of Environment Malaysia for their valuable contribution to this study.

REFERENCES

Alazemi, F. K. A. O. H., Ariffin, M. K. A. M., Mustapha, F., and Supeni, E. E. 2022. A new Fuzzy TOPSIS-Based Machine Learning framework for minimizing completion time in supply chains. *International Journal of Fuzzy Systems*, 24 (3): 1669–1695.

ADB. 2007. Klang River Basin environmental improvement and flood mitigation project. https://www.adb.org/projects/ln1500/main.

Castro-Pardo, M. D., Martínez, P. F., and Zabaleta, A. P. 2022. An initial assessment of water security in Europe using a DEA approach. *Sustainable Technology and Entrepreneurship*, 1: 100002.

Celik, E., & Akyuz, E. 2018. An interval type-2 fuzzy AHP and TOPSIS methods for decision-making problems in maritime transportation engineering: The case of ship loader. *Ocean Engineering*, 155: 371–381.

Chapagain, K., Aboelnga, H. T., Babel, M. S., Ribbe, L., Shinde, V. R., Sharma, D., and Dang, N. M. 2022. Urban water security: A comparative assessment and policy analysis of five cities in diverse developing countries of Asia. *Environmental Development*, 43: 100713.

Chen, S.-M., and Lee, L.-W. 2010. Fuzzy multiple attributes group decision-making based on the interval type-2 TOPSIS method. *Expert Systems with Applications*, 37: 2790–2798.

Chen, Z.-S., Yang, Y., Wang, X.-J., Chin, K.-S., and Tsui, K.-L. 2019. Fostering linguistic decision-making under uncertainty: A proportional interval type-2 hesitant fuzzy TOPSIS approach based on Hamacher aggregation operators and andness optimization models. *Information Sciences*, 500: 229–258.

Chowdhury, N. K., Kabir, M. A., Rahman, M. M., and Islam, S. M. S. 2022. Machine learning for detecting COVID-19 from cough sounds: An ensemble-based MCDM method. *Computers in Biology and Medicine*, 145: 105405.

Cohen, J. 1960. A coefficient of agreement for nominal scales. *Educational and Psychological Measurement*, 20 (1): 37–46.

Demerdash, D. E., Omar, M. E. D., El-Din, M. N., El-Badry, H., Aly, E., and El-Molla, D. A. 2022. Development of a quality-based irrigation water security index. *Ain Shams Engineering Journal*, 13: 101735.

Elleuch, M. A., Anane, M., Euchi, J., and Frikha, A. 2019. Hybrid fuzzy multi-criteria decision making to solve the irrigation water allocation problem in the Tunisian case. *Agricultural System*, 176: 102644.

Evengard, B., Berner, J., Brubaker, M., Mulvard, G., and Revich, B. 2011. Climate change and water security with a focus on the Arctic. *Global Health Action*, 4: 1–4.

Fernández-López, C., García, M. G., and Sánchez-Lozano, J. M. 2021. Analysis of WWTPs technologies based on the removal efficiency of pharmaceutical activated compounds for water reuse purposes. A fuzzy multi-criteria decision making approach. *Journal of Water Process Engineering*, 42: 102098.

Gao, F. and Zhou, D. 2025. A novel decision-making approach for risk assessment of converter steelmaking process based on fuzzy DEMATEL method. *Alexandria Engineering Journal*, 115: 222–237.

GWP. 2000. Towards Water Security: A Framework for Action. Global Water Partnership, Stockholm.

Hoffmann, M. 2005. Numerical control of Kohonen neural network for scattered data approximation. *Numerical Algorithms*, 39: 175–186.

Hubert, L. and Arabic, P. 1985. Comparing partitions. *Journal of Classification,* 2 (1): 193–218.

Jones, J. A. A., Vardianian, T. G., and Hakopian, C. 2009. Threats to Global Water Security: Population Growth, Terrorism, Climate Change or Communication. Springer, The Netherlands.

Kohonen, T. 1984. *Self-Organization and Associative Memory.* Springer, New York.

Liu, S., Ouyang, H., and Peng, F. 2008. Spatial data mining and visualization based on self-organizing map. *Journal of Communication and Computer*, 5 (12): 55–59.

Meniz, B. 2021. An advanced TOPSIS method with new fuzzy metric based on interval type-2 fuzzy sets. *Expert Systems with Applications*, 186: 115770.

Mohsin, M., Ali, S. A., Shamim, S. K., and Ahmad, A. 2022. A GIS based novel approach for suitable sanitary landfill site selection using integrated fuzzy analytic hierarchy process and machine learning algorithms. *Environmental Science and Pollution Research*, 29 (41): 61892–61917.

Noori, A., Bonakdari, H., Salimi, A. H., and Gharabaghi, B. 2021. A group multi-criteria decision-making method for water supply choice optimization. *Socio-Economic Planning Sciences*, 77: 101006.

Onda, K., LoBuglio, J., and Bartram, J. 2012. Global access to safe water: accounting for water quality and the resulting impact on MDG progress. *International Journal of Environmental Research and Public Health*, 9: 880–894.

Richard, A. J. and Dean, W. W. 2002. *Applied Multivariate Statistical Analysis.* Prentice Hall, London, p. 265.

Saha, S. and Mondal, P. 2022. Estimation of the effectiveness of multi-criteria decision analysis and machine learning approaches for agricultural land capability in Gangarampur Subdivision, Eastern India. *Artificial Intelligence in Geosciences*, 3: 179–191.

Sahoo, D., Parida, P. K., and Pati, B. 2024. Efficient fuzzy multi-criteria decision-making for optimal college location selection: A comparative study of min-max fuzzy TOPSIS approach. *Results in Control and Optimization*, 15: 100422.

Torynyeviadzi, H. M., Neba, F. A., Mohammed, H., and Seidu, R. 2021. Nodal vulnerability assessment of water distribution networks: An integrated Fuzzy AHP-TOPSIS approach. *International Journal of Critical Infrastructure Protection*, 34: 100434.

Tu, Y., Wang, H., Zhou, X., Shen, W., and Lev, B. 2021. Comprehensive evaluation of security, equity, and efficiency on regional water resources coordination using a hybrid multi-criteria decision-making method with different hesitant fuzzy linguistic term sets. *Journal of Cleaner Production*, 310: 127447.

UNESCO. 2009. Water in a Changing World. Earthscan, Paris.

Veettil, A. V., Mishra, A. K., and Green, T. R. 2022. Explaining water security indicators using hydrologic and agricultural systems models. *Journal of Hydrology*, 607: 127463.

Vesanto, J. and Alhoniemi, E. 2000. Clustering of the self-organizing map. *IEEE Transactions on Neural Networks*, 11 (3): 586–600.

Vörösmarty, C. J., McIntyre, P. B., Gessner, M. O., Dudgeon, D., Prusevich, A., Green, P., Glidden, S., Bunn, S. E., Sullivan, C. A., Liermann, C. R., and Davies, P. M. 2010. Global threats to human water security and river biodiversity. *Nature*, 467: 555–561.

Ward, J. H. 1963. Hierarchical grouping to optimize an objective function. *Journal of the American Statistical Association*, 58 (301): 236–244.

Warrens, M. J. 2008. On the equivalence of Cohen's kappa and the Hubert-Arabie adjusted Rand index. *Journal of Classification*, 25 (2): 177–183.

Wu, T., Liu, X., and Liu, F. 2018. An interval type-2 fuzzy TOPSIS model for large scale group decision making problems with social network information. *Information Sciences*, 432: 392–410.

WWC. 2000. World Water Vision. Earthscan, Paris.

Yazici, I., Beyca, O. F., Gurcan, O. F., Zaim, H., Delen, D., and Zaim, S. 2022. A comparative analysis of machine learning techniques and fuzzy analytic hierarchy process to determine the tacit knowledge criteria. *Annals of Operations Research*, 308: 753–776.

Yilmaz, I., Polat, L., and Yildizbaş. 2023. Celiac disease multi-purpose diet plan through integrated goal programming and interval type-2 fuzzy TOPSIS method. *Expert Systems with Applications,* 226: 120173..

Zamri, N. and Abdullah, A. 2014. A new positive and negative linguistic variable of interval triangular type-2 fuzzy sets for MCDM. In: Herawan, T., Ghazali, R., & Deris, M. (eds). *Recent Advances on Soft Computing and Data Mining*. Advances in Intelligent Systems and Computing, vol 287. Springer, Cham.

Zamri, N., Azman, W. N. A. W., Pairan, M. A., Abas, S. S., and Gao, M. 2023. An analysis of finding the best strategies of water security for water source areas using an integrated IT2FVIKOR with machine learning. *Frontiers in Environmental Science*, 10: 971129. doi: 10.3389/fenvs.2022.971129.

Zamri, N., Naim, S., and Abdullah, L. 2015. A new linguistic scale for Interval Type-2 Trapezoidal Fuzzy Number based Multiple Criteria Decision Making method. *2015 IEEE International Conference on Fuzzy Systems (FUZZ-IEEE)*, Istanbul, Turkey, pp. 1–9. doi: 10.1109/FUZZ-IEEE.2015.7337870.

6 A Group Decision-Making by Hesitant Fuzzy Set: Determination of Criterion Weights in Biomass Power Plant Investment

Özer Eroğlu and Cevriye Temel Gencer

6.1 INTRODUCTION

There is no area left where energy is not used, from the smallest task to the largest activities. Energy utilization and demand, from individuals to industry and public institutions, are increasing rapidly because of the individual and industrial utilization of technological devices, production, distribution, disposal of food, clothing, vehicle equipment, heating, transportation, new discoveries, public and nonpublic activities, lighting, etc., within the scope of the increasing world population. Energy is used in many activities. This situation has led to the finding of cleaner and sustainable/renewable energy resources instead of decreasing fossil-based resources (Abas et al., 2015). According to the global energy demand forecast, it has been stated that the demand for electrical energy in 2025 will increase by 4% compared to the previous year. Despite the increase in renewable energy utilization, global power generation from coal increases due to strong growth in demand, especially in China and India (IEA World Energy Outlook, 2024).

The situation in Türkiye is not different from the information in the report. Due to economic and population growth, there was not only strong growth in energy demand but also an increase in import dependency over the past two decades. Despite significant renewable energy investments, fossil fuels continue to drive Türkiye's economy, with a heavy dependency on imports, especially oil and gas (IEA, 2024). According to the Ministry of Energy and Natural Resources, the distribution of installed capacity by resources at the end of August 2024 is 28.3% hydraulic energy, 21.7% natural gas, 19.2% coal, 10.8% wind, 16.2% solar, 1.5% geothermal, and 2.4% from other sources in Türkiye. A total of 59.2% of energy power plants are from renewable

 DOI: 10.1201/9781003529088-6

resources. Turkey's installed power in electricity increased to 113,750 MW by the end of 2023 but only 65,309 MW (57%) comes from the renewable energy sources (Ministry of Energy and Natural Resources, 2024). Oil and natural gas-based power plants are externally dependent. Due to advantageous geographical location, many power plant investments are made in Türkiye in order to avoid foreign dependency and increasing energy demands, to ensure energy security, and to meet energy production diversity. Türkiye has an advantageous geographical position in the world in terms of renewable energy resources. Although this situation provides benefits in energy production, the location where the power plant will be built stands out as a problem. The main problems are agricultural lands, tourist facilities, public land status, proximity to electricity transmission lines, etc.

In this study, location selection for a biomass power plant is examined by two multi-criteria decision-making methods. Due to the existence of decision makers, alternatives, and criteria, this study is considered a multi-criteria decision-making problem. Decision makers used linguistic terms to express themselves more easily and to convey the issues they were hesitant about. The main source of motivation for this study was hesitant fuzzy sets, which allow decision makers to avoid mistakes and to evaluate all possible sets in linguistic expressions in which they are undecided.

The theoretical amount of energy that Türkiye can obtain from biomass is 33 million TEP/year, which is determined by the Ministry of Energy and Natural Resources. The Ministry of Energy and Natural Resources has made Turkey's biomass energy potential available to researchers with an interface called Türkiye's Biomass Energy Potential Atlas (BEPA). In this software, reports are received on a provincial basis or on a product basis (Ministry of Energy and Natural Resources, 2024). BEPA software forms the basis of the main data of this study.

The main purpose of this chapter is the implementation of hesitant fuzzy set in a multi-criteria decision-making problem in triangular form. This study is organized as a brief literature review in Section 6.2, explanation of multi-criteria decision-making methods in the Sections 6.3 and 6.4, a case study in the Section 6.5, and conclusions in Section 6.6.

6.2 A BRIEF LITERATURE REVIEW

Fuzzy set theory, which was first presented by L. A. Zadeh in 1965, allows partial membership of an element in a set. In this theory, if the value of the membership function, called membership degree, is equal to 1, the x element completely belongs to the fuzzy set. If this value is 0, x does not belong to the fuzzy set. If the degree of membership is between 0 and 1, x is a partial member of the fuzzy set (Zadeh, 1965; Babuška, 2012). After the proposal of fuzzy set theory, which has a wide range of utilization in solving real-life problems, many studies have been carried out based on membership functions. According to membership functions, utilization methods such as interval sets, triangular sets, trapezoidal sets, neutrosophic sets, Pythagorean fuzzy sets, spherical fuzzy sets, hesitant fuzzy sets, interval sets, etc., are treated to resolve imprecision, inaccuracy, vagueness, uncertainty, and vulnerability (Dubois and Prade, 2012).

Hesitant fuzzy sets, which were proposed by Torra, are an extension of fuzzy sets. In Torra's study, it was stated that when defining the membership of an element, the difficulty in determining the degree of membership is not due to errors and probabilities but due to the possibility of possible sets. In this context, hesitant fuzzy sets enable decision-making with all possible values, rather than considering them just as an addition operator (Torra, 2010). A hesitant fuzzy set is defined in terms of a function that returns a set of membership values for each element in the domain. Let X be a reference set, a hesitant fuzzy set on X is a function h, that returns a subset of values in [0, 1]. Therefore, given a set of fuzzy sets, a hesitant fuzzy set is defined as the union of their membership functions. This set has its lower and upper bounds. Let S be a linguistic term set, S = {s_0: nothing, s_1: very low, s_2: low, s_3: medium, s_4: high, s_5: very high, s_6: perfect}; a different hesitant fuzzy linguistic set might be HS_a = {s_1: very low, s_2: low, s_3: medium} and HS_b = {s_3: medium, s_4: high, s_5: very high, s_6: perfect} (Rodríguez et al., 2011). Hesitant fuzzy sets were introduced to define hesitant feelings that decision makers face during the evaluation process. Hesitant fuzzy linguistic term sets, which were developed by Rodriguez et al., help decision makers/experts to express their hesitant feelings using a context-free grammar and provide the opportunity to use compound terms defined by context-independent grammar, where the fuzzy linguistic approach restricts experts to express their thoughts in a single term. The opportunity to use multiple linguistic terms for expressions is the main advantage over other fuzzy approaches (Ecer, 2021).

Considering the extent of the literature, this section is limited to multi-criteria decision-making, facility/location selection within the scope of energy investment, and biomass energy plants articles. Prominent studies in this field are as follows: Cebi et al. (2016); Uyan (2017); Rodríguez et al. (2017); Jeong et al. (2017); Li (2018); Babalola (2018); Davtalab and Alesheikh (2018); Yuruk and Erdogmus (2018); Woo et al. (2018); Díaz-Cuevas et al. (2019); Wang et al. (2019); Costa et al. (2020); Aydin and Sarptas (2020); Settou et al. (2021); Jayarathna et al. (2020); Famoso et al. (2020); Yücenur et al. (2020); Yalcinkaya (2020); Durmaz and Bilgen (2020); Chukwuma et al. (2021); Klimek et al. (2021); Bingöl et al. (2023); Narayanamoorthy et al. (2023). In these studies, in general, criteria for biomass facility location selection were determined by one of the multi-criteria decision-making methods and then alternative areas were determined by using a geographic information system (GIS). The most used criteria are distance to raw material, land use, distance to road/railway, slope, distance to village centers/settlement areas, distance to faults/earthquake risk, distance to wetlands, distance to stream, population density, estimated biomass, productivity of the land, costs (investment, labor, transportation), risk of terror, and impact on environment. In the literature review, the locations of candidate lands were generally determined, but no ranking study was found. In this study, candidate places will be ranked without using GIS. Additionally, no study has been found on this subject with the CoCoSo method. In this context, it is aimed to contribute to the literature.

6.3 F-DEMATEL METHOD

The DEMATEL method was developed between 1972 and 1976 by the researchers of the Geneva Research Centre of the Battelle Memorial Institute (Tzeng et al., 2007). This method identifies both the importance order and relations among the criteria. In addition, with this graph-based method, cause and effect groups can also be determined (Wu and Lee, 2007; Aksakal and Dağdeviren, 2010; Shieh et al., 2010). Over time, f-DEMATEL methods have been accepted in the literature by using linguistic expressions that enable decision makers or experts to express themselves better during pairwise comparisons by integrating triangular fuzzy numbers, interval type-2, trapezoidal fuzzy numbers, etc., such as Chang et al. (2011); Abdullah and Zulkifli (2015); etc. In this study, the weights of criteria will be determined by integrating the hesitant fuzzy set into the DEMATEL method. The phases are explained below:

Phase 1: Defining the criteria and the fuzzy scale Criteria can be defined by questionnaires, face-to-face conversations with experts/groups, or by referring to the literature depending on the type of problem. In this study, criteria are defined by experts. Moreover, the hesitant fuzzy set is preferred due to the idea that experts could express themselves better in pairwise comparisons. The hesitant fuzzy set also provides a larger scale and interval preference when experts are unsure of their decisions. This hesitant fuzzy scale, which is considered to be advantageous to decision makers, is presented in Table 6.1 (Candan and Toklu, 2021). In this study, l_{ij} is the first and the smallest value; m_{ij} is the middle value, and r_{ij} is the largest value of the triangular fuzzy numbers.

Phase 2: Determining the fuzzy direct relation matrix (FDRM): In this phase, decision makers/experts are asked to make pairwise comparisons in order to understand the relations between the criteria. The direct influence between any two criteria is evaluated by linguistic expressions, an example given in Table 6.2. The obtained

TABLE 6.1
Hesitant fuzzy linguistic scale

S	Linguistic expression	Abbreviation	Triangular fuzzy scale (*l*; *m*; *r*)
10	Absolutely High Importance	AHI	7; 9; 9
9	Very High Importance	VHI	5; 7; 9
8	Essentially High Importance	ESHI	3; 5; 7
7	Weakly High Importance	WHI	1; 3; 5
6	Equally High Importance	EHI	1; 1; 3
5	Exactly Equal	EE	1; 1; 1
4	Equally Low Importance	ELI	0.33; 1; 1
3	Weakly Low Importance	WLI	0.2; 0.33; 1
2	Essentially Low Importance	ESLI	0.14; 0.2; 0.33
1	Very Low Importance	VLI	0.11; 0.14; 0.2
0	Absolutely Low Importance	ALI	0.11; 0.11; 0.14

TABLE 6.2
FDRM

Criterion	C-1	C-2	C-3	C-4
C-1	*0*	AHI	Between ELI and EE	ALI
C-2	Between ESLI and EE	*0*	VHI	EE
C-3	EHI	EE	*0*	ESLI
C-4	Between EHI and VHI	Between ELI and EE	ESHI	*0*

matrix is experts' consolidated evaluations and is not symmetrical, and its diagonal elements are 0.

Phase 3: Determining the normalized fuzzy direct relation matrix (NFDRM): Before determining the NFDRM, interval values obtained from linguistic expressions must be converted into a single triangular fuzzy set. The following equations are converted into a single triangular fuzzy set (Başar, 2017). An example is given, which is obtained from Table 6.2. The terms of triangular fuzzy numbers are expressed as "a_i" in the formulas.

$$a_l = \min\left\{a_l^i, a_m^i, a_m^{i+1}, \dots\dots\dots a_m^j, a_r^j\right\} = a_l^i \tag{6.1}$$

$$a_r = \max\left\{a_l^i, a_m^i, a_m^{i+1}, \dots\dots\dots a_m^j, a_r^j\right\} = a_r^j \tag{6.2}$$

$$a_m = \begin{cases} a_m^i, & \text{if } i+1=j \\ OWA_w\left(a_m^i, \dots\dots\dots a_m^j\right), & \text{in other case} \end{cases} \tag{6.3}$$

$$w_1 = \propto^{(n-1)}, w_2 = (1-\propto)\propto^{(n-1)} \dots\dots\dots w_n = (1-\propto) \tag{6.4}$$

$$\propto = \frac{g-j+1}{g-1} \tag{6.5}$$

"w" is the weight vector for ordered weighted averaging (OWA) operation (Filev and Yager, 1998); "g" is number of the terms ($S_g = 10$); "i" is the lowest evaluation order; "j" is the highest evaluation in Table 6.1 (S values are sorted from the largest to the smallest, S_0 is the lowest and S_g is the highest, thus, if the expert opinion varies between S_i and S_j, the relationship $S_0 \le S_i < S_j \le S_g$). According to the sample evaluations in Table 6.2, the process is explained below.

Sample: In this evaluation, decision makers think C4's importance is between "Equally High Importance" and "Very High Importance" compared to C1.

[C4-C1 Between EHI and VHI Between (1; 1; 3) and (5; 7; 9)].

According to Equations (6.1) and (6.3), (1; 3; 5; 7; 9); l = 1, r = 9. To obtain m, Equations (6.3)–(6.5) are utilized. $i = S_6 = 6; j = S_9 = 9$ and g = 10. According to Table 6.1, $i + 1 = 7 \neq j$. OWA operators must be utilized. "n" is the number of remaining elements. In this sample, the remaining elements are (3, 5, 7) so that n = 3. We need w_1, w_2, and w_3.

$$\propto = \frac{g-j+i}{g-1} = \frac{10-9+6}{10-1} = 0,778; \quad w_1 = 0,778^{(3-1)} = 0,605,$$

$$w_2 = (1-0,778) * 0,778^{(3-2)} = 0,173; \quad w_3 = (1-\propto) = 0,222$$

Middle element; m= 0,605*7+0,222*5+0,173*3= 5,864.

After interval linguistic values transformed into a single triangular form, the NFDM is formed by using Equations (6.6) and (6.7). Elements of NFDRM are expressed as "X_{ij}".

$$x_{ij} = \frac{C}{\partial} = \left(\frac{l_{ij}}{\partial_l}; \frac{m_{ij}}{\partial_m}; \frac{r_{ij}}{\partial_r}\right) \tag{6.6}$$

$$\partial_l = \max(\sum_{i=1}^{n} l_{ij}); \quad \partial_m = \max(\sum_{i=1}^{n} m_{ij}); \quad \partial_r = \max(\sum_{i=1}^{n} r_{ij}) \tag{6.7}$$

Phase 4: Determining the fuzzy total relation matrix (FTRM): FTRM is obtained by Equation (6.10). Equations (6.8) and (6.9) are mathematical processes of Equation (6.10). "I" is the identity matrix, and "C" is the element of the NFDRM, which is obtained in Phase 3.

$$\lim_{H \to \infty} c + c^2 + c^3 + \quad + c^H \tag{6.8}$$

$$NFDM = c^1 + c^2 + c^3 + \quad + c^H \tag{6.9}$$

$$NFDM = C * (I - C)^{-1} \tag{6.10}$$

Phase 5: Defining Causer and Receiver groups: After obtaining FTRM, Causer and Receiver groups can be determined. The sum of the ith row of the FTRM shows the sum of the direct and indirect effects sent by criterion i to the other criteria D_i. The sum of the column shows the sum of the effects coming from the other criteria to the criterion R_i. The sum of the D_i and R_i index, determined by the sums of the row and column for each criterion, shows the total sent and received value, and the subtraction of the D_i and R_i index shows the total net effect that the i factor has on the system. If this value is positive, it shows that the *i*th criterion is a "Net Causer", and when it is negative, it means it is a "Net Receiver". The $D_i + R_i$ value shows the degree of criterion *i* within the total system. D_i is the sent effects; R_i is the received effects; $D_i + R_i$ is the degree of central role, and $D_i - R_i$ is the degree of effect (Çınar, 2013; Eroğlu and Gencer, 2017).

Phase 6: Defuzzification process: An appropriate defuzzification method is given in Equation (6.11) (Organ, 2013).

$$D_i + R_i = \frac{1}{4}(l + 2m + r) D_i - R_i = \frac{1}{4}(l + 2m + r) \tag{6.11}$$

Phase 7: Defining criterion weights: By utilizing Equations (6.12) and (6.13), weights can be found (Organ, 2013).

$$W_i = \sqrt{(D_i + R_i)^2 + (D_i - R_i)^2} \tag{6.12}$$

$$w_i = \frac{W_i}{\sum_{k=1}^{n} W_k} \tag{6.13}$$

6.4 COCOSO METHOD

Based on integrated simple additive weighting and exponentially weighted product model, CoCoSo is a kind of multi-criteria decision-making method that orders the alternatives by experts' preference. Phases are given below (Yazdani et al., 2019; Gül et al., 2022).

Phase 1: Defining initial decision matrix (IDM): Including the initial information, the data (X_{ij}) and weights (W_j) about criteria (C_j) and alternatives (A_i) are shown in Table 6.3 (Eroğlu and Gencer, 2022).

Phase 2: Normalizing decision matrix and determining comparability sequence measures (φ_i): To obtain φ_i, Equation (6.14) must be utilized according to cost/benefit value.

$$\varphi_{ij\_benefit} = \frac{x_{ij} - min\ x_{ij}}{max\ x_{ij} - min\ x_{ij}}; \varphi_{ij\_cost} = \frac{max\ x_{ij} - x_{ij}}{max\ x_{ij} - min\ x_{ij}} \tag{6.14}$$

Phase 3: Determining the sum of the weighted comparability sequence for each alternative (S_i) and sum of the exponentially weighted comparability sequence for each alternative (P_i): These measures can be determined by utilizing Equation (6.15) according to cost/benefit value, respectively (Eroğlu and Gencer, 2022).

$$S_i = \sum_{j=1}^{n} (W_J \varphi_{ij}); P_i = \sum_{j=1}^{n} (\varphi_{ij})^{W_j} \tag{6.15}$$

TABLE 6.3
Initial decision matrix

Weights	W_1	W_2	.	.	W_j
Optimal value	Max	Min	.	.	Max
Criterion	$C1$	$C2$	.	.	CJ
A1	X_{11}	X_{12}	.	.	X_1j
A2	X_{21}	X_{22}	.	.	X_2j
.	.	.	.	.	.
.	.	.	.	.	.
A_i	Xi_1	Xi_2			X_{ij}

***Phase 4: Determining relative weights*:** To determine the relative weights of alternatives, three relative score strategies, of which mathematical notations are given below, can be used. In Equation (6.16), λ is chosen by experts and usually determined (Eroğlu and Gencer, 2022).

$$k_{ia} = \frac{P_i + S_i}{\sum_{i=1}^{n}(P_i + S_i)}; \; k_{ib} = \frac{P_i}{\min P_i} + \frac{S_i}{\min S_i}; \; k_{ic} = \frac{\lambda\, S_i + (1-\lambda)P_i}{\lambda\, max\, S_0 + \lambda\, max\, P_i}, \quad 0 < \lambda \leq 1 \tag{6.16}$$

***Phase 5: Obtaining final ranks*:** Final ranks can be calculated by Equation (6.17).

$$k_i = \sqrt[3]{k_{ia}k_{ib}k_{ic}} + \frac{1}{3}(k_{ia} + k_{ib} + k_{ic}) \tag{6.17}$$

6.5 CASE STUDY

Biomass power plants obtain energy from biological materials such as forest products, agricultural products, household waste, or animal waste. They convert biological waste, which would otherwise decompose in nature, into energy.

Biomass both enables the use of urban waste in energy production and provides a solution to the problem of stacking waste. According to the Türkiye Biomass Energy Potential Atlas (BEPA) software, from which the data of this study were obtained, the theoretical and economic potential of biomass in Türkiye is shown in Table 6.4 (Ministry of Energy and Natural Resources, 2024).

In this study, site selection for biomass power plant installation was handled in Gelibolu District of Çanakkale Province. Three candidate places and the five criteria for selecting these places were evaluated by taking five experts' opinions.

The criteria were determined by obtaining expert opinions and information in the literature and were weighted with the f-DEMATEL method, and the candidates were ranked with the CoCoSo method.

TABLE 6.4
Biomass potential in Türkiye

Resource	Potential (TOE/year)	
	Theoretic	Economic
Animal waste	4.385.371	1.084.506
Herbal	25.384.268	1.462.159
Municipal waste	3.373.011	485.858
Forest and forestry product	N/A	859.899

Information about the experts is given in Table 6.5.

Biomass potential of the region is given in Table 6.6.

Criteria: Distance to high-voltage (HV) substation center (C1) refers to the distance of the power plant to the HV substation centers. There are two candidate HV substation centers in the Gelibolu region that can be connected to this new power

TABLE 6.5
Experts information

Education	Age	Profession	Status	Experience (year)
MsC	>45	Energy systems	Manager	19
Doctor	>40	Op. research	Manager	12
Graduate (university)	43	Energy systems	Manager	13 (public)
Graduate (university)	42	Energy systems	Employee	15 (private sec.)
Graduate (high school)	47	Energy systems	Employee	22 (municipality)

TABLE 6.6
Biomass potential

Resource	Waste (tonnes/year)	Potential energy equivalent (TOE/year)	
		Theoretic	Economic
Animal waste	139.946,6	1139,2	279,1
Herbal	152.004,5	61.480	4065,2
Municipal waste	18.972,2	1736,4	252,4
Forest and forestry product	85.718[a]	38.203,5	38.203,5

[a] 1 ster = 500 kg.

plant. The capacity of these two HV substation centers is available for new connections. In this study, the nearest candidate HV substation center is preferred due to reduced connection costs. ***Transportation costs (C2)*** refers to the pick-up and transportation costs, including maintenance costs and labor expenses. (There will be a continuous flow in shifts on a 24-hour basis, especially for municipal waste.) It is estimated that the facility, which is planned to be operated by the local municipality, will consume 2 liters of fuel per kilometer for the collection of waste and raw materials ($2.64 per km). In addition, three drivers and six workers are planned per vehicle for three shifts in a day, and it is evaluated that today's minimum wage will be paid: $4454.6 for three drivers and two pick-up and delivery workers per month. The transportation costs vary depending on distance, pick-up and delivery points/hubs, and road conditions. ***Investment costs (C3):*** Since the candidate sites are in the public domain, only the installation cost of the facility is included. It has been evaluated that an investment cost of $1.3 million will be required for a 1 MW power plant within the scope of construction, roads, and connection lines. The investment costs vary depending on distance and ground conditions. ***Population density (C4):*** Gallipoli is an important region both historically and touristically. While the population density increases in places close to the sea during the summer, the historical site is used almost every day of the year. This situation periodically increases both the energy need and the collection of municipal waste. For this reason, for sustainability to meet energy need is of greater importance. However, it was preferred not to spoil the texture of the historical site and to be away from touristic places. ***Economic potential (C5)*** refers to the economic potential of obtainable products/waste. The potential that can be obtained from the region is expressed in Table 6.6.

Alternatives: Cevizli region (A1): There is public land in the region, 13 km from Gallipoli District. Distance and transportation network is available. There is no need for an additional road for the facility. It is 5 km away from the nearest HV facility. There is additional cost for line connection. The region is away from historical and touristic places. In this region, which is very close to the Çanakkale Bridge, raw materials that can be converted into energy can be easily supplied by Anatolia, because it is 30 km on the opposite side (Çanakkale—many settlements and agricultural areas). ***Bolayır region (A2):*** There is public land in the region, 15 km from Gallipoli District. Distance and transportation network is available. But there is need for a 1 -km new connection road to the facility. It is 5 km away from the nearest HV facility. There is additional cost for line connection. The region is away from historical places but very near to the sea. There are some available places that can be used as a beach in the future. ***Koruköy region (A3):*** There is public land in the region, 13 km from Gallipoli District. Distance and transportation network is available. There is no need for an additional road for the facility. It is 15.99 km away from the nearest HV facility. There is additional cost for line connection. The region is away from historical places, but the region is very near to the touristic and military facilities. It is considered that this situation will be suitable to meet the energy needs due to the increase in holiday sites in the future. Additionally, if the facility is built in this area, there may be a need for an HV facility in the future. The alternatives and HV substations are visualized in Figure 6.1.

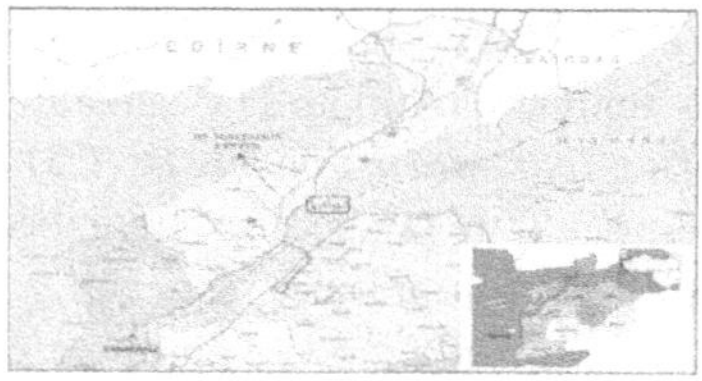

FIGURE 6.1 The alternatives and HV substations

To determine the criteria weights, a survey and pairwise comparisons were implemented with the experts whose information is given above. The individual evaluations of the experts and the combined group pairwise comparison matrix, FDRM, according to Table 6.1, are presented in Table 6.7.

After taking the individual pairwise comparisons, the individual matrices are combined and presented in Table 6.8 with both linguistic and numerical values. Pairwise comparison values that decision makers did not use/evaluate the same linguistic expressions/ideas were designated as "Between" and evaluated as an interval. This is the advantage of hesitant fuzzy set. In the next step, the interval values are reduced to a single triangular fuzzy number set by OWA operators, and this new triangular number set is processed until defuzzification.

After determining the FDRM by applying Equations (6.1)–(6.7), NFDRM is obtained and shown in Table 6.9.

By implementing Equations (6.8)–(6.10), the FTRM is obtained and given in Table 6.10.

According to Equations (6.11)–(6.13), obtained criteria weights are, respectively, w_1: 0.1414, w_2: 0.0815, w_3: 0.2321, w_4: 0.2478, and w_5: 0.2972. The sum is equal to 1.

When the data obtained from the DEMATEL method are seen, it is clear that the economic potential criterion is the most important criterion, but transportation costs and the distance of the power plant to the HV center are not considered more important by decision makers.

Alternatives are evaluated and ordered by the CoCoSo method, which the initial matrix of the method is identified in Table 6.11.

According to Equation (6.14), the normalized decision matrix is obtained and shown in Table 6.12.

After obtaining the normalized decision matrix, by implementing Equation (6.15), the sum of the weighted comparability sequence and the sum of the exponentially weighted comparability sequence for each alternative are obtained and given in Table 6.13.

Relative weights and final ranks are obtained by Equations (6.16) and (6.17) and presented in Table 6.14.

According to the final ranks calculated by the CoCoSo method within the framework of the criteria and criterion weights among the existing alternatives, Alternative 1 (Cevizli region) has become the best available place to establish a biomass power

TABLE 6.7
Individual pairwise comparisons

	C1	C2	C3	C4	C5
C1	0	ALI-ESLI	WLI	EE	VLI
C2	EE	0	WLI	VLI	EE
C3	ALI	ELI	0	AHI	VHI-AHI
C4	WHI	EE	ALI	0	AHI
C5	WHI	ELI	WHI	WHI	0

Expert 1

	C1	C2	C3	C4	C5
C1	0	ALI	EE	EE	AHI
C2	EE	0	WLI	VLI	EE
C3	ALI	ELI	0	AHI	VHI
C4	ESLI	WHI	ALI	0	AHI
C5	WHI	ELI	WHI	WHI	0

Expert 2

	C1	C2	C3	C4	C5
C1	0	ALI-ESLI	WLI	EE	VLI
C2	EE	0	WLI	VLI	EE
C3	ALI	ELI	0	AHI	AHI
C4	ELI	EHI	VLI	0	AHI
C5	WHI	ELI	WLI	WHI	0

Expert 3

	C1	C2	C3	C4	C5
C1	0	ALI-ESLI	WLI	EE	VLI
C2	EHI	0	WLI	ALI	EE
C3	ALI	ELI	0	AHI	EE
C4	WHI	EE	ESLI	0	AHI
C5	WHI	ELI	WHI	WHI	0

Expert 4

	C1	C2	C3	C4	C5
C1	0	ALI-ESLI	WLI	EE	VLI
C2	EE	0	WLI	VLI	EE
C3	ALI	ELI	0	AHI	VHI-AHI
C4	WHI	EE	ALI	0	AHI
C5	ESHI	ELI	WHI	WHI	0

Expert 5

TABLE 6.8
FDRM

	C1	C2	C3	C4	C5
C1	0	Btween ALI and ESLI (0.11; 0.11; 0.14) (0.14; 0.2; 0.33)	Between WLI and EE (0.2; 0.33; 1) -(1; 1; 1)	EE (1; 1; 1)	Between VLI and AHI (0.11; 0.14; 0.2) -(7; 9; 9)
C2	Between EE and EHI (1; 1; 1) -(1; 1; 3)	0	WLI (0.2; 0.33; 1)	Between ALI and VLI (0.11; 0.11; 0.14) -(0.11; 0.14; 0.2)	EE (1; 1; 1)
C3	ALI (0.11; 0.11; 0.14)	ELI (0.33; 1; 1)	0	AHI (7; 9; 9)	Between EE and AHI (1; 1; 1) -(7; 9; 9)
C4	Between ESLI and WHI (0.14; 0.2; 0.33) -(1; 3; 5)	Between EE and WHI (1; 1; 1) -(1; 3; 5)	Between ALI and ESLI (0.11; 0.11; 0.14) -(0.14; 0.2; 0.33)	0	AHI (7; 9; 9)
C5	WHI (1; 3; 5)	ELI (0.33; 1; 1)	WHI (1; 3; 5)	WHI (1; 3; 5)	0

plant. However, it should not be forgotten that the ranking may also change when the criteria, criterion weights, data, and one or more of the alternatives change.

6.6 CONCLUSIONS

People want the settlements to have qualities that can meet their needs in many aspects. Settlements with these qualities are mostly security, health facilities, natural opportunities (existence of water, sea, forests, etc.) and also meeting the energy needs. Additionally, governments serve citizens to meet all the needs in a sustainable manner and make investments accordingly. Energy needs are one of them.

As the main aim of this chapter is the implementation of a hesitant fuzzy set in a multi-criteria decision-making problem in triangular form, DEMATEL and CoCoSo methods are preferred. In order not to deviate from the main goal, the problem area of the chapter is kept limited and small.

Decision makers do not feel comfortable or cannot fully express what is on their mind due to a lack of knowledge in most studies using mathematical notations and methods. Therefore, using linguistic expressions can make decision makers/experts more accurate in explaining their ideas. This is why fuzzy logic and the methods developed for this purpose are very useful.

TABLE 6.9
NFDRM

C1			C2			C3			C4			C5		
0			0,012	0,010	0,012	0,022	0,044	0,036	0,110	0,050	0,036	0,012	0,310	0,321
0,110	0,050	0,107	0			0,022	0,016	0,036	0,012	0,005	0,007	0,110	0,050	0,036
0,012	0,005	0,005	0,036	0,050	0,036	0			0,768	0,447	0,321	0,110	0,193	0,321
0,015	0,094	0,179	0,110	0,133	0,179	0,012	0,010	0,012	0			0,768	0,447	0,321
0,110	0,149	0,179	0,036	0,050	0,036	0,110	0,149	0,179	0,110	0,149	0,179	0		

TABLE 6.11
Initial decision matrix

Weights	**0.1414**	**0.0815**	**0.2321**	**0.2478**	**0.2972**
Optimal value	Min	Min	Min	Max	Max
Criterion	C1	C2	C3	C4	C5
	Distance to HV substation (km)	Unit transportation costs ($)	Unit investment costs ($)	Population density	Economic potential (TOE/year)
A1 Cevizli region	5	5.854,6	1,3	3	59.500
A2 Bolayır region	15	7.814,7	1,37	3	42.800
A3 Koruköy region	15,99	7.914,2	1,45	1	43.500

TABLE 6.10
FTRM

C1			C2			C3			C4			C5		
0,023	0,085	0,117	0,037	0,061	0,063	0,041	0,116	0,125	0,160	0,171	0,161	0,144	0,438	0,453
0,134	0,068	0,140	0,019	0,012	0,018	0,046	0,034	0,063	0,082	0,038	0,054	0,182	0,095	0,119
0,146	0,145	0,211	0,182	0,158	0,148	0,117	0,092	0,123	0,975	0,574	0,474	0,893	0,521	0,587
0,147	0,210	0,323	0,176	0,191	0,232	0,129	0,115	0,126	0,225	0,154	0,147	0,976	0,613	0,522
0,149	0,218	0,300	0,080	0,111	0,115	0,143	0,199	0,248	0,262	0,285	0,320	0,228	0,239	0,283

TABLE 6.12
Normalized decision matrix

	C1	C2	C3	C4	C5
A1	1.000	1.000	1.000	0.000	1.000
A2	0.090	0.048	0.533	0.000	0.000
A3	0.000	0.000	0.000	1.000	0.042

TABLE 6.13
Si and Pi values

	C1	C2	C3	C4	C5	*Si*	C1	C2	C3	C4	C5	*Pi*
A1	0.1414	0.0815	0.2321	0.0	0.2972	0.7522	1.0	1.0	1.0	0.0	1.0	4.0
A2	0.0127	0.0039	0.1238	0.0	0.0	0.1405	0.7115	0.7812	0.8642	0.0	0.0	2.3569
A3	0.0	0.0	0.0	0.2478	0.0125	0.2603	0.0	0.0	0.0	1.0	0.389	1.3896

TABLE 6.14
Relative weights and final ranks

	Relative weights			Final weight	
Alternatives	k_{ia}	k_{ib}	k_{ic}	k_i	**Ranks**
A1	0.534	8.2338	1	4.8942	1
A2	0.2806	7.0514	0.5255	3.6323	2
A3	0.1854	2.8529	0.3472	1.6968	3

Hesitant fuzzy set, on the other hand, offers decision makers a wider range of options that they can evaluate interval. The main reason for using hesitant fuzzy set in this study is the opportunity for indecisive experts to make interval evaluations.

The main reason for choosing the DEMATEL method is that it has more advantages in pairwise comparisons compared to the analytical hierarchy process (AHP). To explain with an example, in AHP, if an expert/decision maker thinks C1 criterion is more important than C3 criterion, it means 7 in its scale, and in the pairwise comparisons process, just the opposite 1/7. This is mandatory. But in the DEMATEL method, the expert/decision maker has no obligation. If the expert/decision maker thinks C1 criterion is more important than C3 criterion, it means 7; just the opposite can be 3. Another reason is that the DEMATEL method provides the opportunity to explain conflicting criteria based on the cause and effect relationship. The CoCoSo method was preferred because of its simplicity, understandability, and little mathematical notation. Also, decision makers can be involved and contribute to the solution.

When DEMATEL method results are examined in the scope of criterion weights, the most important criterion is Economic Potential. The causer criterion is Economic Potential (C5) with the $D + R$ index (3.108); the receiver criterion is Transportation Costs (C2) with the $D - R$ index (−0.195). According to the obtained criterion weights, decision makers want the power plant to be established without paying much attention to the costs in the presence of energy potential.

As a result, hesitant fuzzy set provided greater comfort to decision makers when making their pairwise comparisons. Since this study is limited, the number of alternatives can be increased, the number of criteria can be increased, can be developed using different linguistic terms and decision-making methods, can be tried in different regions, or the results of a previous study can be repeated and compared with the same data.

REFERENCES

Abas, N., Kalair, A. and Khan, N. (2015). Review of fossil fuels and future energy technologies. *Futures*, 69, 31–49 (http://dx.doi.org/10.1016/j.futures.2015.03.003).

Abdullah, L. and Zulkifli, N. (2015). Integration of fuzzy AHP and interval type-2 fuzzy DEMATEL: An application to human resource management. *Expert Systems with Applications*, 42(9), 4397–4409 (https://doi.org/10.1016/j.eswa.2015.01.021).

Aksakal, E. and Dağdeviren, M. (2010). An integrated approach for personel selection with DEMATEL and ANP methods. *Gazi Üniversitesi Mühendislik Mimarlık Fakültesi Dergisi*, 25(4), 905–913.

Aydin, F. and Sarptas, H. (2020). Spatial assessment of site suitability for solar desalination plants: A case study of the coastal regions of Turkey. *Clean Technologies and Environmental Policy*, 22, 309–323 (https://doi.org/10.1007/s10098-019-01783-0).

Babalola, M. A. (2018). Application of GIS-based multi-criteria decision technique in exploration of suitable site options for anaerobic digestion of food and biodegradable waste in Oita City, Japan. *Environments*, 5(7), 77 (https://doi.org/10.3390/environments5070077).

Babuška, R. (2012). *Fuzzy Modeling for Control* (Vol. 12). Springer Science & Business Media (https://doi.org/10.1007/978-94-011-4868-9)

Başar, A. (2017). Klasik ve sezgisel bulanık ikili karşılaştırma ile yazılım geliştirme projelerinin maliyet tahmini: uygulama örneği. *Bilişim Teknolojileri Dergisi*, 10(2), 129–137.

Bingöl, F., Altıner, F. and Kelkit, A. (2023). Analyzing the most eligible site selection for biomass energy facilities through weighted overlay analysis: Case of Balikesir (Turkey) Province. *Ege Üniversitesi Ziraat Fakültesi Dergisi*, 60(1), 19–35 (https://doi.org/10.20289/zfdergi.1136651).

Candan, G. and Toklu, M. C. (2021). Determining solar power plant location using hesitant fuzzy AHP method. *Alphanumeric Journal*, 9(1), 25–34.

Cebi, S., Ilbahar, E. and Atasoy, A. (2016). A fuzzy information axiom based method to determine the optimal location for a biomass power plant: A case study in Aegean Region of Turkey. *Energy*, 116, 894–907 (https://doi.org/10.1016/j.energy.2016.10.024).

Chang, B., Chang, C. W. and Wu, C. H. (2011). Fuzzy DEMATEL method for developing supplier selection criteria. *Expert systems with Applications*, 38(3), 1850–1858 (https://doi.org/10.1016/j.eswa.2010.07.114).

Chukwuma, E. C., Okey-Onyesolu, F. C., Ani, K. A. and Nwanna, E. C. (2021). GIS bio-waste assessment and suitability analysis for biogas power plant: A case study of Anambra state of Nigeria. *Renewable Energy*, 163, 1182–1194 (https://doi.org/10.1016/j.renene.2020.09.046).

Costa, F. R., Ribeiro, C. A. A. S., Marcatti, G. E., Lorenzon, A. S., Teixeira, T. R., Domingues, G. F. and de Carvalho, J. R. (2020). GIS applied to location of bioenergy plants in tropical agricultural areas. *Renewable Energy*, 153, 911–918 (https://doi.org/10.1016/j.renene.2020.01.050).

Çınar, Y. (2013). Kariyer Tercihi Probleminin Yapısal Bir Modeli ve Riske Karşı Tutumlar: Olasılıklı DEMATEL Yöntemi Temelli Bütünleşik Bir Yaklaşım. *Sosyoekonomi*, 1, 158–186.

Davtalab, M. and Alesheikh, A. A. (2018). Spatial optimization of biomass power plant site using fuzzy analytic network process. *Clean Technologies and Environmental Policy*, 20, 1033–1046 (https://doi.org/10.1007/s10098-018-1531-5).

Díaz-Cuevas, P., Domínguez-Bravo, J. and Prieto-Campos, A. (2019). Integrating MCDM and GIS for renewable energy spatial models: Assessing the individual and combined potential for wind, solar and biomass energy in Southern Spain. *Clean Technologies and Environmental Policy*, 21, 1855–1869 (https://doi.org/10.1007/s10098-019-01754-5).

Dubois, D. and Prade, H. (Eds.). (2012). *Fundamentals of Fuzzy Sets* (Vol. 7). Springer Science & Business Media (https://doi.org/10.1007/978-1-4615-4429-6).

Durmaz, Y. G. and Bilgen, B. (2020). Multi-objective optimization of sustainable biomass supply chain network design. *Applied Energy*, 272, 115259 (https://doi.org/10.1016/j.apenergy.2020.115259).

Ecer, B. (2021). Hesitant fuzzy linguistic decision making approach for stroller selection. *Bitlis Eren Üniversitesi Fen Bilimleri Dergisi*, 10(4), 1464–1473 (https://doi.org/10.17798/bitlisfen.960814).

Eroğlu, Ö. and Gencer, C. (2017). Integrating fuzzy DEMATEL and SMAA-2 for maintenance expenses. *International Journal of Engineering Science Invention*, 6(2), 60–71.

Eroğlu, Ö. and Gencer, C. (2022). Prioritizing post-disaster debris for recycling. In M. Gul, M. Yucesan, & M. Erdogan (Eds.), *Multi-Criteria Decision Analysis* (pp. 223–234). CRC Press (https://doi.org/10.1201/9781003212904-15).

Famoso, F., Prestipino, M., Brusca, S. and Galvagno, A. (2020). Designing sustainable bioenergy from residual biomass: Site allocation criteria and energy/exergy performance indicators. *Applied Energy*, 274, 115315 (https://doi.org/10.1016/j.apenergy.2020.115315).

Filev, D. and Yager, R. R. (1998). On the issue of obtaining OWA operator weights. *Fuzzy Sets and Systems*, 94(2), 157–169.

Gul, M., Yucesan, M. and Erdogan, M. (Eds.). (2022). *Multi-Criteria Decision Analysis: Case Studies in Disaster Management* (1st ed.). CRC Press (https://doi.org/10.1201/9781003212904).

IEA (International Energy Agency), 2024, World Energy Outlook, 2024. https://www.iea.org/events/world-energy-outlook-2024, 22.09.2024.

IEA (International Energy Agency), Report 2024. https://www.iea.org/countries/turkiye, 22.09.2024.1

Jayarathna, L., Kent, G., O'Hara, I. and Hobson, P. (2020). A Geographical Information System based framework to identify optimal location and size of biomass energy plants using single or multiple biomass types. *Applied Energy*, 275, 115398 (https://doi.org/10.1016/j.apenergy.2020.115398).

Jeong, J. S. and Ramírez-Gómez, Á. (2017). A multicriteria GIS-based assessment to optimize biomass facility sites with parallel environment—A case study in Spain. *Energies*, 10(12), 2095 (https://doi.org/10.3390/en10122095).

Klimek, K., Kapłan, M., Syrotyuk, S., Bakach, N., Kapustin, N., Konieczny, R. and Wałowski, G. (2021). Investment model of agricultural biogas plants for individual farms in Poland. *Energies*, 14(21), 7375 (https://doi.org/10.3390/en14217375).

Li, Z. (2018). Study of site suitability assessment of regional wind resources development based on multi-criteria decision. *Clean Technologies and Environmental Policy*, 20, 1147–1166 (https://doi.org/10.1007/s10098-018-1538-y).

Ministry of Energy and Natural Resources. (2024a). https://enerji.gov.tr/haber-detay?id=21234, 22.09.2024.

Ministry of Energy and Natural Resources. (2024b). https://bepa.enerji.gov.tr/, 22.09.2024.

Ministry of Energy and Natural Resources. (2024c). https://bepa.enerji.gov.tr/, 10.10.2024.

Narayanamoorthy, S., Ramya, L., Gunasekaran, A., Kalaiselvan, S. and Kang, D. (2023). Selection of suitable biomass conservation process techniques: A versatile approach to normal wiggly interval-valued hesitant fuzzy set using multi-criteria decision making. *Complex & Intelligent Systems*, 9(6), 6681–6695 (https://doi.org/10.1007/s40747-023-01097-1).

Organ, A. (2013). Bulanık Dematel Yöntemiyle Makine Seçimini Etkileyen Kriterlerin Değerlendirilmesi. *Çukurova Üniversitesi Sosyal Bilimler Enstitüsü Dergisi*, 22(1), 157–172.

Rodríguez, R. M., Martínez, L. and Herrera, F. (2011). Hesitant fuzzy linguistic term sets. In: Wang, Y., Li, T. (eds) *Foundations of Intelligent Systems. Advances in Intelligent and Soft Computing*, vol 122. Springer, Berlin, Heidelberg (https://doi.org/10.1007/978-3-642-25664-6_34).

Rodríguez, R., Gauthier-Maradei, P. and Escalante, H. (2017). Fuzzy spatial decision tool to rank suitable sites for allocation of bioenergy plants based on crop residue. *Biomass and Bioenergy*, 100, 17–30 (https://doi.org/10.1016/j.biombioe.2017.03.007).

Settou, B., Settou, N., Gouareh, A., Negrou, B., Mokhtara, C. and Messaoudi, D. (2021). A high-resolution geographic information system-analytical hierarchy process-based method for solar PV power plant site selection: A case study Algeria. *Clean Technologies and Environmental Policy*, 23, 219–234 (https://doi.org/10.1007/s10098-020-01971-3).

Shieh, J. I., Wu, H. H. and Huang, K. K. (2010). A DEMATEL method in identifying key success factors of hospital service quality. *Knowledge-Based Systems*, 23(3), 277–282 (https://doi.org/10.1016/j.knosys.2010.01.013).

Torra, V. (2010). Hesitant fuzzy sets. *International Journal of Intelligent Systems*, 25(6), 529–539 (https://doi.org/10.1002/int.20418).

Tzeng, G. H., Chiang, C. H. and Li, C. W. (2007). Evaluating intertwined effects in e-learning programs: A novel hybrid MCDM model based on factor analysis and DEMATEL. *Expert Systems with Applications*, 32(4), 1028–1044.

Uyan, M. (2017). Optimal site selection for solar power plants using multi-criteria evaluation: A case study from the Ayranci region in Karaman, Turkey. *Clean Technologies and Environmental Policy*, 19, 2231–2244 (https://doi.org/10.1007/s10098-017-1405-2).

Wang, C. N., Tsai, T. T. and Huang, Y. F. (2019). A model for optimizing location selection for biomass energy power plants. *Processes*, 7(6), 353 (https://doi.org/10.3390/pr7060353).

Woo, H., Acuna, M., Moroni, M., Taskhiri, M. S. and Turner, P. (2018). Optimizing the location of biomass energy facilities by integrating Multi-Criteria Analysis (MCA) and Geographical Information Systems (GIS). *Forests*, 9(10), 585 (https://doi.org/10.3390/f9100585).

Wu, W. W. and Lee, Y. T. (2007). Developing global managers' competencies using the fuzzy DEMATEL method. *Expert Systems with Applications*, 32(2), 499–507 (https://doi.org/10.1016/j.eswa.2005.12.005).

Yalcinkaya, S. (2020). A spatial modeling approach for siting, sizing and economic assessment of centralized biogas plants in organic waste management. *Journal of Cleaner Production*, 255, 120040 (https://doi.org/10.1016/j.jclepro.2020.120040).

Yazdani, M., Zarate, P., Kazimieras Zavadskas, E. and Turskis, Z. (2019). A combined compromise solution (CoCoSo) method for multi-criteria decision-making problems. *Management Decision*, 57(9), 2501–2519 (https://doi.org/10.1108/MD-05-2017-0458).

Yuruk, F. and Erdogmus, P. (2018). Finding an optimum location for biogas plant: A case study for Duzce, Turkey. *Neural Computing and Applications*, 29, 157–165 (https://doi.org/10.1007/s00521-016-2424-3).

Yücenur, G. N., Çaylak, Ş., Gönül, G. and Postalcıoğlu, M. (2020). An integrated solution with SWARA&COPRAS methods in renewable energy production: City selection for biogas facility. *Renewable Energy*, 145, 2587–2597 (https://doi.org/10.1016/j.renene.2019.08.011).

Zadeh, L. A. (1965). Fuzzy sets. *Information and Control*, 8, 338–353 (https://doi.org/10.1016/S0019-9958(65)90241-X).

7 Fuzzy Set Theory Applications in Smart Cities and IoT

Akram Elomiya, Jiří Křupka, Libor Švadlenka, Stefan Jovčić, and Marko Radovanović

7.1 INTRODUCTION

The proliferation of urban centers and the imperative for sustainable growth have given rise to the smart city paradigm as a pivotal approach to contemporary urban challenges. These technologically enhanced urban environments harness the power of advanced information and communication technologies (ICTs) to achieve multiple objectives: enhancing residents' well-being, optimizing city operations, and fostering environmentally responsible practices (Albino et al., 2015). Central to this urban transformation is the Internet of Things (IoT), a network of interconnected devices that collect and exchange data. This technological infrastructure enables sophisticated decision-making processes and automates various aspects of urban management across multiple sectors (Zanella et al., 2014).

As cities become more complex and data centric, the need for advanced methods to handle decision-making uncertainties increases. For instance, city planners face challenges like predicting traffic flow with fluctuating demand, managing waste collection routes with dynamic bin fill levels, or allocating resources for public services with imprecise population data. Traditional binary logic and crisp set theory often prove inadequate in capturing the complexities of real-world scenarios, where boundaries are often unclear, and elements may belong to multiple sets to varying degrees (Zadeh, 1965). Fuzzy set theory (FST) provides a robust framework for managing uncertainty and imprecision, making it particularly relevant for the dynamic and complex environments of smart cities and IoT (Ross, 2005). In the context of smart cities and IoT, FST offers significant advantages, such as the ability to represent linguistic variables, which align closely with human intuition, and to model complex, nonlinear relationships between variables. This makes fuzzy systems ideal for applications in urban management and control, where these characteristics are prevalent (Castillo and Melin, 2007).

This chapter aims to examine the fundamentals and applications of FST in smart cities and IoT, bridging theoretical concepts with practical implementations. The chapter structure is organized as follows: We begin by discussing the essential

DOI: 10.1201/9781003529088-7

concepts of FST within the context of smart cities and IoT (Section 7.2). Next, we explore various applications of FST in smart cities and IoT (Section 7.3). The chapter concludes with a detailed case study on optimizing waste management using the spherical fuzzy analytic hierarchy process (SFAHP) and fuzzy evaluation based on distance from average solution (FEDAS) (Section 7.4). Finally, we have summarized the key findings, discussed the practical implications of our research, and provided recommendations for future work (Section 7.5).

7.2 ESSENTIAL CONCEPTS OF FUZZY SET THEORY IN SMART CITIES AND IOT CONTEXT

FST provides a robust framework for managing the ambiguity and uncertainty inherent in complex, real-world systems (Zadeh, 1965). By extending classical set theory, FST allows elements to have partial membership in sets, quantified through a membership function assigning values between 0 and 1. This capability is particularly well-suited for addressing the vagueness and imprecision common in human reasoning and decision-making processes (Zimmermann, 2001). Such a nuanced approach is especially valuable in smart cities and IoT applications, where data often exhibit noise and lack clear-cut boundaries (Wu et al., 2020; Alghassab, 2024).

In a universe of discourse X , a fuzzy set A is characterized by its membership function $\mu_A(x)$, which assigns to each element $x \in X$ a degree of membership ranging from 0 to 1. This is mathematically represented as:

$$A = \{(x, \mu_A(x)) \;\; x \in X, \mu_A(x) \in [0,1]\} \tag{7.1}$$

This formulation enables a flexible representation of data, accommodating elements that partially belong to multiple sets to varying degrees. For instance, in applications such as assessing traffic congestion levels or air quality indices, measurements are not strictly binary but can vary continuously, making fuzzy sets an effective modeling tool (Ilyas, 2021).

Building upon FST, fuzzy logic (FL) utilizes linguistic variables to translate natural language terms into mathematical expressions. This approach facilitates the quantitative handling of qualitative concepts, aligning computational processes more closely with human cognition. For instance, the term "traffic density" can be categorized into fuzzy sets such as "low," "medium," or "high," each defined by specific membership functions (Balaji and Srinivasan, 2011). This enables systems to interpret and act upon imprecise information more effectively, enhancing decision-making processes in smart city and IoT contexts.

7.2.1 MECHANISM OF FUZZY INFERENCE SYSTEM

Fuzzy inference system (FIS) plays a vital role in numerous FL applications within smart cities and IoT. FIS translates inputs into outputs through a series of steps: fuzzification, rule evaluation, aggregation, and defuzzification (Castillo and Melin,

2007). The commonly used models include Mamdani and Sugeno, each tailored to different types of application requirements.

The fundamental steps involved in FIS are:

1. Fuzzification: Transform precise inputs into fuzzy sets.
2. Rule evaluation: Use fuzzy operators to assess the antecedents and determine the strength of each rule.
3. Aggregation: Merge the outcomes of all rules.
4. Defuzzification: Convert the combined fuzzy output back into a specific crisp value.

7.2.2 Fuzzy Control Systems

Fuzzy control systems extend FIS to control engineering, offering a method for designing controllers based on FL principles (Pedrycz, 1993). A typical fuzzy control system includes fuzzification, a knowledge base, an inference engine, and defuzzification. The control action $u(t)$ at time t can be expressed as:

$$u(t) = F\left(\mu_{A_1}\left(e(t)\right), \mu_{A_2}\left(\dot{e}(t)\right), \ldots, \mu_{A_n}\left(x_n(t)\right)\right) \tag{7.2}$$

where $e(t)$ is the error, (t) is its derivative, $x_n(t)$ are other variables relevant to the control context, and μ_{A_i} represents the fuzzy membership functions corresponding to fuzzy sets A_i in the system.

7.2.3 Relevance to Smart Cities and IoT Applications

FST, FIS, and fuzzy control systems are highly relevant in smart cities and IoT due to their ability to handle vague and imprecise data, often collected via IoT sensors (Dionova et al., 2020; Ilyas, 2021; Vivas et al., 2022).

Smart cities have emerged as a strategic response to the challenges posed by rapid urbanization and the imperative for sustainable development. Although definitions can vary, the core concept of a smart city revolves around the innovative application of ICT to enhance urban services, improve quality of life, and promote sustainability (Albino et al., 2015). Giffinger et al. (2007) proposed a comprehensive framework for smart cities, identifying six critical domains: economic competitiveness, advanced mobility systems, environmental stewardship, human capital development, quality of life, and participatory governance.

The main components of smart cities include:

1. Intelligent infrastructure: Buildings, roads, and utilities equipped with sensors for real-time data collection.
2. Data analytics platforms: Systems that process and analyze urban data to generate actionable insights.

3. Communication networks: High-speed, reliable networks like 5G and Wi-Fi for seamless data exchange.
4. Citizen engagement platforms: Tools that facilitate public participation in urban governance.

7.2.4 Role of IoT in Smart Cities

IoT plays a pivotal role in the realization of smart city initiatives. At its core, IoT is a wide network of interconnected devices equipped with sensors and software, enabling seamless data exchange over internet protocols (Atzori et al., 2010). In the context of smart cities, IoT acts as a digital nervous system, facilitating the collection and dissemination of information across various urban subsystems.

Several key IoT technologies are fundamental to smart city implementations:

1. **Sensor technologies:** These include environmental sensors, motion detectors, and smart meters, which gather crucial data about the urban environment.
2. **Radio-frequency identification (RFID) and near-field communication (NFC):** These technologies enable object tracking and transaction identification, streamlining logistics and payment systems within the city.
3. **Low-power, wide-area network (LPWAN):** Technologies such as Long-range, wide-area network (LoRaWAN) and Narrowband Internet of Things (NB-IoT) facilitate long-distance, low-power communication, essential for city-wide IoT deployments.
4. **Edge computing:** By moving data processing closer to the source, edge computing improves response times and reduces bandwidth requirements, enhancing the efficiency of IoT systems.

7.2.5 Key Challenges in Implementing Smart Cities and IoT

Implementing smart cities and IoT technologies faces several significant challenges:

- **Data management and scalability:** The high volume of data from IoT networks requires robust solutions for storage, processing, and analysis. Innovations in cloud and edge computing are essential for managing scalability (Puliafito et al., 2019).
- **Security and privacy:** IoT systems introduce critical risks in security and raise privacy concerns, necessitating robust security protocols and privacy-preserving technologies to protect urban infrastructure and manage personal data (Zhang et al., 2017).
- **Interoperability:** Diverse devices and protocols can lead to integration issues, impacting the effectiveness of urban solutions (Atzori et al., 2017).
- **Uncertainty and decision-making:** The complexity and dynamic nature of urban environments complicate decision-making, requiring models that accommodate multiple stakeholders and incomplete data (Chen et al., 2021).

- **Energy efficiency:** IoT devices must be designed for energy efficiency to support sustainable urban management (Hammi et al., 2018).
- **Social and digital divide:** There is a risk of exacerbating social inequalities through uneven access to smart technologies, highlighting the need for inclusive deployment strategies (Neirotti et al., 2014).

Addressing these challenges requires an interdisciplinary approach that combines technological innovation with policy development and social integration to enhance the resilience, efficiency, and equity of urban environments.

7.3 FUZZY SET THEORY APPLICATIONS IN SMART CITIES AND IOT

FST has been widely applied in smart cities and IoT systems, offering robust solutions to address the inherent uncertainties and complexities of urban environments. Figure 7.1 illustrates the broad range of domains where FL is effectively employed, from traffic management to healthcare monitoring. These applications integrate FL with various methods, relying on different sensors, as detailed in Table 7.1, to collect and process data, thereby enabling intelligent decision-making.

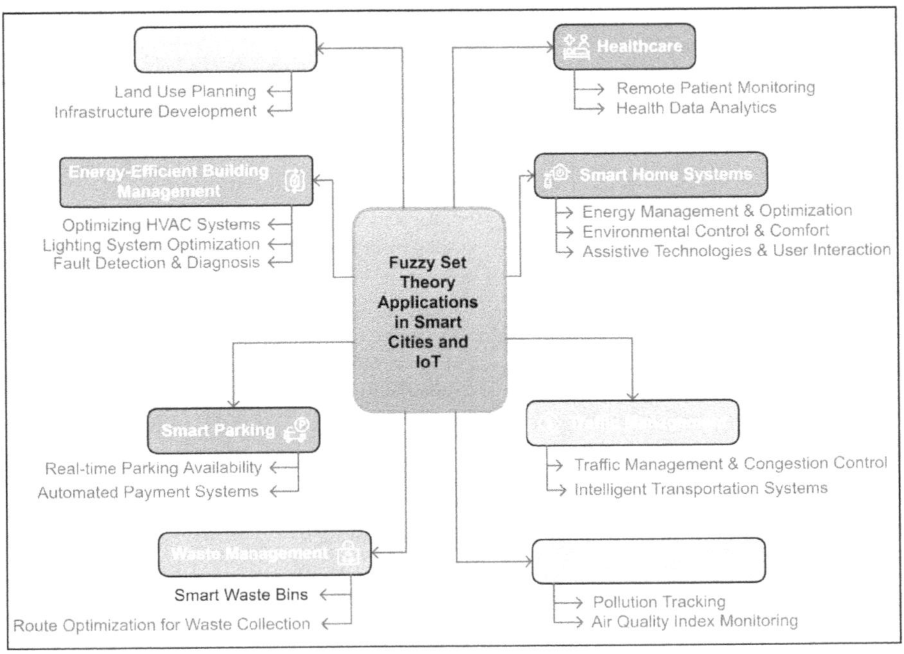

FIGURE 7.1 Taxonomy of FST applications in smart cities and IoT

TABLE 7.1
Fuzzy logic integration and devices in smart city applications

Area of focus	Applications	Devices used	Fuzzy logic integration
Traffic management and congestion control	Traffic signal optimization, congestion prediction	Traffic sensors, cameras, connected vehicles	FL combined with deep learning for improved accuracy of traffic congestion prediction (Berrouk et al., 2022)
Intelligent transportation systems	Emergency vehicle routing, autonomous vehicle control, traffic flow prediction	Wireless communication devices, autonomous vehicles, traffic sensors	Machine learning (ML) with a deep belief network with FL for traffic prediction (George and Santra, 2020)
Energy-efficient building management systems	Heating, ventilation, and air conditioning (HVAC) optimization, lighting control, fault detection	HVAC systems, lighting systems, energy meters	ML with adaptive neuro fuzzy inference system (ANFIS) for fault detection (Martínez Viol et al., 2020)
Smart parking systems	Parking space detection, occupancy prediction	Parking area proximity sensors, surveillance cameras	ML with ANFIS for occupancy prediction (Elomiya et al., 2024)
Waste management systems	Bin fill level monitoring, collection route optimization	Smart bins, GPS-enabled collection vehicles	Fuzzy analytic hierarchy process (FAHP) vcombined with geographic information System (GIS) for facility locations (Aghad et al., 2024)
Air quality monitoring systems	Real-time air quality measurement, pollution level prediction	Weather stations, air quality monitoring devices	Neuro-fuzzy techniques for anomaly detection (Jain and Shah, 2021)
Urban planning and development	Infrastructure performance evaluation, investment decision support	GIS systems, environmental monitoring stations	FL combined with AHP for MCDM in infrastructure assessment (Khatri et al., 2011)
Smart home systems	Energy management, environmental control, assistive technologies	Smart thermostats, lighting systems, appliances, wearable devices	ML with FL for energy management (Shah et al., 2020)
Healthcare and monitoring systems	Remote disease diagnosis, heart disease monitoring, COVID-19 monitoring	Wearable health devices, IoT-enabled medical devices	FL integrated with mathematical modeling for disease diagnosis (Yan et al., 2024)

7.3.1 Traffic Management and Congestion Control

Efficient traffic management and congestion mitigation are essential for enhancing circulation, minimizing delays, and improving overall transportation system performance in smart cities.. Elkari et al. (2024) developed FL-driven frameworks for optimizing traffic signals at intersections, resulting in reduced vehicle waiting times compared to traditional methods. Similarly, Bi et al. (2017) implemented a type-2 FL control mechanism across multiple junctions, fine-tuning green light durations to improve traffic flow and reduce delays based on real-time data. Furthermore, Hao and Hsieh (2024) utilized a fuzzy system based on the Greenshields model for accurate congestion forecasting on highways using vehicle velocity and flow data. Additionally, Berrouk et al. (2022) combined fuzzy inference with deep learning to create a congestion index minimizing prediction inaccuracies.

7.3.2 Intelligent Transportation Systems

Intelligent transportation systems (ITSs) utilize advanced technologies to enhance the safety, efficiency, and environmental sustainability of transportation. FL plays a pivotal role in various ITS applications such as vehicle navigation, self-driving vehicle management, and traffic flow prediction. For instance, Rout et al. (2020) implemented an IoT-based system with FL-driven data fusion for optimizing emergency vehicle routing, thereby improving response times by avoiding congested paths. Djahel et al. (2015) extended this approach by developing an adaptive traffic management system adjusting traffic controls to expedite emergency responses. In autonomous vehicle technologies, Milanés et al. (2009) created an FL-based control system for urban crossroads, enabling self-driving vehicles to adjust speeds according to traffic conditions. Aslam and Bilal (2024) explored broader applications with a Takagi-Sugeno (T-S) FL framework for an unmanned aerial vehicle (UAV). Moreover, FST enhances ITS by improving traffic flow forecasting and data integrity. George and Santra (2020) developed a fuzzy-inspired deep belief network (DBN) for accurate traffic predictions, while Tang et al. (2021) integrated fuzzy neural networks (FNNs) with fuzzy rough set (FRS) theory to correct data discrepancies and increase reliability of traffic projections.

7.3.3 Energy-Efficient Building Management Systems

FST application in smart cities enhances building management systems (BMSs), focusing on energy optimization and occupant comfort. FL improves HVAC systems by integrating occupant preferences and environmental parameters to achieve energy efficiency. For instance, Escobar et al. (2020) and Turhan et al. (2021) demonstrated that FL-based controllers optimize HVAC operations, reducing energy usage. Baniyounes et al. (2022) and Chojecki et al. (2023) further confirm these findings, showing reductions in energy consumption while enhancing comfort metrics. FL is pivotal also in optimizing lighting systems within BMS. Chiesa et al. (2020) and Baniyounes et al. (2020) utilize FL to manage artificial and natural lighting effectively, cutting energy costs and adapting to changing conditions. Cılasun Kunduracı and Kazanasmaz (2019) and Chojecki et al. (2020) explore FL in simulating human-controlled lighting behaviors and integrating renewable energy sources, respectively, leading to energy savings and reduced peak demand. FL has been used in fault detection and diagnosis in BMS, enhancing system reliability and performance. Martínez Viol et al. (2020) employed an adaptive neuro-fuzzy inference system (ANFIS) for HVAC fault detection, achieving high accuracy and reducing false alarms, maintaining effective operation within smart buildings.

7.3.4 Smart Parking Systems

FST enhances smart parking systems within smart cities and IoT frameworks. For instance, Elomiya et al. (2023) utilized FL to improve parking suggestions using data from surveillance systems, considering factors like proximity, trip duration, and space availability. Elomiya et al. (2024) further extended this by combining adaptive neuro-fuzzy inference systems (ANFISs) with machine learning techniques such as recurrent neural networks (RNNs), long short-term memory (LSTM), and gated recurrent units (GRUs) to enhance predictive accuracy and decision-making in parking management. Alyas et al. (2019) also developed a cloud-based, IoT-integrated parking management system using the Mamdani fuzzy inference system (MFIS). This system leverages sensor data and cloud computing to provide real-time parking insights, optimizing space utilization and vehicle guidance, validated through MATLAB simulations.

7.3.5 Waste Management Systems

FST and IoT technologies advance waste management in smart urban environments by optimizing processes and decision-making. For example, Soleymani et al. (2022) and Aghad et al. (2024) employed FST integrated with geographic information systems (GISs) and multi-criteria analysis to identify optimal locations for waste management facilities in Iran and Morocco, respectively. On the other side, Ma et al. (2017) and Li and Huang (2010) developed models to manage municipal solid waste under uncertainty, using fuzzy chance-constrained and dual-interval fuzzy stochastic programming methods, respectively. Regarding intelligent waste management systems, Bano et al. (2020) and Abuga and Raghava (2021) introduced smart bin mechanisms optimizing trash bin placement and collection routes using FL, reducing costs and enhancing urban sanitation. FST has been combined with other optimization techniques like genetic algorithms (GAs) to enhance waste management efficiency. Ikram et al. (2023) integrated IoT with a GA-fuzzy inference system (GA-FIS) for solid waste management, while Oyebode and Abdulazeez (2023) used a hybrid GA and FL approach to optimize waste management supply chains.

7.3.6 Air Quality Monitoring Systems

Smart city initiatives use IoT technologies integrated with FST to improve air quality monitoring, crucial for public health in urban environments. Fahim et al. (2023)

developed IoT-based weather station measuring pollutants like CO_2 and NO_2, using FIS, to categorize air quality into six levels and create an air quality index (AQI). Saleem et al. (2024) advanced this approach by employing a fuzzy inference-based model to monitor a broader range of pollutants, including SO_2, NO_2, CO, O_3, and PM_{10}, providing a detailed AQI. Enhancing predictive capabilities, Praveenchandar et al. (2024) used ANFIS optimized with GA to forecast pollutant concentrations, surpassing other models in precision and effectiveness. Jain and Shah (2021) tackled anomaly detection in air pollution data using neuro-fuzzy techniques, and Gufroni et al. (2024) applied FL to classify environmental parameters in green open spaces, expanding FST applications in environmental monitoring.

7.3.7 Urban Planning and Development

FST has contributed to urban planning by enhancing infrastructure performance evaluations and investment decision-making. Khatri et al. (2011) utilized a fuzzy synthetic evaluation (FSE) approach combined with the analytic hierarchy process (AHP) to assess urban infrastructure, providing a hierarchical evaluation of performance indicators.

In the context of financing urban development, Mortazavi et al. (2023) developed an intuitionistic fuzzy weighting and ranking model to address the complexities of public-private partnerships (3P). This model leverages expert weight calculations and criteria importance determination to manage uncertainty in evaluating 3P projects.

7.3.8 Smart Home Systems

FST enhances various aspects of smart home systems within smart cities and IoT infrastructures, focusing on energy management, user comfort, and assistive technologies. For instance, de Bruyn et al. (2021) integrated FL with a hidden Markov model (HMM) to enhance solar heating efficiency and water usage in geyser systems, achieving significant energy savings. Similarly, Jiménez-Bravo et al. (2020) used a multiagent system coupled with FL to regulate central heating, reducing energy consumption and costs. Abusharia and Al-Aubidy (2022) developed an embedded control unit that applies FL in photovoltaic systems, optimizing energy from various sources.

In addition to energy management, FL enhances user comfort and the quality of experience (QoE) within smart homes. Techniques such as the context-aware FL approach by Ming et al. (2024) automate IoT device coordination, improving comfort and energy use. Shah et al. (2020) integrated a deep extreme learning machine (DELM) with FL to dynamically adjust temperature and lighting, optimizing environmental control.

Furthermore, FL is instrumental in assistive technologies, improving accessibility for individuals with disabilities. Tayyaba et al. (2020) introduced an FL-based system that assists visually impaired users by using IoT devices to detect obstacles and guide navigation. Ahvar et al. (2016) designed a system for precise user location tracking, enhancing interaction and ensuring privacy within smart home environments.

7.3.9 Healthcare and Monitoring Systems

FST enhances smart healthcare and monitoring systems by managing uncertainty and imprecision in medical data, supporting advanced clinical decision-making within smart cities and IoT infrastructures. For example, Yan et al. (2024) implemented a fuzzy-based remote disease diagnosis system utilizing fuzzy clustering to improve diagnostic accuracy for conditions such as heart disease and diabetes. Dewangan et al. (2024) also introduced an IoT-enabled healthcare system using ANFIS to deliver intelligent care in remote areas. In disease monitoring, Munagala et al. (2022) developed a system using a Fuzzy-LSTM model for continuous heart disease monitoring, enhancing predictive accuracy with IoT-generated patient data. Rahman et al. (2023) focused on real-time cardiac arrhythmia detection in COVID-19 patients using IoT and FL for ECG signal processing and autonomous diagnosis. FL is crucial in infectious disease monitoring and risk assessment. Panja et al. (2023) designed an Internet of Medical Things (IoMT) framework using FL for real-time COVID-19 risk assessment, improving patient management during pandemics. Ilyas et al. (2021) developed a system integrating IoT data with FL to analyze COVID-19 symptoms and support clinical decisions. FL facilitates chronic disease management. Maniam et al. (2022) created a monitoring system for chronic kidney disease (CKD) employing IoT and FL for early detection of related conditions, using sensors to collect and process data for real-time decision-making, thus enhancing patient care and quality of life.

7.4 CASE STUDY: OPTIMIZATION OF WASTE MANAGEMENT IN A SMART CITY USING THE SFAHP-FEDAS MODEL

This study employed a hybrid approach using SFAHP to determine the weights of evaluation criteria and fuzzy EDAS to rank the alternatives. For the AHP component, we chose spherical fuzzy sets (SFSs) to handle the complexity of expert judgments during pairwise comparisons, especially when there is a degree of uncertainty or hesitation. SFS allows us to incorporate this hesitation as an independent parameter, making it ideal for capturing nuanced expert opinions. Meanwhile, for the FEDAS calculations, we used trapezoidal fuzzy numbers (TFNs) to evaluate alternatives based on weighted values due to its simplicity, computational efficiency, and suitability for data handling, ensuring reliable rankings for our waste management solutions.

Application of the SFAHP-FEDAS hybrid model for ranking different smart waste management solutions, based on defined criteria, with the goal of optimizing waste collection, processing, and recycling systems, while reducing environmental impact and operational costs, is shown in Figure 7.2.

7.4.1 Spherical Fuzzy AHP Method

Typically, pairwise comparison algorithms rely on precise real numbers. However, expert evaluations in pairwise comparisons are often subjective and imprecise.

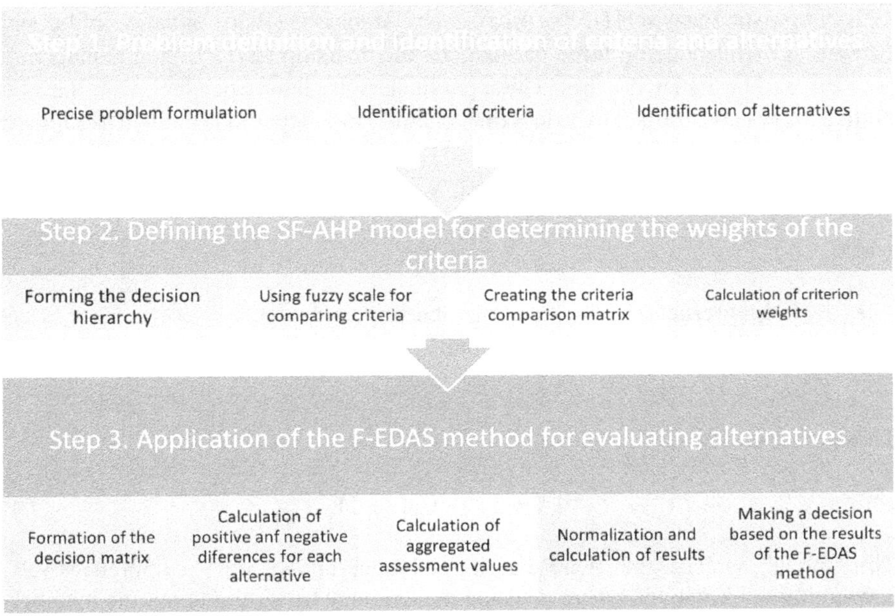

FIGURE 7.2 SFAHP with FEDAS model

Therefore, incorporating fuzzy approaches within the AHP method seems to be a viable solution for handling expert assessments (Abdulkareem and Erzaij, 2022).

In addition to using a fuzzy formulation for pairwise comparisons, the proposed FAHP model follows the same phases as the standard AHP model. The steps of the AHP model are as follows (Radovanović et al., 2021):

1. Define the problem by creating a model that represents the key aspects and relationships of the issue (Radovanovic et al., 2020).
2. Compare items in pairs, providing judgments based on knowledge or opinion.
3. Quantify the judgments using appropriate numerical values.
4. Use these values to determine the priorities of the components in the hierarchy.

7.4.2 Spherical Fuzzy Sets

The development of SFS represents a significant advancement in FST, building upon foundational concepts introduced by intuitionistic fuzzy sets (IFSs) and Pythagorean fuzzy sets (PFSs). These predecessors have provided the groundwork for addressing uncertainties in decision-making processes by extending the capabilities of classical fuzzy sets.

Intuitionistic fuzzy sets (IFSs), proposed by Atanassov (1986), enhance traditional fuzzy sets by introducing three parameters: membership ($\mu_A(x)$), nonmembership ($\nu_A(x)$), and hesitancy degrees. These parameters are interdependent, with the sum of their values constrained to be less than or equal to 1. The IFS is defined as follows:

$$A = \{(x, \mu_A(x),\ \nu_A(x)) \mid x \in X\} \tag{7.3}$$

where

- $\mu_A(x)$ represents the degree of membership, and
- $\nu_A(x)$ represents the degree of nonmembership of the element x to the set A.

These must satisfy

$$0 \leq \mu_A(x) + \nu_A(x) \leq 1. \tag{7.4}$$

This model enables the representation of uncertainty more comprehensively, accounting for situations where decision makers may be unsure about the membership of an element.

Pythagorean fuzzy sets (PFSs), introduced by (Yager, 2013), further extend the flexibility of fuzzy sets by relaxing the constraints imposed by IFS. Instead of the linear summation of membership and nonmembership degrees, PFS requires that the sum of their squares remain less than or equal to 1. The PFS is defined as follows:

$$A = \{(x, \mu_A(x),\ \nu_A(x)) \mid x \in X\} \tag{7.5}$$

where

- $\mu_A(x)$ represents the degree of membership, and
- $\nu_A(x)$ represents the degree of nonmembership of the element x to the set A.

These must satisfy

$$0 \leq (\mu_A(x))^2 + (\nu_A(x))^2 \leq 1. \tag{7.6}$$

This modification provides greater adaptability in modeling complex real-world scenarios where higher degrees of uncertainty or hesitancy are present.

Building on these principles, spherical fuzzy sets (SFSs) incorporate an additional parameter, the hesitancy degree (θ), which is independent of the membership degree (λ) and nonmembership degree (ω). This independence allows for a more flexible and comprehensive framework for decision-making, as illustrated in Figure 7.3. SFS

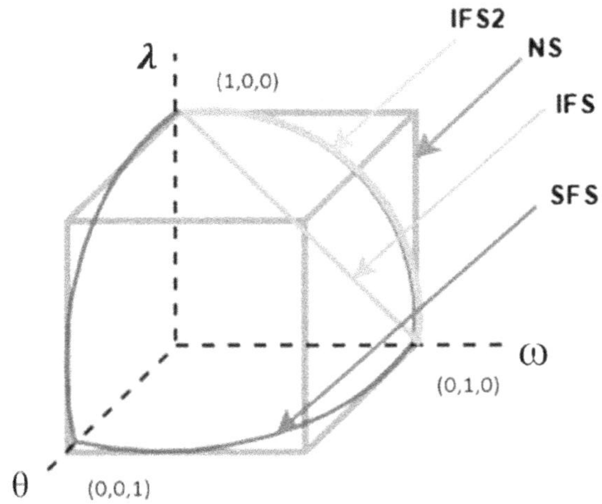

FIGURE 7.3 Geometric representation of spherical fuzzy sets (SFSs), intuitionistic fuzzy sets (IFSs), and neutrosophic fuzzy sets (NFSs). The diagram illustrates how SFSs extend beyond traditional fuzzy sets by incorporating an additional degree of hesitancy, providing a more flexible framework for decision-making (Kutlu Gündoğdu and Kahraman, 2020; Kahraman and Gündogdu, 2021).

ensures that the sum of the squares of λ , ω , and θ does not exceed 1, encapsulating the relationships within a spherical geometric structure. This approach not only enhances the descriptive power of fuzzy sets but also provides decision makers with a robust tool to manage uncertainties across diverse applications (Kahraman and Gündogdu, 2021). This is expressed in the following equation:

$$0 \leq \lambda_{\tilde{A}}^2(u) + \omega_{\tilde{A}}^2(u) + \theta_{\tilde{A}}^2(u) \leq 1, \quad \forall u \in U \tag{7.7}$$

On the surface of the sphere, this relationship simplifies to the following equation:

$$\lambda_{\tilde{A}}^2(u) + \omega_{\tilde{A}}^2(u) + \theta_{\tilde{A}}^2(u) = 1, \quad \forall u \in U \tag{7.8}$$

Spherical fuzzy sets offer decision makers the ability to extend various forms of fuzzy sets, providing a more flexible and comprehensive approach (Kahraman and Gündogdu, 2021). A spherical fuzzy set $\tilde{A}_S$ in the universe U is defined as follows:

$$\tilde{A}_S = \left\{ \left\langle u, \lambda_{\tilde{A}_S}(u), \omega_{\tilde{A}_S}(u), \quad \theta_{\tilde{A}_S}(u) \right\rangle \mid u \in U \right\}, \tag{7.9}$$

where

$$\begin{aligned} \lambda_{\tilde{A}_S}(u) \quad &: U \to [0,1], \\ \omega_{\tilde{A}_S}(u) \quad &: U \to [0,1], \\ \theta_{\tilde{A}_S}(u) \quad &: U \to [0,1]. \end{aligned}$$

For each $u \in U$, the rejection degree of the element is determined using the membership degree $\lambda_{\tilde{A}_S}(u)$, nonmembership degree $\omega_{\tilde{A}_S}(u)$, and hesitation degree $\theta_{\tilde{A}_S}(u)$, as described in Kutlu Gündoğdu and Kahraman (2020) and Kahraman and Gündogdu (2021). The refusal degree $X_{\tilde{A}_S}(u)$ is calculated as follows:

$$X_{\tilde{A}_S}(u) = \sqrt{1 - \lambda^2_{\tilde{A}_S}(u) - \omega^2_{\tilde{A}_S}(u) - \theta^2_{\tilde{A}_S}(u)} \tag{7.10}$$

7.4.3 Zadeh's SFS Extension Principle

Let $i = 1,\ldots,n$ and U_i be a universe with $V \neq 0$. Let $f : \prod_{i=1}^{n} U_i \rightarrow V$ be a mapping, where $y = f(z_1,\ldots,z_n)$. Let z_i be linguistic variables on U_i for $i = 1,\ldots,n$. Assume that $(\tilde{A}_{S_i})$ is a SFS on U_i . Then, the output of the mapping f is $\tilde{B}_S$. For $y \in V$, the set $\tilde{B}_S$ is an SFS on V and is defined as follows:

$$\tilde{B}_S(y) = \left\{ \begin{array}{c} \max\limits_{Z(y)} \left(\min\limits_{i=1}^{n} \lambda_{\tilde{A}_{S_i}}(u_i) \right), \\ \min\limits_{Z(y)} \left(\max\limits_{i=1}^{n} \omega_{\tilde{A}_{S_i}}(u_i) \right), \\ \min\limits_{Z(y)} \left(\min\limits_{i=1}^{n} \theta_{\tilde{A}_{S_i}}(u_i) \right) \end{array} \right\}, \quad \forall u_i \in U_i, i = 1,\ldots,n \tag{7.11}$$

if $f^{-1}(y) \neq 0$, where $Z(y) = f^{-1}(y)$.

For operations such as addition and multiplication (Konyalıoğlu et al., 2025), the SFS addition is defined as follows:

$$\tilde{A}_S \oplus \tilde{B}_S = \left\{ \begin{array}{c} z, \left(\max\limits_{Z=x+y} \min\left(\lambda_{\tilde{A}_S}(x), \lambda_{\tilde{B}_S}(y) \right) \right), \\ \left(\min\limits_{Z=x+y} \max\left(\omega_{\tilde{A}_S}(x), \omega_{\tilde{B}_S}(y) \right) \right), \\ \left(\min\limits_{Z=x+y} \min\left(\theta_{\tilde{A}_S}(x), \theta_{\tilde{B}_S}(y) \right) \right) \end{array} \right\} \tag{7.12}$$

The geometric mean of spherical weights for $w = (w_1, w_2,\ldots,w_n)$, where $w_i \in [0,1]$ and $\sum_{i=1}^{n} w_i = 1$, is defined as follows (Abdulkareem and Erzaij, 2022):

$$\mathrm{SWGM}_w\left(\tilde{A}_{S_1}^{w_1},\ldots,\tilde{A}_{S_n}^{w_n}\right)=\left\{\begin{array}{r}\prod_{i=1}^{n}\lambda_{\tilde{A}_{S_i}}^{w_i},\\ \left[1-\prod_{i=1}^{n}\left(1-\omega_{\tilde{A}_{S_i}}^2\right)^{w_i}\right]^{\frac{1}{2}},\\ \left[\prod_{i=1}^{n}\left(1-\lambda_{\tilde{A}_{S_i}}^2\right)^{w_i}-\prod_{i=1}^{n}\left(1-\omega_{\tilde{A}_{S_i}}^2-\theta_{\tilde{A}_{S_i}}^2\right)^{w_i}\right]^{\frac{1}{2}}\end{array}\right\} \tag{7.13}$$

Table 7.2 presents the linguistic evaluation scale used in the SFAHP for pairwise comparisons.

The consistency of pairwise comparison matrices within the AHP is measured using the consistency index (CI) and consistency ratio (CR). These matrices are evaluated to ensure consistent judgments, with all matrices found to be consistent. In AHP, $\lambda_{\max}$ refers to the largest eigenvalue of the comparison matrix, and n represents its size. The random index (RI) is derived based on the matrix size and is used as an average consistency index (CI) from several randomly generated comparison matrices (Saaty, 1980). The expressions for CI and CR are as follows:

$$CI=\frac{\lambda_{\max}-n}{n-1},\quad CR=\frac{CI}{RI} \tag{7.14}$$

7.4.4 Fuzzy EDAS

EDAS method, initially proposed by Keshavarz Ghorabaee et al. (2015), is commonly used for multi-criteria inventory classification. In its fuzzy version, decision

TABLE 7.2
Linguistic scales applied in pairwise comparison evaluations

SF (λ,ω,θ)	Linguistic expression	Scale index
(0.9, 0.1, 0.0)	Extremely high importance (EHI)	9
(0.8, 0.2, 0.1)	Strong importance (SI)	7
(0.7, 0.3, 0.2)	High importance (HI)	5
(0.6, 0.4, 0.3)	Slightly higher importance (SHI)	3
(0.5, 0.4, 0.4)	Equal importance (EI)	1
(0.4, 0.6, 0.3)	Slightly lower importance (SLI)	1/3
(0.3, 0.7, 0.2)	Low importance (LI)	1/5
(0.2, 0.8, 0.1)	Very low importance (VLI)	1/7
(0.1, 0.9, 0.0)	Extremely low importance (ELI)	1/9

makers assign criteria weights and evaluate alternatives for each criterion using linguistic terms. These linguistic evaluations are transformed into positive TFNs, as outlined in Table 7.3.

The extended fuzzy EDAS methodology comprises the following steps, adapted from Ghorabaee et al. (2016):

Step 1: Construction of the average decision matrix ($\dot{Y}$): The average decision matrix $\boldsymbol{Y}$ is formed as:

$$\boldsymbol{Y} = [\tilde{y}_{ij}]_{n\times m} \tag{7.15}$$

where $\tilde{y}_{ij} = \frac{1}{h} \oplus_{r=1}^{h} \tilde{y}_{ij}^{r}$. Here, $\tilde{y}_{ij}^{r}$ represents the performance rating of alternative A_i for criterion c_j , as provided by the rth decision maker.

Step 2: Determining the weight matrix ($\boldsymbol{V}$): The weight matrix $\boldsymbol{V}$ for the criteria is calculated as:

$$\boldsymbol{V} = [\tilde{v}_j]_{1\times m} \tag{7.16}$$

where $\tilde{v}_j = \frac{1}{h} \oplus_{r=1}^{h} \tilde{v}_j^{r}$ and $\tilde{v}_j^{r}$ denotes the weight of criterion $\boldsymbol{c}_j$ as assigned by the $\boldsymbol{r}^{th}$ decision maker.

Step 3: Generation of the average solution matrix: The average solution matrix is expressed as:

$$\boldsymbol{AS} = [\widetilde{as}_j]_{1\times m} \tag{7.17}$$

TABLE 7.3
Linguistic terms and their corresponding TFNs

Linguistic term	TFN
Very high (VH)	(0, 0, 1, 2)
High (H)	(1, 2, 2, 3)
Medium high (MH)	(2, 3, 4, 5)
Medium (M)	(4, 5, 5, 6)
Medium low (ML)	(5, 6, 7, 8)
Low (L)	(7, 8, 8, 9)
Very low (VL)	(8, 9, 10, 10)

where $\widetilde{as}_j = \frac{1}{n} \oplus_{i=1}^{n} \tilde{y}_{ij}$. This matrix provides the average solutions for each criterion.

Step 4: Calculation of the positive and negative distance matrices (PDA and NDA)

For the beneficial criteria set ***B*** and nonbeneficial set ***N***, the matrices for positive distance from average (PDA) and negative distance from average (NDA) are computed as:

$$PDA = [\widetilde{pda}_{ij}]_{n\times m} \tag{7.18}$$

$$NDA = [\widetilde{nda}_{ij}]_{n\times m} \tag{7.19}$$

The values $\widetilde{pda}_{ij}$ and $\widetilde{nda}_{ij}$ are calculated as:

$$\widetilde{pda}_{ij} = \begin{cases} \dfrac{\phi\left(\tilde{y}_{ij} \ominus \widetilde{as}_j\right)}{h\cdot \widetilde{as}_j}, & \text{if } j \in B \\ \dfrac{\phi\left(\widetilde{as}_j \ominus \tilde{y}_{ij}\right)}{h\cdot \widetilde{as}_j}, & \text{if } j \in N \end{cases} \tag{7.20}$$

$$\widetilde{nda}_{ij} = \begin{cases} \dfrac{\phi\left(\widetilde{as}_j \ominus \tilde{y}_{ij}\right)}{h\cdot \widetilde{as}_j}, & \text{if } j \in B \\ \dfrac{\phi\left(\tilde{y}_{ij} \ominus \widetilde{as}_j\right)}{h\cdot \widetilde{as}_j}, & \text{if } j \in N \end{cases} \tag{7.21}$$

Step 5: Calculation of weighted sums for positive and negative distances

The weighted sum of positive and negative distances for each alternative is computed as:

$$\widetilde{sp}_i = \oplus_{j=1}^{m} \left(\tilde{v}_j \otimes \widetilde{pda}_{ij}\right) \tag{7.22}$$

$$\widetilde{sn}_i = \oplus_{j=1}^{m} \left(\tilde{v}_j \otimes \widetilde{nda}_{ij}\right) \tag{7.23}$$

Step 6: Normalization of positive and negative distances

The normalized positive and negative distances are calculated as:

$$\widetilde{nsp}_i = \frac{\widetilde{sp}_i}{\max_i h\left(\widetilde{sp}_i\right)} \tag{7.24}$$

$$\widetilde{nsn_i} = \frac{\widetilde{sn_i}}{\max_i h\left(\widetilde{sn_i}\right)} \tag{7.25}$$

Step 7: Appraisal score calculation: Finally, the appraisal score ($\widetilde{aps_i}$) for each alternative is determined as:

$$\widetilde{aps_i} = \frac{1}{2}\left(\widetilde{nsp_i} \otimes \widetilde{nsn_i}\right) \tag{7.26}$$

Alternatives are ranked based on their appraisal scores, with the highest score representing the best alternative.

7.4.5 Identification of Criteria and Alternatives

In this section, seven criteria (Table 7.4) and seven alternatives (Table 7.5) are defined, which were used in the case study addressing the waste management problem in a smart city. The criteria and alternatives were established based on an analysis of the available literature and the involvement of experts.

7.4.6 Implementation of the Proposed Model

The weight coefficients for the criteria were computed using the SFAHP method, as determined by three experts, and the results are presented in Table 7.6.

Following this, the initial decision matrix was established using data from Table 7.3, and the average decision matrix was derived by applying Equations (7.12) to (7.14), as shown in Table 7.7. Subsequently, Equations (7.15) to (7.23) were utilized to select the optimal alternative based on the seven predefined criteria. The results of this analysis are provided in Table 7.8.

The results shown in Table 7.8 indicate that alternative A1, represented by "smart bins with sensors," ranks as the most favorable choice based on the stability and consistency of the findings.

7.5 CONCLUSION

This chapter explored the diverse applications of FST within smart cities and IoT, highlighting its effectiveness in managing uncertainties and complexities in urban environments. We reviewed various applications of FST, including traffic management, intelligent transportation systems, energy-efficient building management, smart parking systems, waste management, air quality monitoring, urban planning, smart home systems, and healthcare monitoring. Each application showed how FST enhances decision-making processes and improves system performance in areas that need to handle imprecise or uncertain information.

TABLE 7.4
Identify the criteria

Criteria	Description of criteria
(**C1**) Efficiency of waste collection	This criterion measures how effectively the waste collection system operates in terms of territorial coverage and collection frequency. It encompasses aspects such as route optimization, reduction of the time required for collection, and minimization of missed bins that are full. Additionally, it considers how quickly and efficiently the collection trucks can empty the bins and transport the waste to processing centers.
(**C2**) Environmental impact	This criterion assesses the impact of the solution on the environment. It includes aspects such as the reduction of carbon dioxide emissions during waste transportation, the environmental benefits of waste recycling, and the reduction of landfills. An important factor is also how much the technology contributes to minimizing the negative impact on natural resources and how it helps in preserving biodiversity.
(**C3**) Energy efficiency	It evaluates the amount of energy required for the operation of the solution, whether it is energy used for waste collection, processing, or recycling. The focus is on reducing energy consumption through optimized processes and the use of renewable energy sources, as well as on balancing energy consumption with gains, especially in solutions that utilize energy from waste.
(**C4**) Economic efficiency	This criterion assesses the costs of implementing, maintaining, and operating the solution over the long term. It includes an analysis of capital investments, operational expenses, and the financial sustainability of the solution in the long run. Economic efficiency is also evaluated based on the potential for cost reduction for the city, as well as the ability to generate additional revenue through recycling or selling energy from waste.
(**C5**) Sustainable recycling	This criterion measures how much the solution contributes to increasing recycling rates and sustainable resource management. It includes the ability to sort waste into recyclable materials, minimize waste that ends up in landfills, and achieve targets related to the circular economy. It also assesses how easily the solution can be adapted for different types of waste, such as paper, plastic, glass, metal, and organic waste.
(**C6**) Integration of smart technologies	It evaluates how well the solution utilizes modern technologies such as sensors, IoT, artificial intelligence (AI), and data analytics to enhance waste management operations. The focus is on the technological advancement and interoperability of the system, as well as the capability for automatic collection, processing, and transmission of real-time waste data.
(**C7**) Citizen engagement	This criterion measures the level of citizen engagement in the waste management process. It assesses how well the solution enables citizens to actively participate in recycling, proper waste disposal, and monitoring their own environmental habits. Additionally, it includes aspects such as the ease of use of applications or platforms for citizens, as well as the effectiveness of promoting environmentally responsible behavior through education or incentives.

TABLE 7.5
Definition and description of alternatives

Alternatives	Description of alternatives
(**A1**) Smart bins with sensors	Bins equipped with sensors that detect fill levels and automatically send information to a central system. This system enables real-time monitoring of bin status and optimizes the timing for collection trucks to empty them. Additionally, the sensors can detect specific types of waste, facilitating sorting.
(**A2**) Optimization of waste collection truck routes	A software solution that utilizes data from sensors in bins and real-time traffic information to optimize waste collection truck routes. The aim is to reduce travel time, fuel consumption, and harmful gas emissions, while also improving the overall efficiency of waste collection.
(**A3**) Automated waste sorting	The introduction of artificial intelligence and robotics-based technologies in recycling centers for the automatic sorting of waste. These systems can quickly and accurately classify waste by material type (plastic, metal, glass, etc.) without the need for human labor, thereby accelerating the process and increasing recycling efficiency.
(**A4**) Illegal waste dumping detection system	The use of smart cameras, sensors, and AI technology to identify illegal waste dumping in urban areas. This system automatically reports incidents to the relevant authorities, enabling faster response and reducing improper waste disposal.
(**A5**) Digital platforms for citizen engagement	Applications or digital platforms that allow citizens to track their waste and recycling habits. Citizens can receive feedback on their ecological behavior, rewards for responsible waste management, and guidance for better recycling practices.
(**A6**) Smart recycling centers powered by waste energy	Centers that utilize advanced technologies for waste separation and energy production from organic waste. These centers generate renewable energy from waste, reduce the amount of waste sent to landfills, and contribute to the energy sustainability of smart cities.
(**A7**) Infrastructure for waste separation at home	Solutions that enable households to presort waste using provided equipment and tracking applications. Citizens receive support in the form of apps that monitor proper waste separation and offer rewards for contributing to environmental goals.

TABLE 7.6
Weight coefficients assigned to criteria using the SFAHP method

Criteria	Spherical fuzzy weights	Crisp weights
C1 Efficiency of waste collection	0.770, 0.228, 0.181	0.203
C2 Environmental impact	0.680, 0.331, 0.208	0.177
C3 Energy efficiency	0.692, 0.324, 0.203	0.180
C4 Economic efficiency	0.714, 0.287, 0.218	0.186
C5 Sustainable recycling	0.404, 0.583, 0.283	0.098
C6 Integration of smart technologies	0.343, 0.661, 0.263	0.082
C7 Citizen engagement	0.317, 0.690, 0.259	0.075

TABLE 7.7
Average decision matrix for evaluating alternatives

	v1				v2				v3				v4			
	0.77	0.228	0.181	0.203	0.68	0.331	0.208	0.177	0.692	0.324	0.203	0.18	0.714	0.287	0.218	0.186
	C1				**C2**				**C3**				**C4**			
A1	8.000	9.000	10.000	10.000	6.333	7.333	7.667	8.667	8.000	9.000	10.000	10.000	8.000	9.000	10.000	10.000
A2	8.000	9.000	10.000	10.000	3.333	4.333	4.667	5.667	7.667	8.667	9.333	9.667	8.000	9.000	10.000	10.000
A3	6.333	7.333	7.667	8.667	1.667	2.667	3.333	4.333	7.333	8.333	8.667	9.333	7.667	8.667	9.333	9.667
A4	2.333	3.333	3.667	4.667	4.667	5.667	6.333	7.333	2.333	3.333	3.667	4.667	0.667	1.333	1.667	2.667
A5	3.667	4.667	5.333	6.333	3.333	4.333	4.667	5.667	2.667	3.667	4.333	5.333	4.333	5.333	5.667	6.667
A6	4.333	5.333	5.667	6.667	4.667	5.667	6.333	7.333	5.667	6.667	7.333	8.333	6.333	7.333	7.667	8.667
A7	1.667	2.667	3.333	4.333	2.333	3.333	3.667	4.667	2.667	3.667	4.333	5.333	4.667	5.667	6.333	7.333
	v5				**v6**				**v7**							
	0.404	0.583	0.283	0.098	0.343	0.661	0.263	0.082	0.317	0.69	0.259	0.075				
	C5				**C6**				**C7**							
A1	2.667	3.667	4.333	5.333	7.667	8.667	9.333	9.667	3.667	4.667	5.333	6.333				
A2	2.667	3.667	4.333	5.333	8.000	9.000	10.000	10.000	2.333	3.333	3.667	4.667				
A3	7.333	8.333	8.667	9.333	8.000	9.000	10.000	10.000	1.000	1.667	2.333	3.333				
A4	0.333	0.667	1.333	2.333	4.333	5.333	5.667	6.667	3.667	4.667	5.333	6.333				
A5	1.333	2.333	2.667	3.667	6.333	7.333	7.667	8.667	5.333	6.333	6.667	7.667				
A6	8.000	9.000	10.000	10.000	7.667	8.667	9.333	9.667	3.667	4.667	5.333	6.333				
A7	3.667	4.667	5.333	6.333	4.667	5.667	6.333	7.333	8.000	9.000	10.000	10.000				

TABLE 7.8
Results obtained from the proposed model application

Alternatives	aps_i				$K(aps_i)$	Rank
A1	0.556	0.918	1.151	1.491	1.028	1
A2	0.410	0.676	0.812	1.410	0.848	2
A3	0.192	0.953	0.761	1.131	0.710	4
A4	−0.360	−0.151	−0.210	0.336	−0.063	7
A5	−0.068	0.240	0.268	1.114	0.431	6
A6	−0.173	1.034	1.143	1.358	0.769	3
A7	−0.122	0.603	0.524	0.899	0.442	5

Our case study focused on optimizing waste management based on the SFAHP-FEDAS model. SFS was utilized in the AHP component to better account for expert hesitation, while TFN was employed in the FEDAS method to ensure optimal evaluation of weighted values. By ranking various smart waste management solutions based on predefined criteria, the analysis identified smart bins equipped with sensors as the most effective option. These findings indicate the practical potential of FST-based models in creating more efficient and sustainable urban management systems.

From a practical perspective, our study highlights how FST can significantly improve resource allocation, enhance system efficiency, and optimize decision-making in smart cities. The implementation of FST-based methods can help city planners and stakeholders to develop and deploy urban management strategies with a higher level of resilience and adaptability.

Future research should focus on further integrating FST with emerging technologies such as artificial intelligence, machine learning, and edge computing to enhance the scalability and adaptability of smart city systems. Additionally, exploring the application of FST in other urban domains, such as water management, public safety, and disaster response, could provide valuable insights. Furthermore, addressing the challenges of data security, privacy, and interoperability will be crucial for the widespread adoption of FST-based solutions in smart cities.

ACKNOWLEDGMENT

This chapter is supported by the project SGS_2025 "Modelling of Selected Aspects of Transportation Technology and Control V" at the University of Pardubice. Furthermore, this work was supported by the European Regional Development Fund (ERDF) through the project "Cross-Sectoral and Interdisciplinary Collaboration in Research and Development of Communication, Information, and Detection Technologies for Control and Signalling Systems (CIDET)" (Grant No. CZ.02.01.01/00/23_021/0008402).

REFERENCES

Abdulkareem, H.G., Erzaij, K.R., 2022. A spherical fuzzy AHP model for contractor assessment during project life cycle. *Journal of the Mechanical Behavior of Materials* 31, 369–380.

Abuga, D., Raghava, N., 2021. Real-time smart garbage bin mechanism for solid waste management in smart cities. *Sustainable Cities and Society* 75, 103347.

Abusharia, R.M., Al-Aubidy, K.M., 2022. Embedded control unit design for energy management in smart homes. *Bulletin of Electrical Engineering and Informatics* 11, 2537–25462537–2546.

Aghad, M., Manaouch, M., Sadiki, M., Pham, Q.B., Al karkouri, J., 2024. Integrating fuzzy-AHP and GIS for solid waste disposal site selection in Kenitra province, NW Morocco. *Environmental Monitoring and Assessment* 196, 537.

Ahvar, E., Lee, G.M., Han, S.N., Crespi, N., Khan, I., 2016. Sensor network-based and user-friendly user location discovery for future smart homes. *Sensors* 16, 969.

Albino, V., Berardi, U., Dangelico, R.M., 2015. Smart cities: Definitions, dimensions, performance, and initiatives. *Journal of Urban Technology* 22, 3–21.

Alghassab, M.A., 2024. Fuzzy-based smart energy management system for residential buildings in Saudi Arabia: A comparative study. *Energy Reports* 11, 1212–1224.

Alyas, T., Ahmad, G., Saeed, Y., Asif, M., Farooq, U., Kanwal, A., 2019. Cloud and IoT based smart car parking system by using Mamdani fuzzy inference system (MFIS). *ISeCure: The ISC International Journal of Information Security* 11, 153–160.

Aslam, M.S., Bilal, H., 2024. Modeling a Takagi-Sugeno (T-S) fuzzy for unmanned aircraft vehicle using fuzzy controller. *Ain Shams Engineering Journal*, 15(10), 102984.

Atanassov, K.T., 1986. Intuitionistic fuzzy sets. *Fuzzy Sets and Systems* 20, 87–96.

Atzori, L., Iera, A., Morabito, G., 2010. The internet of things: A survey. *Computer Networks* 54, 2787–2802787–2805.

Atzori, L., Iera, A., Morabito, G., 2017. Understanding the internet of things: Definition, potentials, and societal role of a fast evolving paradigm. *Ad Hoc Networks* 56, 122–140.

Balaji, P., Srinivasan, D., 2011. Type-2 fuzzy logic based urban traffic management. *Engineering Applications of Artificial Intelligence* 24, 12–22.

Baniyounes, A.M., Ghadi, Y.Y., Alnabulsi, M., 2020. Exergy analysis of day light using fuzzy logic controllers of Jordanian commercial buildings. *International Journal of Power Electronics and Drive Systems (IJPEDS)* 11, 1686–1692.

Baniyounes, A.M., Ghadi, Y.Y., Radwan, E., Al-Olimat, K.S., 2022. Functions of fuzzy logic-based controllers used in smart building. *International Journal of Electrical and Computer Engineering (IJECE)* 12, 3061–30713061–3071.

Bano, A., Din, I.U., Al-Huqail, A.A., 2020. AIoT-based smart bin for real-time monitoring and management of solid waste. *Scientific Programming* 2020, 13.

Berrouk, S., Fazziki, A.E., Sadgal, M., 2022. Hybrid deformable convolutional with recurrent neural network for optimal traffic congestion prediction: A fuzzy logic congestion index system. *International Journal of Advanced Computer Science and Applications*, 13(5), 635–651.

Bi, Y., Lu, X., Srinivasan, D., Sun, Z., Sun, Z., 2017. Optimal type-2 fuzzy system for arterial traffic signal control. *IEEE Transactions on Intelligent Transportation Systems*, 19(9), 3009–3027.

de Bruyn, D.N., Kotze, B., Hurst, W., 2021. A hidden Markov model and fuzzy logic forecasting approach for solar geyser water heating. *Infrastructures* 6, 67.

Castillo, O., Melin, P., 2007. *Type-2 Fuzzy Logic: Theory and Applications*. Springer.

Chen, J., Li, P., Fang, W., Zhou, N., Yin, Y., Zheng, H., Xu, H., Wang, R., 2021. Fuzzy frequent pattern mining algorithm based on weighted sliding window and type-2 fuzzy sets over medical data stream. *Wireless Communications and Mobile Computing* 2021, 6662254.

Chiesa, G., Di Vita, D., Ghadirzadeh, A., Muñoz Herrera, A.H., Rodriguez, J.C.L., 2020. A fuzzy-logic IoT lighting and shading control system for smart buildings. *Automation in Construction* 120, 103397.

Chojecki, A., Ambroziak, A., Borkowski, P., 2023. Fuzzy controllers instead of classical PIDS in HVAC equipment: Dusting off a well-known technology and today's implementation for better energy efficiency and user comfort. *Energies* 16, 2967.

Chojecki, A., Rodak, M., Ambroziak, A., Borkowski, P., 2020. Energy management system for residential buildings based on fuzzy logic: Design and implementation in smart-meter. *IET Smart Grid* 3, 254–266.

Cılasun Kunduracı, A., Kazanasmaz, Z.T., 2019. Fuzzy logic model for the categorization of manual lighting control behaviour patterns based on daylight illuminance and interior layout. *Indoor and Built Environment* 28, 584–598.

Dewangan, N.K., Pandey, N., Gautam, R., Goswami, A.K., Mitkari, S.R., Singh, A., Kopare, A., Gobi, N., 2024. A novel healthcare decision support system using IoT and ANFIS. *International Journal of Information Technology*, 16, 5421–5427.

Dionova, B.W., Mohammed, M., Al-Zubaidi, S., Yusuf, E., 2020. Environment indoor air quality assessment using fuzzy inference system. *ICT Express* 6, 185–194.

Djahel, S., Smith, N., Wang, S., Murphy, J., 2015. Reducing emergency services response time in smart cities: An advanced adaptive and fuzzy approach, in: *2015 IEEE First International Smart Cities Conference (ISC2)*, IEEE. pp. 1–8.

Elkari, B., Ourabah, L., Sekkat, H., Chaibi, Y., 2024. Minimizing intersection delays: A novel fuzzy logic-based architecture for traffic signal control. *International Journal of Intelligent Transportation Systems Research*, 22, 542–560.

Elomiya, A., Křupka, J., Jovčić, S., 2023. A smart parking system using surveillance cameras and fuzzy logic: A case study at Pardubice university's campus. *Procedia Computer Science* 225, 4881–48904881–4890.

Elomiya, A., Křupka, J., Jovčić, S., Simic, V., 2024. Enhanced prediction of parking occupancy through fusion of adaptive neuro-fuzzy inference system and deep learning models. *Engineering Applications of Artificial Intelligence* 129, 107670.

Escobar, L.M., Aguilar, J., Garces-Jimenez, A., De Mesa, J.A.G., Gomez-Pulido, J.M., 2020. Advanced fuzzy-logic-based context-driven control for HVAC management systems in buildings. *IEEE Access* 8, 16111–1612616111–16126.

Fahim, M., Mhouti, A.E., Boudaa, T., Jakimi, A., 2023. Modeling and implementation of a low-cost IoT-smart weather monitoring station and air quality assessment based on fuzzy inference model and MQTT protocol. *Modeling Earth Systems and Environment* 9, 4085–41024085–4102.

George, S., Santra, A.K., 2020. Fuzzy inspired deep belief network for the traffic flow prediction in intelligent transportation system using flow strength indicators. *Big Data* 8, 1–17.

Ghorabaee, M.K., Zavadskas, E.K., Amiri, M., Turskis, Z., 2016. Extended EDAS method for fuzzy multi-criteria decision-making: An application to supplier selection. *International Journal of Computers Communications & Control* 11, 358–371.

Giffinger, R., Fertner, C., Kramar, H., Kalasek, R., Pichler-Milanovic, N., Meijers, E., 2007. Smart cities: Ranking of European medium-sized cities. *Centre of Regional Science, Vienna University of Technology* 16, 13–18.

Gufroni, A.I., Rachman, A.N., Mubarok, H., Hiron, N., Ramdani, C.M.S., Sambas, A., 2024. Classification of temperature and humidity in green open spaces by implementing internet of things (IoT) using Mamdani fuzzy logic. *Journal of Advanced Research in Applied Sciences and Engineering Technology* 49, 205–217.

Hammi, B., Khatoun, R., Zeadally, S., Fayad, A., Khoukhi, L., 2018. IoT technologies for smart cities. *IET Networks* 7, 1–13.

Hao, M.J., Hsieh, B.Y., 2024. Greenshields model-based fuzzy system for predicting traffic congestion on highways. *IEEE Access*, 12, 115868–115882.

Ikram, S.T., Mohanraj, V., Ramachandran, S., Balakrishnan, A., 2023. An intelligent waste management application using IoT and a genetic algorithm–fuzzy inference system. *Applied Sciences* 13, 3943.

Ilyas, M., 2021. IoT applications in smart cities, in: *2021 International Conference on Electronic Communications, Internet of Things and Big Data (ICEIB)*, IEEE, pp. 44–47.

Ilyas, T., Mahmood, D., Ahmed, G., Akhunzada, A., 2021. Symptom analysis using fuzzy logic for detection and monitoring of covid-19 patients. *Energies* 14, 7023.

Jain, R., Shah, H., 2021. Anomaly detection in smart cities modeled as wireless sensor networks. *International Journal of Smart City Applications* 10, 123–135.

Jiménez-Bravo, D.M., Murciego, A.L., de la Iglesia, D.H., Paz, J.F.D., González, G.V., 2020. Central heating cost optimization for smart-homes with fuzzy logic and a multi-agent architecture. *Applied Sciences* 10, 4057.

Kahraman, C., Gündogdu, F., 2021. Decision making with spherical fuzzy sets. *Studies in Fuzziness and Soft Computing* 392, 3–25.

Keshavarz Ghorabaee, M., Zavadskas, E.K., Olfat, L., Turskis, Z., 2015. Multi-criteria inventory classification using a new method of evaluation based on distance from average solution (edas). *Informatica* 26, 435–451.

Khatri, K.B., Vairavamoorthy, K., Akinyemi, E., 2011. Framework for computing a performance index for urban infrastructure systems using a fuzzy set approach. *Journal of Infrastructure Systems* 17, 163–175.

Konyalıoğlu, A.K., Bereketli, I., Ozcan, T., 2025. A novel spherical fuzzy AHP method to managing waste from face masks and gloves: An Istanbul-based case study. *International Journal of Environmental Science and Technology*, 22, 4229–4248.

Kutlu Gündoğdu, F., Kahraman, C., 2020. A novel spherical fuzzy analytic hierarchy process and its renewable energy application. *Soft Computing* 24, 4607–46214607–4621.

Li, Y., Huang, G., 2010. Dual-interval fuzzy stochastic programming method for long-term planning of municipal solid waste management. *Journal of Computing in Civil Engineering* 24, 188–202.

Ma, X., Ma, C., Wan, Z., Wang, K., 2017. A fuzzy chance-constrained programming model with type 1 and type 2 fuzzy sets for solid waste management under uncertainty. *Engineering Optimization* 49, 1040–1056.

Maniam, G., Sampe, J., Jaafar, R., Ibrahim, M.F., 2022. Smart monitoring system for chronic kidney disease patients based on fuzzy logic and IoT. *International Journal of Advanced Computer Science and Applications*, 13(2), 324–333.

Martínez Viol, V., Urbano González, E.M., Delgado Prieto, M., Romeral Martínez, J.L., 2020. HVAC early fault detection using a fuzzy logic based approach, in: *Renewable Energy and Power Quality Journal (RE&PQJ)*, Vol. 18, June 2020, AEDERMACP (European Association for the Development of Renewable Energies), pp. 1–6.

Milanés, V., Pérez, J., Onieva, E., González, C., 2009. Controller for urban intersections based on wireless communications and fuzzy logic. *IEEE Transactions on Intelligent Transportation Systems* 11, 243–248.

Ming, C., Kadry, S., Dasel, A.A., 2024. Automating smart internet of things devices in modern homes using context-based fuzzy logic. *Computational Intelligence* 40, e12370.

Mortazavi, S.M., Adlparvar, M.R., Shahbazi, M., 2023. An intuitionistic fuzzy weighting and ranking model to evaluate the financing challenges of urban development with public-private partnership projects. *Scientia Iranica* 30, 877–890.

Munagala, N.K., Langoju, L.R.R., Rani, A.D., Reddy, D.R.K., 2022. A smart IoT-enabled heart disease monitoring system using meta-heuristic-based fuzzy-LSTM model. *Biocybernetics and Biomedical Engineering* 42, 1183–1204.

Neirotti, P., De Marco, A., Cagliano, A.C., Mangano, G., Scorrano, F., 2014. Current trends in smart city initiatives: Some stylised facts. *Cities* 38, 25–36.

Oyebode, O., Abdulazeez, Z., 2023. Optimization of supply chain network in solid waste management using a hybrid approach of genetic algorithm and fuzzy logic: A case study of Lagos state. *Nature Environment & Pollution Technology*, 22 (4), 1707–1722.

Panja, S., Chattopadhyay, A.K., Nag, A., Singh, J.P., 2023. Fuzzy-logic-based IoMT framework for COVID19 patient monitoring. *Computers & Industrial Engineering* 176, 108941.

Pedrycz, W., 1993. *Fuzzy Control and Fuzzy Systems.* Research Studies Press Ltd.

Praveenchandar, J., Venkatesh, K., Mohanraj, B., Prasad, M., Udayakumar, R., 2024. Prediction of air pollution utilizing an adaptive network fuzzy inference system with the aid of genetic algorithm. *Natural and Engineering Sciences* 9, 46–56.

Puliafito, C., Mingozzi, E., Longo, F., Puliafito, A., Rana, O., 2019. Fog computing for the internet of things: A survey. *ACM Transactions on Internet Technology (TOIT)* 19, 1–41.

Radovanović, M., Petrovski, A., Žindrašič, V., Ranđelović, A., 2021. Application of the fuzzy AHP-VIKOR hybrid model in the selection of an unmanned aircraft for the needs of tactical units of the armed forces. *Scientific Technical Review* 71, 26–35.

Radovanovic, M., Ranđelović, A., Jokić, Ž., 2020. Application of hybrid model fuzzy AHP-VIKOR in selection of the most efficient procedure for rectification of the optical sight of the long-range rifle. *Decision Making: Applications in Management and Engineering* 3, 131–148.

Rahman, M.Z., Akbar, M.A., Leiva, V., Tahir, A., Riaz, M.T., Martin-Barreiro, C., 2023. An intelligent health monitoring and diagnosis system based on the internet of things and fuzzy logic for cardiac arrhythmia covid-19 patients. *Computers in Biology and Medicine* 154, 106583.

Ross, T.J., 2005. *Fuzzy Logic with Engineering Applications.* John Wiley & Sons.

Rout, R.R., Vemireddy, S., Raul, S.K., Somayajulu, D., 2020. Fuzzy logic-based emergency vehicle routing: An IoT system development for smart city applications. *Computers and Electrical Engineering* 88, 106839.

Saaty, T., 1980. *The Analytic Hierarchy Process: Planning, Priority Setting, Resource Allocation.* Advanced book program, McGraw-Hill International Book Company.

Saleem, M., Shingari, N., Farooq, M.S., Mago, B., Khan, M.A., 2024. Real-time air quality monitoring model using fuzzy inference system. *International Journal of Advanced Computer Science and Applications* 15, 838–846.

Shah, A.S., Nasir, H., Fayaz, M., Lajis, A., Ullah, I., Shah, A., 2020. Dynamic user preference parameters selection and energy consumption optimization for smart homes using deep extreme learning machine and bat algorithm. *IEEE Access* 8, 204744–204762204744–204762.

Soleymani, M., Asakereh, A., Safieddin Ardebili, S.M., 2022. A gis-based multi-criteria fuzzy approach to select a suitable location for a MSW-based power plant and landfill: A case study, Khuzestan province, Iran. *Environmental Monitoring and Assessment* 194, 174.

Tang, J., Zhang, X., Yin, W., Zou, Y., Wang, Y., 2021. Missing data imputation for traffic flow based on combination of fuzzy neural network and rough set theory. *Journal of Intelligent Transportation Systems* 25, 439–454.

Tayyaba, S., Ashraf, M.W., Alquthami, T., Ahmad, Z., Manzoor, S., 2020. Fuzzy-based approach using IoT devices for smart home to assist blind people for navigation. *Sensors* 20, 3674.

Turhan, C., Simani, S., Akkurt, G.G., 2021. Development of a personalized thermal comfort driven controller for HVAC systems. *Energy* 237, 121568.

Vivas, F.J., Segura, F., Andújar, J., 2022. Fuzzy logic-based energy management system for grid-connected residential dc microgrids with multi-stack fuel cell systems: A multi-objective approach. *Sustainable Energy, Grids and Networks* 32, 100909.

Wu, T.Y., Lin, J.C.W., Yun, U., Chen, C.H., Srivastava, G., Lv, X., 2020. An efficient algorithm for fuzzy frequent itemset mining. *Journal of Intelligent & Fuzzy Systems* 38, 5787–57975787–5797.

Yager, R.R., 2013. Pythagorean fuzzy subsets, in: *2013 Joint IFSA World Congress and NAFIPS Annual Meeting (IFSA/NAFIPS)*, IEEE, pp. 57–61.

Yan, F., Huang, H., Pedrycz, W., Hirota, K., 2024. A disease diagnosis system for smart healthcare based on fuzzy clustering and battle royale optimization. *Applied Soft Computing* 151, 111123.

Zadeh, L.A., 1965. Fuzzy sets. *Information and Control* 8, 338–353.

Zanella, A., Bui, N., Castellani, A., Vangelista, L., Zorzi, M., 2014. Internet of things for smart cities. *IEEE Internet of Things Journal* 1, 22–32.

Zhang, K., Ni, J., Yang, K., Liang, X., Ren, J., Shen, X.S., 2017. Security and privacy in smart city applications: Challenges and solutions. *IEEE Communications Magazine* 55, 122–129.

Zimmermann, H.J., 2001. *Fuzzy Set Theory—and its Applications.* Springer Science & Business Media.

8 Smart Campus Process Automation

Process Prioritization through Triangular Fuzzy AHP

Onur Dogan, Ali Mert Erdogan, and Bayram Kose

8.1 INTRODUCTION

In recent years, automation and digital transformation have emerged as fundamental drivers of efficiency and innovation across various sectors, including education. Higher education institutions are increasingly adopting smart campus applications to optimize administrative processes, enhance student experiences, and improve decision-making capabilities. A smart campus leverages digital technologies such as artificial intelligence (AI), the Internet of Things (IoT), cloud computing, and robotic process automation (RPA) to create interconnected and intelligent systems that enhance institutional performance. These advancements not only streamline operations but also contribute to sustainability by reducing resource consumption and operational costs.

RPA has become a focal point for organizations aiming to boost operational efficiency, lower costs, and increase productivity. By automating repetitive tasks and workflows with software robots, RPA allows human employees to focus on more strategic and value-driven activities. However, determining which processes are best suited for RPA deployment is a complex decision-making process, as it involves evaluating numerous criteria and accounting for the diverse nature of various processes.

To address this challenge, organizations are increasingly turning to multi-criteria decision-making (MCDM) methods to support the selection of processes for RPA implementation. MCDM approaches offer a structured framework for evaluating and prioritizing alternatives based on multiple criteria, considering the interdependencies and trade-offs among these criteria. In this study, the intuitionistic fuzzy originated interval type-2 fuzzy AHP method has been employed as the primary analytical tool.

The analytic hierarchy process (AHP) is widely recognized as an effective tool for determining the relative weights of criteria within the decision-making framework.

 DOI: 10.1201/9781003529088-8

It allows decision makers to systematically assess the significance of each criterion, thereby deriving priority weights that facilitate a more structured and objective decision-making approach. Through hierarchical structuring and pairwise comparisons, AHP captures the nuanced importance of each criterion, ensuring a thorough and well-rounded evaluation.

While previous studies have leveraged the combined strengths of various MCDM techniques for RPA process selection, this study employs the intuitionistic fuzzy originated interval type-2 fuzzy AHP method as an advanced alternative. This approach addresses the inherent complexities of decision-making under uncertainty, offering a robust analytical framework that accommodates the ambiguity and imprecision associated with evaluating RPA processes. By integrating this method, decision makers can comprehensively assess and prioritize RPA processes, considering their intrinsic characteristics, alignment with organizational objectives, and operational needs.

The application of this sophisticated methodology empowers organizations to make well-informed decisions regarding RPA process selection, aligning these choices with strategic goals, optimizing resource distribution, and enhancing the overall efficacy of RPA implementations. By refining the decision-making process and facilitating a more precise evaluation, the utilization of the intuitionistic fuzzy originated interval type-2 fuzzy AHP contributes to decision quality and increases the likelihood of successful RPA deployments.

The following sections of this study explore the methodology in detail, present a case study that validates the effectiveness of the proposed approach, discuss the results and their analysis, and offer insights into future research directions and practical implications.

8.2 LITERATURE REVIEW

The use of RPA has drawn significant interest in a variety of industries, including the education sector, in recent years (Oluçoğlu et al. 2023). Various studies have examined the challenges and opportunities linked to RPA implementation, especially with regard to process selection (Wanner et al. 2019).

Numerous studies examine the key drivers and obstacles to the adoption of RPA. Lacity and Willcocks (2016) point out that cost savings, adherence to compliance, and scalability are the main motivating factors, while governance challenges and complications with integration present significant hurdles. Similarly, Aguirre and Rodriguez (2017) investigate the social and technical aspects of RPA implementation, highlighting that resistance from the workforce and inadequate training are crucial elements hindering successful adoption. In addition, Madakam et al. (2019) present a classification of RPA technologies, stressing the incorporation of machine learning (ML) and artificial intelligence to broaden its usability beyond merely repetitive tasks.

Selecting the most appropriate processes for automation is a significant challenge in RPA projects, as the effectiveness of RPA implementation largely hinges on selecting the correct processes to automate (Lindström et al. 2018). Organizations function within diverse frameworks, where processes vary significantly in their characteristics. These traits are crucial in assessing the automation potential of each process. Given the influence of different process characteristics on automation potential, RPA cannot be applied universally; it demands careful evaluation and well-informed decision-making (Geyer-Klingeberg et al. 2018). Therefore, it is essential for organizations to thoroughly understand their processes to guarantee the overall success of their RPA initiatives.

A major challenge in implementing RPA is the inability to realize the expected cost savings (Lacity et al. 2015). This problem frequently stems from incorrect predictions of the economic value of RPA at the process level. To overcome this challenge, it is essential to develop a system that can precisely assess the economic benefits of automating individual processes. This necessitates the creation of a decision-support system that can aid in selecting processes for automation and facilitate a comprehensive cost-benefit analysis (Akyol et al. 2022). Such a system would offer valuable insights into potential financial benefits and assist organizations in making informed decisions regarding which processes to automate, ensuring that RPA initiatives align with realistic cost-saving expectations (Schmitz et al. 2019).

Lacity et al. (2015) examined the dynamic characteristics of processes and their changing automation potential over time. By continuously evaluating the appropriateness of processes for automation, organizations can recognize shifts in requirements and adapt their RPA strategies accordingly.

A primary difficulty in adopting RPA is found in the phase of selecting processes, where organizations need to identify which processes are appropriate for automation. Various studies have introduced decision-making frameworks to tackle this issue (Abdel-Basset et al. 2019). Recently, more advanced methods like Fuzzy MCDM techniques have been utilized to address the limitations of traditional approaches (Çalık 2021; Ghaleb et al. 2020). These methods enhance decision-making by incorporating fuzzy logic, allowing for a more nuanced evaluation of criteria weights and process characteristics. This integration helps decision makers navigate the inherent uncertainties and ambiguities in complex decision problems, providing a more comprehensive and precise assessment framework.

For example, Costa et al. (2023) offer a method that merges AHP and TOPSIS for the selection of processes for automation. Their model assesses processes based on multiple criteria, including cost, complexity, and improvement potential, providing a systematic approach to pinpoint processes that will gain the most from automation. By employing AHP to prioritize each criterion's significance and using TOPSIS to rank the processes, this method guarantees a thorough and impartial selection process. The authors illustrate the practical use of their model in a case study, outlining a clear path for process selection in RPA initiatives. In a similar vein, Erdogan and Dogan (2023) explore the application of MCDM techniques for selecting processes in RPA initiatives at Izmir Bakircay University. They utilize a modified MCDM method, which includes aspects such as process volume, time consumption, and human resource usage. The chapter emphasizes the need to take into account the organization's unique context when choosing processes for automation. The authors stress that a customized approach, which considers both operational efficiency and the organization's strategic goals, is crucial for maximizing RPA's advantages.

Focusing on supply chain management, John and Jeyapaul (2024) use a fuzzy MCDM method to choose RPA software packages for a reverse supply chain firm. This method tackles the inherent uncertainty in decision-making by applying fuzzy logic to represent the ambiguity in the criteria evaluation. Their research demonstrates how fuzzy MCDM techniques can be utilized for RPA software selection, considering critical factors like software compatibility, scalability, and cost-effectiveness. By implementing fuzzy logic, the authors offer a more adaptable and resilient framework for decision-making, which is particularly beneficial in uncertain and rapidly evolving environments. Another crucial aspect of RPA implementation is selecting a vendor, where various criteria need to be assessed to find the most suitable vendor for the organization's requirements. Banerjee and Parkhi (2020) examine the evaluation and selection of RPA vendors by utilizing a combination of AHP and the Kano model. The study presents a hybrid model that assesses vendors based on both their technical attributes and customer satisfaction, with the Kano model pinpointing elements that enhance customer delight. By merging these two approaches, the authors contend that organizations can reach a more comprehensive and thoughtful conclusion when choosing an RPA vendor. Their method is especially advantageous for organizations aiming to balance technical expertise with user satisfaction and support services.

While process and vendor selection are vital components of RPA adoption, it is also essential to evaluate the risks related to utilizing digital robots in process automation. Dogan et al. (2024) tackle this concern by performing a risk assessment of digital robots employed in RPA. They point out possible risks such as cybersecurity threats, operational interruptions, and employee pushback, proposing strategies to alleviate these risks. The research underscores the importance of organizations adopting a proactive stance on risk management, ensuring that RPA implementation is both safe and sustainable.

In the procurement sector, RPA has demonstrated considerable potential for automation. van Hoek et al. (2022) assess RPA technologies for automating procurement processes. The findings reveal that RPA can greatly decrease processing times and errors, especially in repetitive tasks such as order processing, invoice handling, and inventory monitoring. The authors use an MCDM approach to evaluate various RPA tools and technologies, taking into account criteria such as cost, functionality, and ease of integration with existing systems. Their results indicate that RPA can lead to significant efficiency improvements in procurement, particularly when combined with other technologies like AI and machine learning (ML).

Wanner et al. (2019) further the conversation on process selection in RPA initiatives by creating a quantifiable decision-making method. Their framework integrates performance metrics like automation potential, complexity, and scalability to assist organizations in objectively determining which processes should be automated. The authors highlight the necessity for a structured and quantitative decision-making approach, especially when navigating complex organizational landscapes with multiple decision makers.

Although previous studies have employed simpler methods, there is a notable lack of research using more complex methodologies to support the selection of appropriate processes for RPA implementation in specific domains (Marrella 2018; Wanner et al. 2019). This study aims to address this gap by developing a comprehensive framework based on the intuitionistic fuzzy originated interval type-2 Fuzzy AHP method. The objective is to utilize this advanced approach to select RPA processes within the context of a smart university case. By employing intuitionistic fuzzy originated interval type-2 fuzzy AHP, the study provides a systematic methodology that incorporates multiple criteria and evaluates the automation potential of various processes. This research seeks to contribute to the existing literature by presenting a robust and practical methodology that can guide organizations in making well-informed decisions regarding RPA process selection in the smart university domain.

8.3 METHODOLOGY

Triangular Fuzzy AHP is an advanced extension of the traditional AHP, incorporating fuzzy logic to handle uncertainty and imprecision in decision-making. However, traditional AHP assumes precise numerical judgments, which may not always be practical in complex decision-making environments. To address this limitation, triangular Fuzzy AHP integrates fuzzy set theory to account for the inherent uncertainty in expert evaluations. In triangular Fuzzy AHP, linguistic variables are used to express subjective judgments, which are then represented by triangular fuzzy numbers (TFNs). A TFN is defined by a triplet (l, m, u), where:

- l (lower bound) represents the minimum possible value,
- m (modal value) represents the most likely or expected value, and
- u (upper bound) represents the maximum possible value.

These fuzzy numbers provide a more flexible approach to capturing expert opinions, allowing for a range of possible values rather than fixed single values. The methodology described by Kahraman et al. (2018) is employed in this study, with steps detailed as follows:

Step 1: Define evaluation criteria and alternatives

The initial phase involves identifying the assessment criteria and the alternative options pertinent to the decision-making problem. Subsequently, a hierarchical structure of the problem should be developed to serve as the analytical framework for the evaluation process.

Step 2: Construct pairwise comparison matrices

The second step involves constructing pairwise comparison matrices using a triangular intuitionistic fuzzy linguistic assessment scale, as presented in Table 8.1. Each linguistic term reflects the strength of comparison between elements in the pairwise comparison matrix. The parameters μ and υ represent the triangular fuzzy numbers associated with the linguistic terms, encompassing their lower, middle, and upper values. The term "reciprocal" denotes the inverse values of the corresponding fuzzy numbers. These linguistic terms and their associated values are applied to evaluate

TABLE 8.1
Triangular intuitionistic fuzzy scale

Linguistic terms	μ			υ			Reciprocal
Absolutely strong (AS)	8	9	10	7	9	11	AW
Very strong (VS)	6	7	8	5	7	9	VW
Fairly strong (FS)	4	5	6	3	5	7	FW
Slightly strong (SS)	2	3	4	1	3	5	SW
Exactly equal (E)	1	1	1	1	1	1	E
Slightly weak (SW)	0.5	0.3	0.25	1	0.33	0.2	SS
Fairly weak (FW)	0.3	0.2	0.17	0.3	0.2	0.14	FS
Very weak (VW)	0.2	0.1	0.13	0.2	0.14	0.11	VS
Absolutely weak (AW)	0.1	0.1	0.1	0.1	0.11	0.09	AS

and compare various criteria or alternatives within a fuzzy environment, enabling more refined decision-making. Following this, the linguistic terms are converted into triangular intuitionistic fuzzy numbers (TIFNs) based on the given scale.

Step 3: Aggregation of intuitionistic fuzzy pairwise comparison matrices

In the third step, the individual intuitionistic fuzzy pairwise comparison matrices are aggregated. The aggregation operator employed in this study plays a crucial role in consolidating expert judgments in the group decision-making process. It is achieved using the aggregation operator, resulting in a single aggregated matrix that reflects the collective judgments of all experts involved.

Let $\tilde{A}_j = \left(\left(a_j^L, a_j^M, a_j^U\right), \left(a_j'^L, a_j'^M, a_j'^U\right)\right), j = 1, 2, \ldots, n$ be a set of triangular fuzzy number intuitionistic fuzzy numbers (*TFNIFN*s), where a_j^L, a_j^M, a_j^U, $a_j'^L$, $a_j'^M, a_j'^U \geq 1$. We propose the aggregation operator in Equation (8.1) for these numbers:

$$f_m\left(I_1, I_2, \ldots, I_n\right) = \left(\begin{array}{c} \left[\prod_{j=1}^{n} \left(a_j^L\right)^{w_j}, \prod_{j=1}^{n} \left(a_j^M\right)^{w_j}, \prod_{j=1}^{n} \left(a_j^U\right)^{w_j} \right], \\ \left[\prod_{j=1}^{n} \left(a_j'^L\right)^{w_j}, \prod_{j=1}^{n} \left(a_j'^M\right)^{w_j}, \prod_{j=1}^{n} \left(a_j'^U\right)^{w_j} \right] \end{array} \right) \tag{8.1}$$

where $w = \left(w_1, w_2, \ldots, w_n\right)^T$ is the weight vector of $I_j\left(j = 1, 2, \ldots, n\right), w_j \in [0,1]$,

$\sum_{j=1}^{n} w_j = 1$.

Step 4: Transformation of aggregated matrix to type-2 fuzzy pairwise comparison matrix

The transformation of an aggregated matrix into a type-2 fuzzy pairwise comparison matrix is a process used to incorporate higher levels of uncertainty and imprecision in decision-making, especially when expert opinions are vague or uncertain. Type-2 fuzzy sets extend the standard type-1 fuzzy sets by allowing uncertainty in the membership functions themselves, making them useful for more complex and uncertain environments.

Once the lower and upper membership functions are defined, each element of the aggregated matrix is replaced by an interval of fuzzy numbers to form the type-2 fuzzy matrix. This matrix now contains type-2 fuzzy numbers that account for both the comparison between alternatives (or criteria) and the associated uncertainty in that comparison.

For example, a comparison between two alternatives A1 and A2 with respect to a subcriterion C11 in a type-1 matrix may have a fuzzy value of 0.7. After transformation into a type-2 fuzzy pairwise comparison matrix, this value may now be represented as an interval, like [0.65, 0.75], indicating the range of possible membership values.

The fourth step entails transforming the aggregated triangular intuitionistic fuzzy pairwise comparison matrix into a triangular type-2 fuzzy (TT2F) pairwise comparison matrix, as illustrated in Figure 8.1.

Step 5: Calculation of geometric means and fuzzy weights

In the fifth step, the geometric mean of each row in the TT2F matrix is computed using equation 19 from Kahraman et al. (2018). The fuzzy weights are subsequently obtained by normalizing these geometric means, following the procedure outlined in equation 20 of the same study.

Step 6: Aggregation of fuzzy weights and performance scores

The sixth step involves aggregating the fuzzy weights and performance scores of the alternatives, in accordance with equation 21 of Kahraman et al. (2018). This step integrates the criteria weights with performance scores to calculate the overall fuzzy utility of each alternative.

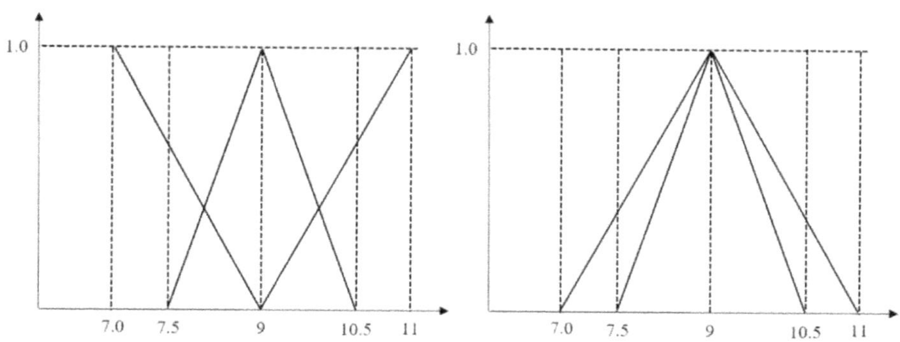

FIGURE 8.1 The transformation from TIFN to TT2F number

Step 7: Selection of the optimal alternative

The seventh step is the selection of the best alternative. The classical AHP method is employed for this purpose. Using the defuzzification formula provided in equation 22 of Kahraman et al. (2018), the fuzzy values are converted into crisp scores, enabling the identification of the optimal choice.

8.4 CASE STUDY

Izmir Bakircay University's Seyrek Campus covers an area of 58,898 m². It houses a computer and internet-equipped library of 1,350 m², capable of serving up to 350 students. The campus also includes meeting and conference rooms totaling 2,170 m², sports facilities of 1,151 m², and an exhibition hall of 700 m² that can accommodate up to 500 individuals. Additionally, there are 41 classrooms and lecture halls of various sizes. The university employs 586 academic and administrative staff members and caters to 6,573 students. These characteristics emphasize the strengths of Izmir Bakircay University as described in the mentioned source. Comparable advantages may be realized by other universities based on their unique attributes. Implementing and integrating technological infrastructure can yield substantial economic and social benefits for public institutions and organizations.

The university has set up a dedicated smart university and Digital Transformation Coordination Office to manage the Smart University Project. This office focuses on achieving sustainability in ecological, financial, data management, and digitalization aspects. A smart reporting system is being created utilizing centralized data management and cloud computing technologies to support sustainability. This system enables authorized users to promptly access necessary reports in their areas of responsibility, enhancing the efficiency of decision-making processes (Dogan and Cengiz Tirpan 2022). Additionally, one of the office's key roles is to digitize and automate university operations.

Step 1: Define evaluation criteria and alternatives

In the first step of the RPA process selection, it is essential to clearly define the evaluation criteria and identify the alternatives. Five evaluation criteria and 16 subcriteria serve as the foundation for decision-making, providing measurable factors as given below. These criteria ensure that the selected process aligns with the strategic goals of the smart campus initiative.

- Process suitability (C1)
  - Configurability (C11): The process's ability to be adjusted or customized for automation.
  - Repeatability (C12): How frequently the process is performed, impacting automation efficiency.
  - Clarity of rules (C13): The level of clear, predefined rules governing the process.
  - Standardization (C14): The ease of creating standardized steps for automation.

- Process complexity (C2)
  - Decision-making needs (C21): The complexity of decisions required during the process.
  - Exceptions (C22): The number and complexity of exceptions or unique cases in the process.
  - Manual intervention (C23): The degree of human involvement needed for process completion.
- Technological compatibility (C3)
  - Integration with existing systems (C31): How easily the process can integrate with current technology systems.
  - Data structure (C32): Whether the process uses structured or unstructured data.
  - Input and output formats (C33): The standardization of data inputs and outputs for the process.
- Efficiency potential (C4)
  - Time savings (C41): The potential reduction in process time through automation.
  - Cost savings (C42): The expected cost reductions from automating the process.
  - Reduction of human errors (C43): The potential for reducing mistakes made by manual processes.
- Process priority (C5)
  - Business criticality (C51): How essential the process is to core business operations.
  - Risk reduction (C52): The ability of automation to minimize operational risks.
  - Regulatory compliance (C53): Whether the process meets legal and regulatory requirements.

Alongside the criteria, a list of process alternatives is established, representing different campus operations that could benefit from automation. This structured approach allows for a comprehensive and objective evaluation in subsequent steps.

- Student enrollment and registration (A1): Automating the collection and verification of student information, document submissions, approvals, and registration processes for admissions.
- Course and curriculum management (A2): Automating course scheduling, curriculum updates, course registration, and managing conflicts or waitlists.
- Campus facility management (A3): Automating the booking of rooms and facilities, maintenance requests, and scheduling repairs for campus infrastructure.
- Student support services (A4): Automating common student services such as handling inquiries, issuing documents (e.g., transcripts), and managing service requests.

- Campus event management (A5): Streamlining event logistics, including scheduling, ticketing, communication with attendees, and venue coordination.
- IT help desk operations (A6): Automating the ticketing process for IT issues, including ticket creation, routing, resolution, and providing support through chatbots.

Step 2: Construct pairwise comparison matrices

This step presents the computational results used to prioritize university processes within the framework of smart university implementations. Using the proposed method, weights of 0.30, 0.35, and 0.35 were assigned to the three decision makers based on their level of experience. Tables 8.2 and 8.3 display their respective pairwise comparison matrices for both the main and subcriteria, and Table 8.4 indicates pairwise comparison matrices of the alternatives with respect to C11, as an example representation.

Step 3: Aggregation of intuitionistic fuzzy pairwise comparison matrices

Experts compare these criteria pairwise based on their relative importance. For instance, process suitability (C1) might be compared to process complexity (C2) and so on, using intuitionistic fuzzy numbers. The comparisons reflect how much more important one criterion is over another in terms of achieving automation efficiency.

Similarly, subcriteria under each main criterion are compared pairwise. For example, within process suitability (C1), experts would compare configurability (C11) against repeatability (C12), clarity of rules (C13), and standardization (C14) to determine which subcriterion contributes more to process suitability.

The individual evaluations should be combined into a consensus form. This aggregated matrix is created using various methods like averaging or geometric means applied to the original pairwise comparison matrices provided by experts. This study applies to geometric means for the aggregation. Tables 8.5 and 8.6 present the aggregated triangular intuitionistic fuzzy pairwise comparison matrices for the main criteria and subcriteria, respectively.

In the context of selecting processes for robotic process automation (RPA), comparisons of alternatives with respect to subcriteria using intuitionistic fuzzy pairwise comparison matrices follow a similar process to the criteria evaluation. However, instead of comparing the main criteria or subcriteria, the focus shifts to comparing the alternatives (e.g., A1, A2, A3, etc.) with respect to each subcriterion. Decision makers assess each alternative in relation to each subcriterion. Table 8.7 provides the pairwise comparison matrices for the subcriteria with respect to the other alternatives.

Step 4: Transformation of aggregated matrix to type-2 fuzzy pairwise comparison matrix

Table 8.8 shows transformed to TT2F pairwise comparison matrix for the main criteria as an example of this transformation. The remaining transformation matrix is not given due to the page limitation.

Step 5: Calculation of geometric means and fuzzy weights

Table 8.9 presents the geometric and normalized geometric means of each criterion. The defuzzified scores correspond to the criterion weights.

TABLE 8.2
Pairwise comparison matrices of the main criteria concerning the goal

	DM1					DM2					DM3				
	C1	C2	C3	C4	C5	C1	C2	C3	C4	C5	C1	C2	C3	C4	C5
C1	E	E	E	VS	SW	E	E	AS	FS	VS	E	VW	FW	SW	FW
C2	E	E	FS	AS	SS	E	E	FS	FS	FS	VS	E	SS	VS	FS
C3	E	FW	E	SS	SW	AW	FW	E	AW	SW	FS	SW	E	VS	E
C4	VW	AW	SW	E	AW	FW	FW	AS	E	VS	SS	VW	VW	E	AW
C5	SS	SW	SS	AS	E	VW	FW	SS	VW	E	FS	FW	E	AS	E

TABLE 8.3
Pairwise comparison matrices of the subcriteria concerning the goal

	DM1				DM2				DM3			
	C11	**C12**	**C13**	**C14**	**C11**	**C12**	**C13**	**C14**	**C11**	**C12**	**C13**	**C14**
C11	E	VS	E	SS	E	AS	E	VS	E	FW	AW	SW
C12	VW	E	FW	AW	AW	E	AW	SW	FS	E	FW	SW
C13	E	FS	E	SS	E	AS	E	AS	AS	FS	E	SS
C14	SW	AS	SW	E	VW	SS	AW	E	SS	SS	SW	E
	C21	C22	C23		C21	C22	C23		C21	C22	C23	
C21	E	VW	FW		E	SW	AW		E	SS	FW	
C22	VS	E	E		SS	E	FW		SW	E	AW	
C23	FS	E	E		AS	FS	E		FS	AS	E	
	C31	C32	C33		C31	C32	C33		C31	C32	C33	
C31	E	E	FS		E	AS	VS		E	VS	VS	
C32	E	E	FS		AW	E	E		VW	E	E	
C33	FW	FW	E		VW	E	E		VW	E	E	
	C41	C42	C43		C41	C42	C43		C41	C42	C43	
C41	E	E	VW		E	E	AW		E	AW	AW	
C42	E	E	FW		E	E	AW		AS	E	SS	
C43	VS	FS	E		AS	AS	E		AS	SW	E	
	C51	C52	C53		C51	C52	C53		C51	C52	C53	
C51	E	E	AS		E	E	VW		E	FS	AS	
C52	E	E	VS		E	E	VW		FW	E	E	
C53	AW	VW	E		VS	VS	E		AW	E	E	

TABLE 8.4
Pairwise comparison matrices of the alternatives with respect to C11

	DM1						DM2					
C11	**A1**	**A2**	**A3**	**A4**	**A5**	**A6**	**A1**	**A2**	**A3**	**A4**	**A5**	**A6**
A1	E	AS	AW	AS	VS	AS	E	FS	E	E	SS	SS
A2	AW	E	SW	SS	AS	AW	FW	E	AW	FW	AS	SW
A3	AS	SS	E	FW	E	SS	E	AS	E	AW	SS	FS
A4	AW	SW	FS	E	AW	AS	E	FS	AS	E	SW	AW
A5	VW	AW	E	AS	E	VS	SW	AW	SW	SS	E	E
A6	AW	AS	SW	AW	VW	E	SW	SS	FW	AS	E	E
	DM3											
	A1	A2	A3	A4	A5	A6						
A1	E	FW	VW	FS	FW	SW						
A2	FS	E	E	FW	AW	FW						
A3	VS	E	E	SS	E	VS						
A4	FW	FS	SW	E	VS	SS						
A5	FS	AS	E	VW	E	AS						
A6	SS	FS	VW	SW	AW	E						

TABLE 8.5
Aggregated main criteria weights

	μ			υ		
C1	0.14	0.13	0.14	0.13	0.13	0.15
C2	0.27	0.47	0.72	0.36	0.47	0.63
C3	0.12	0.12	0.13	0.10	0.12	0.13
C4	0.08	0.09	0.10	0.07	0.09	0.10
C5	0.14	0.20	0.26	0.14	0.20	0.25

Step 6: Aggregation of fuzzy weights and performance scores
Table 8.10 represents the fuzzy and crisp scores of the alternatives and their ranking.
Step 7: Selection of the optimal alternative
Criterion C2 has the highest defuzzified value of 0.47, indicating it is the most significant among the criteria. This criterion holds nearly half of the total weight in the decision-making process. C5 follows with a moderate defuzzified value of 0.20, suggesting that it also plays a meaningful role. The combined importance of the top two criteria (C2 and C5) is 0.67, reflecting that the majority of the evaluation weight is concentrated on these criteria. On the other hand, C4 and C3 have the lowest defuzzified values of 0.09 and 0.12, respectively, making them the least significant

TABLE 8.6
Aggregated subcriteria weights

Subcriterion	Local subcriterion weights						Global subcriterion weights					
	μ			υ			μ			υ		
C11	0.25	0.25	0.26	0.25	0.25	0.22	0.04	0.03	0.04	0.03	0.03	0.03
C12	0.12	0.09	0.08	0.10	0.09	0.07	0.02	0.01	0.01	0.01	0.01	0.01
C13	0.39	0.53	0.64	0.57	0.53	0.50	0.06	0.07	0.09	0.08	0.07	0.07
C14	0.14	0.13	0.12	0.20	0.13	0.10	0.02	0.02	0.02	0.03	0.02	0.02
C21	0.11	0.10	0.11	0.10	0.10	0.11	0.03	0.05	0.08	0.04	0.05	0.07
C22	0.15	0.16	0.18	0.15	0.16	0.18	0.04	0.07	0.13	0.06	0.07	0.11
C23	0.49	0.74	1.05	0.55	0.74	0.95	0.13	0.34	0.75	0.20	0.34	0.60
C31	0.56	0.76	1.00	0.66	0.76	0.83	0.07	0.09	0.13	0.06	0.09	0.11
C32	0.16	0.17	0.18	0.16	0.17	0.17	0.02	0.02	0.02	0.02	0.02	0.02
C33	0.08	0.07	0.07	0.12	0.07	0.06	0.01	0.01	0.01	0.01	0.01	0.01
C41	0.07	0.07	0.07	0.07	0.07	0.09	0.01	0.01	0.01	0.00	0.01	0.01
C42	0.15	0.19	0.23	0.17	0.19	0.27	0.01	0.02	0.02	0.01	0.02	0.03
C43	0.56	0.75	0.99	0.43	0.75	1.13	0.04	0.07	0.10	0.03	0.07	0.11
C51	0.49	0.54	0.58	0.51	0.54	0.52	0.07	0.11	0.15	0.07	0.11	0.13
C52	0.27	0.26	0.25	0.26	0.26	0.23	0.04	0.05	0.07	0.04	0.05	0.06
C53	0.21	0.21	0.20	0.28	0.21	0.19	0.03	0.04	0.05	0.04	0.04	0.05

TABLE 8.7
Pairwise comparison matrices of the alternatives concerning the subcriterion C11 configurability

	DM1						DM2						DM3					
	A1	**A2**	**A3**	**A4**	**A5**	**A6**	**A1**	**A2**	**A3**	**A4**	**A5**	**A6**	**A1**	**A2**	**A3**	**A4**	**A5**	**A6**
A1	E	FW	SS	VS	VW	SW	E	SW	FS	FS	FS	FW	E	FS	SW	E	VW	VW
A2	FS	E	FW	SW	SW	FW	SS	E	SW	VS	FS	AS	FW	E	AW	VS	SS	AW
A3	SW	FS	E	FS	VW	E	FW	SS	E	E	E	VW	SS	AS	E	VW	AS	SW
A4	VW	SS	FW	E	FS	FW	FW	VW	E	E	VS	AS	E	VW	VS	E	FS	AS
A5	VS	SS	VS	FW	E	SW	FW	FW	E	VW	E	FW	VS	SW	AW	FW	E	FW
A6	SS	FS	E	FS	SS	E	FS	AW	VS	AW	FS	E	VS	AS	SS	AW	FS	E

TABLE 8.8
Transformed to TT2F pairwise comparison matrix for the main criteria

	μ			υ			μ			υ		
Criterion	C1						C2					
C1	1.00	1.00	1.00	1.00	1.00	1.00	0.49	0.46	0.44	0.53	0.46	0.42
C2	1.90	2.18	2.41	2.05	2.18	2.30	1.00	1.00	1.00	1.00	1.00	1.00
C3	0.84	0.88	0.91	0.79	0.88	0.94	0.33	0.88	0.91	0.79	0.88	0.94
C4	0.46	0.54	0.56	0.52	0.54	0.55	0.18	0.54	0.55	0.46	0.54	0.56
C5	0.88	1.27	1.51	1.11	1.27	1.40	0.30	1.27	1.40	0.88	1.27	1.51
	C3						C4					
C1	1.19	1.13	1.09	1.27	1.13	1.06	1.93	1.84	1.81	2.20	1.84	1.80
C2	1.93	4.08	6.12	3.03	4.08	5.10	4.55	6.63	8.67	5.59	6.63	7.65
C3	1.00	1.00	1.00	1.00	1.00	1.00	0.96	1.33	1.56	1.18	1.33	1.45
C4	0.85	0.75	0.69	1.04	0.75	0.64	1.00	1.00	1.00	1.00	1.00	1.00
C5	1.00	1.93	2.63	1.52	1.93	2.30	2.02	2.11	2.20	2.06	2.11	2.16
	C5						C6					
C1	0.90	0.79	0.71	1.13	0.79	0.66	0.90	0.79	0.71	1.13	0.79	0.66
C2	2.28	4.40	6.44	3.36	4.40	5.42	2.28	4.40	6.44	3.36	4.40	5.42
C3	0.66	0.52	0.44	1.00	0.52	0.38	0.66	0.52	0.44	1.00	0.52	0.38
C4	0.48	0.47	0.46	0.50	0.47	0.45	0.48	0.47	0.46	0.50	0.47	0.45
C5	1.00	1.00	1.00	1.00	1.00	1.00	1.00	1.00	1.00	1.00	1.00	1.00

TABLE 8.9
The geometric means of criteria

Criterion	Geometric means						Normalized geometric means						Defuzzified
	μ			υ			μ			υ			
C1	1.11	0.92	0.84	0.99	0.92	0.87	0.14	0.13	0.14	0.13	0.13	0.15	0.13
C2	2.10	3.23	4.17	2.71	3.23	3.72	0.27	0.47	0.72	0.36	0.47	0.63	0.47
C3	0.92	0.81	0.76	0.72	0.81	0.78	0.12	0.12	0.13	0.10	0.12	0.13	0.12
C4	0.61	0.61	0.59	0.51	0.61	0.60	0.08	0.09	0.10	0.07	0.09	0.10	0.09
C5	1.08	1.37	1.54	1.00	1.37	1.46	0.14	0.20	0.26	0.14	0.20	0.25	0.20

TABLE 8.10
Scores of alternatives and ranking

Alternatives		μ			υ		Score	Rank
A1	0.17	0.19	0.20	0.18	0.19	0.20	0.22	2
A2	0.14	0.14	0.13	0.14	0.14	0.13	0.14	6
A3	0.15	0.16	0.17	0.16	0.16	0.17	0.21	3
A4	0.17	0.18	0.18	0.17	0.18	0.18	0.24	1
A5	0.15	0.16	0.16	0.16	0.16	0.16	0.16	5
A6	0.17	0.19	0.21	0.18	0.19	0.20	0.17	4

criteria in the evaluation. C1, with a defuzzified value of 0.13, falls between the most and least significant criteria.

Given the criteria weights, Alternative A4 stands out as the best option, earning the highest defuzzified score of 0.24, making it the top-ranked alternative. A1 follows closely with a score of 0.22, ranking second, indicating it is also a highly competitive choice. A3 is ranked third with a score of 0.21, showing a solid performance. Alternatives A6 and A5 are ranked fourth and fifth, respectively, with defuzzified scores of 0.17 and 0.16, which suggest that while they are viable options, they are not the top performers. Lastly, A2 ranks lowest, with a defuzzified score of 0.14, indicating it is the least preferred alternative in this analysis.

This ranking and the distribution of scores suggest that A4 and A1 are the strongest candidates, largely due to the significant weight placed on C2, while A2's lower score is likely a result of its weaker performance across multiple criteria.

8.5 RESULTS AND DISCUSSION

The application of the intuitionistic fuzzy originated interval type-2 fuzzy AHP methodology has enabled a structured and systematic evaluation of campus processes for robotic process automation. The results highlight the relative importance of key decision criteria and provide insights into the feasibility of automating various university operations.

The most influential criterion in the decision-making process is process complexity (C2), with a defuzzified weight of 0.47. This indicates that automation efforts should focus on processes characterized by high manual intervention, decision-making requirements, and frequent exceptions, as they present the greatest potential for efficiency improvements. Process priority (C5) follows as the second most significant criterion, reinforcing the importance of selecting processes that are critical to the university's core functions, regulatory compliance, and risk management.

In contrast, technological compatibility (C3) and efficiency potential (C4) received lower weightings, suggesting that while ease of integration and efficiency gains are relevant, they are secondary considerations compared to process complexity and priority. Process suitability (C1) ranked in the mid-range, indicating that factors such as configurability and repeatability play a role but do not outweigh the complexities and strategic significance of a process.

Among the process alternatives, student support services (A4) emerged as the top candidate for automation, achieving the highest defuzzified score of 0.24. This is largely due to its high frequency, structured nature, and potential for reducing administrative workload through automation. Student enrollment and registration (A1) and campus facility management (A3) followed closely, demonstrating strong automation potential due to their repetitive tasks and predefined rules.

On the other hand, course and curriculum management (A2) ranked the lowest, reflecting its high variability, dependency on academic policies, and the need for human decision-making, which make it less suitable for RPA. IT help desk operations (A6) and campus event management (A5) also scored moderately, indicating that while automation could assist in streamlining their workflows, their dynamic and unpredictable nature poses challenges to full automation.

The findings validate the robustness of the intuitionistic fuzzy originated interval type-2 Fuzzy AHP methodology in supporting RPA decision-making under uncertainty. By incorporating expert judgments and handling imprecision effectively, this approach ensures that automation decisions align with both operational feasibility and strategic objectives.

From a practical perspective, these results offer a data-driven framework for universities aiming to optimize administrative efficiency through automation. The prioritization model suggests that institutions should focus on rule-based, high-volume processes such as student support services and registration systems, while leveraging hybrid automation strategies (e.g., combining RPA with AI) for more complex tasks such as curriculum management.

Future research could further refine this approach by integrating process mining techniques to analyze real-time process performance metrics. Additionally, cost-benefit analyses and pilot implementations of selected RPA solutions could provide deeper insights into the long-term impact of automation on university operations.

8.6 CONCLUSION

This study presented a structured methodology for prioritizing processes for RPA in a smart campus environment using the intuitionistic fuzzy originated interval type-2 Fuzzy AHP approach. By integrating MCDM techniques, the study effectively addressed the complexities and uncertainties involved in selecting the most suitable campus processes for automation.

The results indicate that process complexity (C2) and process priority (C5) are the most critical factors influencing RPA adoption in university settings. These findings suggest that highly manual, repetitive, and strategically important processes should be prioritized for automation to maximize efficiency and operational effectiveness. Among the evaluated alternatives, student support services (A4), student enrollment and registration (A1), and campus facility management (A3) emerged as the top candidates for automation, highlighting their potential to reduce administrative burdens, improve service delivery, and enhance institutional efficiency.

The proposed decision support framework provides valuable insights for universities seeking to implement automation strategically. By leveraging fuzzy logic and interval type-2 methods, this approach ensures a more robust and objective evaluation of automation candidates, minimizing the risks associated with subjective decision-making.

The study leveraged triangular fuzzy sets to handle the inherent uncertainty in expert evaluations. They effectively captured the variability in decision makers' judgments by representing criteria weights and pairwise comparisons as triangular fuzzy numbers, which consist of a lower bound, a most likely value, and an upper bound. This approach allowed for a more flexible and realistic assessment of process automation potential, reducing bias and improving the robustness of the decision-making framework. By incorporating fuzzy logic into the AHP structure, this study provided a more granular and nuanced prioritization model, ensuring that automation decisions align with both operational feasibility and strategic objectives.

Future research could enhance this methodology by incorporating process mining techniques to analyze real-time operational data and integrating cost-benefit assessments to refine automation strategies further. Additionally, piloting RPA implementations in selected processes could offer practical validation and insights into the long-term impact of digital transformation in higher education institutions.

This study contributes to the growing body of knowledge on smart campus automation by introducing an advanced decision-making model that can be adapted to other organizational contexts, providing a scalable and flexible framework for RPA adoption in various domains.

REFERENCES

Abdel-Basset, M., Manogaran, G., Gamal, A., & Smarandache, F. (2019). A group decision making framework based on neutrosophic TOPSIS approach for smart medical device selection. *Journal of Medical Systems, 43*, 1–13.

Aguirre, S., & Rodriguez, A. (2017). Automation of a business process using robotic process automation (RPA): A case study. *Applied Computer Sciences in Engineering: 4th Workshop on Engineering Applications*. Cartagena, Colombia.

Akyol, S., Dogan, O., & Er, O. (2022). Process automation with digital robots under smart university concept. *International Conference on Intelligent Systems Design and Applications* (pp. 242–251). Cham: Springer Nature Switzerland.

Banerjee, A., & Parkhi, S. (2020). RPA vendor evaluation and selection using AHP and Kano model. *Psychology and Education Journal, 57*(9), 6230–6243.

Çalık, A. (2021). A novel Pythagorean fuzzy AHP and fuzzy TOPSIS methodology for green supplier selection in the Industry 4.0 era. *Soft Computing, 25*, 2253–2265.

Costa, D. S., Mamede, H. S., & da Silva, M. M. (2023). A method for selecting processes for automation with AHP and TOPSIS. *Heliyon, 9*(3), e13683.

Dogan, O., & Cengiz Tirpan, E. (2022). Process mining methodology for digital processes under smart campus concept. *Bilecik Seyh Edebali University. Journal of Science, 9*(2), 1006–1018.

Dogan, O., Arslan, O., Tirpan, E. C., & Cebi, S. (2024). Risk assessment of employing digital robots in process automation. *Systems, 12*(19), 428.

Erdogan, A. M., & Dogan, O. (2023). Process selection for RPA projects with MDCM: The case of Izmir Bakircay University. *International Conference on Applied Machine Learning and Data Analytics*. Cham: Springer Nature Switzerland.

Geyer-Klingeberg, J., Nakladal, J., Baldauf, F., & Veit, F. (2018). Process mining and robotic process automation: A perfect match. *BPM (Dissertation/Demos/Industry)* (pp. 124–131). Sydney.

Ghaleb, A. M., Kaid, H., Alsamhan, A., Mian, S. H., & Hidri, L. (2020). Assessment and comparison of various MCDM approaches in the selection of manufacturing process. *Advances in Materials Science and Engineering*, 4039253.

John, K. K., & Jeyapaul, R. (2024). RPA software package selection for a reverse supply chain organisation—A fuzzy MCDM approach. *International Journal of Procurement Management, 20*(4), 537–556.

Kahraman, C., Öztaysi, B., Onar, S. Ç., & Dogan, O. (2018). Intuitionistic fuzzy originated interval type-2 fuzzy AHP: An application to damless hydroelectric power plants. *International Journal of the Analytic Hierarchy Process, 10*(2), 266–292.

Lacity, M. C., & Willcocks, L. P. (2016). A new approach to automating services. *MIT Sloan Management Review, 58*(1), 41–49.

Lacity, M., Willcocks, L., & Craig, A. (2015). Robotic process automation: Mature capabilities in the energy sector. *The Outsourcing Unit Working Research Paper Series 15/05, 15*(5), 1–19.

Lindstrom, J., Kyosti, P., & Delsing, J. (2018). *European Roadmap for Industrial Process Automation (2018)*. Sweden: DIVA (Digitala Vetenskapliga Arkivet).

Madakam, S., Holmukhe, R. M., & Jaiswal, D. K. (2019). The future digital work force: Robotic process automation (RPA). *Journal of Information Systems and Technology Management, 16*, e201916001.

Marrella, A. (2018). What automated planning can do for business process management. In E. Teniente & M. Weidlich, *Business Process Management Workshops. Lecture Notes in Business Information Processing* (pp. 7–19). Springer, Cham.

Olucoglu, M., Dogan, O., Akkol, E., & Keskin, B. (2023). Digitized and automated a university process with robotic process automation. *Erzincan University Journal of Science and Technology, 16*(1), 58–66.

Schmitz, M., Dietze, C., & Czarnecki, C. (2019). Enabling digital transformation through robotic process automation at deutsche telekom. In N. Urbach & M. Roglinger (Eds.), *Digitalization Cases: How Organizations Rethink their Business for the Digital Age* (pp. 15–33). Springer.

van Hoek, R., Gorm Larsen, J., & Lacity, M. (2022). Robotic process automation in Maersk procurement–applicability of action principles and research opportunities. *International Journal of Physical Distribution & Logistics Management, 52*(3), 285–298.

Wanner, J., Hofmann, A., Fischer, M., Imgrund, F., Janiesch, C., & Geyer-Klingeberg, J. (2019). Process selection in RPA projects-towards a quantifiable method of decision making. *International Conference on Information Systems*. Munich.

9 Dombi t-Norm and t-Conorm-Based Aggregation Operators in an Interval-Valued Fermatean Fuzzy Framework with Confidence Levels

Utpal Mandal and Mijanur Rahaman Seikh

9.1 INTRODUCTION

Multi-criteria decision-making (MCDM) is a systematic method for assessing and ranking multiple conflicting criteria in decision-making situations. It is crucial for complex problem-solving in various fields. However, collecting real-world data for MCDM can be challenging due to uncertainties, incomplete information, and the subjective nature of human judgments, which complicates the assessment of criteria. In this context, fuzzy sets (FSs) (Zadeh, 1965) are highly relevant as they provide a robust framework for expressing imprecise and ambiguous data, allowing decision makers to capture the inherent vagueness in their assessments and leading to more accurate and meaningful conclusions. Over the years, numerous extensions of the original FSs have been proposed by researchers, including interval-valued FSs (IVFSs), intuitionistic FSs (Atanassov, 1986), Pythagorean FSs (Yager, 2014), and Fermatean FSs (FFSs) (Senapati and Yager, 2020). These fuzzy set extensions have been instrumental in solving numerous MCDM problems. Recently, interval-valued Fermatean fuzzy sets (IVFFSs) (Jeevaraj, 2021) have emerged as a powerful tool for handling uncertainty, extending both IVFS and FFSs. IVFFS allows the membership degree (MD) and nonmembership degree (NMD) to be represented as intervals, where the cubic sum of their upper bounds is constrained to be less than or equal to 1. This feature enables decision makers to express their evaluations within a broader range, making IVFFS more effective in solving and modeling

DOI: 10.1201/9781003529088-9

complex decision-making problems. Recent research has expanded on the applications of IVFFS. Ma et al. (2024) developed Bonferroni mean aggregation operators (AOs) under the IVFF environment and proposed an MCDM model based on these operators. Kirişci (2024) developed a hybrid MCDM model based on AHP, TOPSIS, and MABAC methods under the IVFF environment and applied it to the risk assessment for self-driving vehicles. Görçün et al. (2024) developed an integrated MCDM approach based on PIvot Pairwise Relative Criteria Importance Assessment (PIPRECIA) and Simple Weighted Sum Product (WISP) methods under the IVFF environment and applied it to telescopic forklift selection. Korucuk et al. (2024) developed the IVFF environment-based LOPCOW-RAFSI model to deal with MCDM problems and applied it in warehouse site selection for humanitarian relief organizations.

In FS theory and MCDM, t-norms (TNs), and t-conorms (TCNs) are crucial for modeling and handling inherent uncertainty and imprecision. These mathematical constructs are fundamental in extending classical logic to fuzzy logic, allowing for more nuanced and flexible decision-making processes.

The TNs and TCNs are a type of binary operation used to model the intersection and union of fuzzy sets, respectively. The TNs and TCNs are the functions from $\left[0,1\right]\times\left[0,1\right]$ to $\left[0,1\right]$ that satisfy certain axioms, including monotonicity, associativity, and commutativity, with the specific property that $T\left(a,1\right)=a$ and $S\left(a,0\right)=a$ for any $a\in\left[0,1\right]$. In essence, TNs represent the fuzzy intersection where the MD in the intersection is the minimum of the MD in the individual sets. Common examples include the minimum operator and the product operator. On the other hand, TCNs represent the fuzzy union, where the MD in the union is the maximum of the MD in the individual sets. The maximum operator and the probabilistic sum operator are well-known examples.

TNs and TCNs are crucial for several reasons. They provide the mathematical framework for defining and computing the fuzzy intersection and union, respectively. This is essential in fuzzy logic systems, where precise boundaries between membership and nonmembership are unclear. By employing TNs and TCNs, systems can manage and manipulate fuzzy data more effectively, reflecting real-world scenarios where information is often incomplete or imprecise.

Applications of TNs and TCNs are extensive. In fuzzy control systems, they help design systems that can handle uncertainties in input data and control actions. In pattern recognition and image processing, these operators assist in classifying and processing images where the boundaries between classes are not well-defined. In decision-making processes, particularly MCDM, TNs and TCNs are used to aggregate different criteria and provide a coherent decision strategy when dealing with fuzzy preferences. Various established TNs and TCNs, including algebraic, Einstein, Hamacher, and Frank, are commonly applied to construct aggregation operators across different fuzzy frameworks. However, the Dombi t-norm (DTN) and Dombi t-conorm (DTCN), introduced by Dombi (1982), are regarded as superior to other t-norms, such as the algebraic, Hamacher, and Einstein norms, because of their inherent flexibility and adaptability. This is primarily attributed to the adjustable parameter, which allows users to customize the shape of the t-norm to fit specific application needs. Furthermore, utilizing the Dombi operation makes the aggregation process simpler and easier to use. Numerous scholars developed AOs using DTCN and DTN in various fuzzy settings, including intuitionistic fuzzy environments (Seikh and Mandal, 2021), IVFF environment (Seikh and Mandal, 2023), interval-valued spherical fuzzy environment (Mandal and Seikh, 2023), Pythagorean fuzzy environment (Jana et al., 2019), and Fermatean fuzzy environment (Aydemir and Gunduz, 2020).

Existing Dombi AOs often assume that all sources of information are equally reliable, which can lead to suboptimal decisions, especially when dealing with uncertain, imprecise, or conflicting data. These traditional operators do not account for the varying confidence levels associated with each source, making it difficult to differentiate between reliable and unreliable information. As a result, less reliable data may have an undue influence on the final decision, reducing the robustness and accuracy of the aggregation process. Confidence level-based AOs address this by incorporating the reliability of each source, assigning higher weights to more reliable data. This results in more accurate and robust decision-making, particularly in complex environments where uncertainty is prevalent. Despite the importance of confidence level-based aggregation, there has been no work on developing such operators using DTN and DTCN under the IVFF environment. This chapter aims to address this gap by proposing new confidence level-based AOs that utilize the flexibility and strengths of DTN and DTCN, specifically designed for IVFF sets. Therefore, in this chapter, we develop confidence level-based AOs using DTN and DTCN under the IVFF environment.

This chapter aims to enhance decision-making processes by introducing novel aggregation operators that integrate the DTN and DTCN within the IVFF environment. The specific objectives include:

- To propose Dombi-based aggregation operators that effectively handle the uncertainty and imprecision inherent in real-world decision-making scenarios while incorporating confidence levels to refine the process by weighting the importance of information based on the confidence of decision makers.
- To explore the applicability of these aggregation operators in solving MCDM problems, providing a structured approach to evaluate and rank alternatives in complex decision environments.

The main originality of the chapter lies in introducing confidence levels into aggregation operators based on DTN and DTCN within the IVFF framework. This integration enhances the precision and adaptability of handling uncertainty, allowing for more nuanced aggregation that accounts for the reliability of information, aligning decision-making processes more closely with real-world scenarios.

Motivated by the concept of Dombi operations and confidence intervals, we first extend Dombi operations within the context of confidence intervals and then develop two novel AOs: a weighted averaging operator and a geometric operator, each accompanied by a thorough examination of their properties to aggregate IVFF information.

Building upon these operators, we present a multi-attribute group decision-making (MAGDM) framework tailored to deal with decision-making challenges within the IVFF context. The main contributions of this study are outlined as follows:

- To introduce the concept of confidence-level-based Dombi operations under the IVFF environment, expanding their applicability to settings where data reliability is variable.
- To propose two new aggregation operators: the confidence IVFF Dombi weighted averaging (CIVFFDWA) and the confidence IVFF Dombi weighted geometric (CIVFFDWG) operators, along with a verification of their key properties.
- Using these operators, we develop a novel MAGDM framework tailored to address decision-making problems with numerous conflicting criteria within the context of IVFFSs.
- To showcase the practical utility and efficiency of the proposed approach, we apply the framework to an MCDM for selecting a method of plastic waste management, highlighting its applicability in real-world scenarios.

The structure of the study is as follows: Section 9.2 consists of the preliminaries. Section 9.3 contains the novel confidence level-based Dombi aggregation operator for IVFFSs. In Section 9.4, a MAGDM framework is proposed, utilizing the CIVFFDWA and CIVFFDWG operators to address decision-making challenges. To demonstrate the effectiveness and applicability of the proposed confidence level-based Dombi operations in MAGDM, a numerical example is provided in Section 9.5, along with a comparison of the results with existing methods based on aggregation operators. Finally, Section 9.6 provides the conclusion, discusses limitations, and suggests avenues for future research.

9.2 PRELIMINARIES

Here, we revisit several fundamental definitions.

Definition 2.1 (Jeevaraj, 2021): The IVFFS $\tilde{f}$ over the universal set X is defined as:

$$\tilde{f} = \left\langle \tilde{v},\ \overline{\xi}_{\tilde{f}}(\tilde{v}),\ \overline{\kappa}_{\tilde{f}}(\tilde{v}) \right\rangle | \overline{\xi}_{\tilde{f}}(\tilde{v}) = \left[\overline{\xi}^{-}_{\tilde{f}}(\tilde{v}),\ \overline{\xi}^{+}_{\tilde{f}}(\tilde{v})\right] \in I\left(\left[0,1\right]\right), \overline{\kappa}_{\tilde{f}}(\tilde{v})$$
$$= \left[\overline{\kappa}^{-}_{\tilde{f}}(\tilde{v}),\ \overline{\kappa}^{+}_{\tilde{f}}(\tilde{v})\right] \in I\left(\left[0,1\right]\right), \tilde{v} \in X$$

where $I\left(\left[0,1\right]\right)$ represents the collection of all closed intervals contained within the range $\left[0,1\right]$, $\overline{\xi}_{\tilde{f}}, \overline{\kappa}_{\tilde{f}} : X \to I\left(\left[0,1\right]\right)$ respectively represents the MD and NMD of the element $\tilde{v} \in X$, satisfying $0 \leq \left(\overline{\xi}^{+}_{\tilde{f}}(\tilde{v})\right)^3 + \left(\overline{\kappa}^{+}_{\tilde{f}}(\tilde{v})\right)^3 \leq 1$.

For simplicity, we consider $\tilde{f} = \left(\left[\overline{\xi}^{-}, \overline{\xi}^{+}\right], \left[\overline{\kappa}^{-}, \overline{\kappa}^{+}\right]\right)$ as an IVFF number (IVFFN).

Definition 2.2 (Seikh and Mandal, 2023): The score function for the IVFFN $\tilde{f} = \left(\left[\overline{\xi}^{-}, \overline{\xi}^{+}\right], \left[\overline{\kappa}^{-}, \overline{\kappa}^{+}\right]\right)$ is defined as:

$$\mathcal{S}\left(\tilde{f}\right) = 1 + \frac{\left[\left(\overline{\xi}^{-}\right)^{3} + \left(\overline{\xi}^{+}\right)^{3} - \left(\overline{\kappa}^{-}\right)^{3} - \left(\overline{\kappa}^{+}\right)^{3}\right]}{2}. \tag{9.1}$$

Definition 2.3 (Rani et al., 2022): The accuracy function for the IVFFN $\tilde{f} = \left(\left[\overline{\xi}^{-}, \overline{\xi}^{+}\right], \left[\overline{\kappa}^{-}, \overline{\kappa}^{+}\right]\right)$ is defined as:

$$\Phi\left(\tilde{f}\right) = \frac{\left[\left(\overline{\xi}^{-}\right)^{3} + \left(\overline{\xi}^{+}\right)^{3} + \left(\overline{\kappa}^{-}\right)^{3} + \left(\overline{\kappa}^{+}\right)^{3}\right]}{2}. \tag{9.2}$$

Definition 2.4 (Dombi, 1982): For any pair of real numbers u and v, DTN $\mathcal{D}$ and DTCN $\mathcal{D}'$ are defined as: $\mathcal{D}(u,v) = \dfrac{1}{1 + \left[\left(\frac{1-u}{u}\right)^{\sigma} + \left(\frac{1-v}{v}\right)^{\sigma}\right]^{\frac{1}{\sigma}}}$ and $\mathcal{D}'(u,v) = 1 - \dfrac{1}{1 + \left[\left(\frac{u}{1-u}\right)^{\sigma} + \left(\frac{v}{1-v}\right)^{\sigma}\right]^{\frac{1}{\sigma}}}$

where $\sigma > 0$ and $(u,\ v) \in [0,1] \times [0,1]$. Here, σ is the operational parameter of Dombi operations.

In this chapter, the DTN and DTCN will be utilized to develop aggregation operators in an IVFF environment.

Definition 2.5 (Seikh and Mandal, 2023): The DTN and DTCN operations for the IVFFNS, $\tilde{f} = \left(\left[\overline{\xi}^{-}, \overline{\xi}^{+}\right], \left[\overline{\kappa}^{-}, \overline{\kappa}^{+}\right]\right)$ $\tilde{f}_i = \left(\left[\overline{\xi}_i^{-}, \overline{\xi}_i^{+}\right],\ \left[\overline{\kappa}_i^{-}, \overline{\kappa}_i^{+}\right]\right)$, for $(i = 1, 2)$ and $\lambda (\in \mathbb{R}) > 0$, are defined as follows:

(1)

$$\widetilde{f_1} \oplus \widetilde{f_2} = \langle \left[\sqrt[3]{1 - \frac{1}{1 + \left\{ \left(\frac{\left(\overline{\xi}_1^{-}\right)^{3}}{1 - \left(\overline{\xi}_1^{-}\right)^{3}} \right)^{\sigma} + \left(\frac{\left(\overline{\xi}_2^{-}\right)^{3}}{1 - \left(\overline{\xi}_2^{-}\right)^{3}} \right)^{\sigma} \right\}^{\frac{1}{\sigma}}}}, \right.$$

$$\left. \sqrt[3]{1 - \frac{1}{1 + \left\{ \left(\frac{\left(\overline{\xi}_1^{+}\right)^{3}}{1 - \left(\overline{\xi}_1^{+}\right)^{3}} \right)^{\sigma} + \left(\frac{\left(\overline{\xi}_2^{+}\right)^{3}}{1 - \left(\overline{\xi}_2^{+}\right)^{3}} \right)^{\sigma} \right\}^{\frac{1}{\sigma}}}} \right]$$

$$\left[\frac{1}{\sqrt[3]{1+\left\{\left(\frac{1-\left(k_1^-\right)^3}{\left(k_1^-\right)^3}\right)^\sigma+\left(\frac{\left(1-k_2^-\right)^3}{\left(k_2^-\right)^3}\right)^\sigma\right\}^{\frac{1}{\sigma}}}},\right.$$

$$\left.\frac{1}{\sqrt[3]{1+\left\{\left(\frac{\left(1-k_1^+\right)^3}{\left(k_1^+\right)^3}\right)^\sigma+\left(\frac{\left(1-k_2^+\right)^3}{\left(k_2^+\right)^3}\right)^\sigma\right\}^{\frac{1}{\sigma}}}}\right]\rangle,$$

$$\widetilde{f_1}\otimes\widetilde{f_2}=\langle\left[\frac{1}{\sqrt[3]{1+\left\{\left(\frac{1-\left(\overline{\xi}_1^-\right)^3}{\left(\overline{\xi}_1^-\right)^3}\right)^\sigma+\left(\frac{1-\left(\overline{\xi}_2^-\right)^3}{\left(\overline{\xi}_2^-\right)^3}\right)^\sigma\right\}^{\frac{1}{\sigma}}}},\right.$$

$$\left.\frac{1}{\sqrt[3]{1+\left\{\left(\frac{1-\left(\overline{\xi}_1^+\right)^3}{\left(\overline{\xi}_1^+\right)^3}\right)^\sigma+\left(\frac{1-\left(\overline{\xi}_2^+\right)^3}{\left(\overline{\xi}_2^+\right)^3}\right)^\sigma\right\}^{\frac{1}{\sigma}}}}\right],$$

$$\sqrt[3]{1-\frac{1}{1+\left\{\left(\frac{\left(k_1^-\right)^3}{1-\left(k_1^-\right)^3}\right)^\sigma+\left(\frac{1-\left(k_2^-\right)^3}{\left(k_2^-\right)^3}\right)^\sigma\right\}^{\frac{1}{\sigma}}}},$$

$$\sqrt[3]{1-\frac{1}{1+\left\{\left(\frac{\left(k_1^+\right)^3}{1-\left(k_1^+\right)^3}\right)^\sigma+\left(\frac{1-\left(k_2^+\right)^3}{\left(k_2^+\right)^3}\right)^\sigma\right\}^{\frac{1}{\sigma}}}}$$

$$\lambda\widetilde{f_1} = \left\langle \left[\sqrt[3]{1-\frac{1}{1+\left\{\lambda\left(\frac{(\bar{\xi}^-)^3}{1-(\bar{\xi}^-)^3}\right)^{\sigma}\right\}^{\frac{1}{\sigma}}}}, \sqrt[3]{1-\frac{1}{1+\left\{\lambda\left(\frac{(\bar{\xi}^+)^3}{1-(\bar{\xi}^+)^3}\right)^{\sigma}\right\}^{\frac{1}{\sigma}}}} \right]\right.$$

$$\left. \left[\sqrt[3]{1-\frac{1}{1+\left\{\lambda\left(\frac{1-(k^-)^3}{(k^-)^3}\right)^{\sigma}\right\}^{\frac{1}{\sigma}}}}, \sqrt[3]{1-\frac{1}{1+\left\{\lambda\left(\frac{(1-k^+)^3}{(k^+)^3}\right)^{\sigma}\right\}^{\frac{1}{\sigma}}}} \right]\right.$$

$$\widetilde{f^{\lambda}} = \left\langle \left[\frac{1}{\sqrt[3]{1+\left\{\lambda\left(\frac{1-(\bar{\xi}^-)^3}{(\bar{\xi}^-)^3}\right)^{\sigma}\right\}^{\frac{1}{\sigma}}}}, \frac{1}{\sqrt[3]{1+\left\{\lambda\left(\frac{1-(\bar{\xi}^+)^3}{(\bar{\xi}^+)^3}\right)^{\sigma}\right\}^{\frac{1}{\sigma}}}} \right]\right.$$

$$\left. \left[\sqrt[3]{1-\frac{1}{1+\left\{\lambda\left(\frac{(k^-)^3}{(1-k^-)^3}\right)^{\sigma}\right\}^{\frac{1}{\sigma}}}}, \sqrt[3]{1-\frac{1}{1+\left\{\lambda\left(\frac{(k^+)^3}{(1-k^+)^3}\right)^{\sigma}\right\}^{\frac{1}{\sigma}}}} \right] \right\rangle$$

Here, $\sigma > 0$ represents the operational parameter used in Dombi operations.

Definition 2.6 (Seikh and Mandal, 2023): The aggregated values of the IVFFNs $\tilde{f}_i = ([\bar{\xi}_i^-, \bar{\xi}_i^+], [\bar{\kappa}_i^-, \bar{\kappa}_i^+])$ $(i = 1(1)k)$ using IVFF Dombi weighted averaging/geometric (IVFFDWA/IVFFDWG) operators are defined by:

$$IVFFDWA\left(\tilde{f}_1,\tilde{f}_2,\ldots\tilde{f}_k\right)=\left\langle \left[\sqrt[3]{1-\frac{1}{1+\{\sum_{i=1}^{k}\tau_i\left(\frac{\left(\xi_i^-\right)^3}{1-\left(\xi_i^-\right)^3}\right)\sigma\}^{\frac{1}{\sigma}}}},\sqrt[3]{1-\frac{1}{1+\{\sum_{i=1}^{k}\tau_i\left(\frac{\left(\xi_i^+\right)^3}{1-\left(\xi_i^+\right)^3}\right)\sigma\}^{\frac{1}{\sigma}}}}\right]\right.$$

$$\left.\left[\frac{1}{\sqrt[3]{1+\{\sum_{i=1}^{k}\tau_i\left(\frac{\left(\bar{k}_i^-\right)^3}{1-\left(\bar{k}_i^-\right)^3}\right)\sigma\}^{\frac{1}{\sigma}}}},\frac{1}{\sqrt[3]{1+\{\sum_{i=1}^{k}\tau_i\left(\frac{\left(\bar{k}_i^+\right)^3}{1-\left(\bar{k}_i^+\right)^3}\right)\sigma\}^{\frac{1}{\sigma}}}}\right]\right\rangle$$

$$IVFFDWG\left(\tilde{f}_1,\tilde{f}_2,\ldots\tilde{f}_k\right)=\left\langle \left[\frac{1}{\sqrt[3]{1+\{\sum_{i=1}^{k}\tau_i\left(\frac{1-\left(\xi_i^-\right)^3}{\left(\xi_i^-\right)^3}\right)\sigma\}^{\frac{1}{\sigma}}}},\frac{1}{\sqrt[3]{1+\{\sum_{i=1}^{k}\tau_i\left(\frac{1-\left(\xi_i^+\right)^3}{\left(\xi_i^+\right)^3}\right)\sigma\}^{\frac{1}{\sigma}}}}\right]\right.$$

$$\left.\left[\sqrt[3]{1-\frac{1}{1+\{\sum_{i=1}^{k}\tau_i\left(\frac{\left(\bar{k}_i^-\right)^3}{1-\left(\bar{k}_i^-\right)^3}\right)\sigma\}^{\frac{1}{\sigma}}}},\sqrt[3]{1-\frac{1}{1+\{\sum_{i=1}^{k}\tau_i\left(\frac{\left(\bar{k}_i^+\right)^3}{1-\left(\bar{k}_i^+\right)^3}\right)\sigma\}^{\frac{1}{\sigma}}}}\right]\right\rangle$$

where τ_i be the associated weight of f_i satisfying $\tau_i > 0$ and $\sum_i \mathrm{P}_{\tau_i} = 1$

9.3 CONFIDENCE LEVEL-BASED DOMBI AGGREGATION OPERATOR FOR IVFFS

This section proposes confidence level-based Dombi weighted averaging and geometric aggregation operators under the IVFF environment.

Definition 3.1 The aggregated values of the IVFFNs $\tilde{f}_i = \left(\left[\overline{\xi}_i^-, \overline{\xi}_i^+\right], \left[\overline{\kappa}_i^-, \overline{\kappa}_i^+\right]\right)$ $(i = 1, 2, \ldots, k)$ using CIVFFDWA and CIVFFDWG operators are the functions from $IVFFN^k$ to $IVFFN$ and defined by:

$$CIVFFDWA\left(\tilde{f}_1, \tilde{f}_2, \ldots, \tilde{f}_k\right) = \overset{k}{\underset{i=1}{\oplus}}\left(\tau_i l_i \tilde{f}_i\right)$$

$$CIVFFDWG\left(\tilde{f}_1, \tilde{f}_2, \ldots, \tilde{f}_k\right) = \overset{k}{\underset{i=1}{\otimes}}\left(f_i\right)^{\tau_i l_i}$$

where $l_i\ (0 \le l_i \le 1)$ be the confidence level $\tilde{f}_i$, and τ_i's are given in Definition 2.6.

Theorem 3.1 *If* $\tilde{f}_i = \left(\left[\overline{\xi}_i^-, \overline{\xi}_i^+\right], \left[\overline{\kappa}_i^-, \overline{\kappa}_i^+\right]\right)$ $(i = 1, 2, \ldots, k)$ *be a collection of IVFFNs and* l_i *be the confidence level of* $\tilde{f}_i$ *then the aggregated values using CIVFFDWA and CIVFFDWG operators are given by:*

$$CIVFFDW\,A\left(\tilde{f}_1, \tilde{f}_2, \ldots \tilde{f}_k\right) = \left\langle \left[\sqrt[3]{1 - \frac{1}{1 + \{\sum_{i=1}^{k} \tau_i l_i \left(\frac{\left(\overline{\xi}_i^-\right)^3}{1 - \left(\overline{\xi}_i^-\right)^3}\right)^{\sigma}\}^{\frac{1}{\sigma}}}}, \sqrt[3]{1 - \frac{1}{1 + \{\sum_{i=1}^{k} \tau_i l_i \left(\frac{\left(\overline{\xi}_i^+\right)^3}{1 - \left(\overline{\xi}_i^+\right)^3}\right)^{\sigma}\}^{\frac{1}{\sigma}}}} \right] \right.$$

$$\left. \left[\frac{1}{\sqrt[3]{1 + \{\sum_{i=1}^{k} \tau_i l_i \left(\frac{\left(\overline{k}_i^-\right)^3}{1 - \left(\overline{k}_i^-\right)^3}\right)^{\sigma}\}^{\frac{1}{\sigma}}}}, \frac{1}{\sqrt[3]{1 + \{\sum_{i=1}^{k} \tau_i l_i \left(\frac{\left(\overline{k}_i^+\right)^3}{1 - \left(\overline{k}_i^+\right)^3}\right)^{\sigma}\}^{\frac{1}{\sigma}}}} \right] \right\rangle \tag{9.3}$$

$$CIVFFDWG\left(\tilde{f}_1,\tilde{f}_2,\ldots\tilde{f}_k\right)=\left\langle\left[\frac{1}{\sqrt[3]{1+\{\sum_{i=1}^{k}\tau_i l_i\left(\frac{\left(\bar{\xi}_i^-\right)^3}{1-\left(\bar{\xi}_i^-\right)^3}\right)\sigma\}^{\frac{1}{\sigma}}}},\frac{1}{\sqrt[3]{1+\{\sum_{i=1}^{k}\tau_i l_i\left(\frac{\left(\bar{\xi}_i^+\right)^3}{1-\left(\bar{\xi}_i^+\right)^3}\right)\sigma\}^{\frac{1}{\sigma}}}}\right]\right.$$

$$\left.\left[\sqrt[3]{1-\frac{1}{1+\{\sum_{i=1}^{k}\tau_i l_i\left(\frac{\left(\bar{k}_i^-\right)^3}{1-\left(\bar{k}_i^-\right)^3}\right)\sigma\}^{\frac{1}{\sigma}}}},\sqrt[3]{1-\frac{1}{1+\{\sum_{i=1}^{k}\tau_i l_i\left(\frac{\left(\bar{k}_i^+\right)^3}{1-\left(\bar{k}_i^+\right)^3}\right)\sigma\}^{\frac{1}{\sigma}}}}\right]\right\rangle \tag{9.3}$$

Proof: For two IVFFNs $\tilde{f}_1$ and $\tilde{f}_2$, we have:

$$CIVFFDWA\left(\tilde{f}_1,\tilde{f}_2\right)=\left\langle\left[\sqrt[3]{1+\frac{1}{1+\{\tau_1 l_1\left(\frac{\left(\bar{\xi}_1^-\right)^3}{1-\left(\bar{\xi}_1^-\right)^3}\right)^{\sigma}\}^{\frac{1}{\sigma}}}},\sqrt[3]{1+\frac{1}{1+\{\tau_i l_i\left(\frac{\left(\bar{\xi}_1^+\right)^3}{1-\left(\bar{\xi}_1^+\right)^3}\right)^{\sigma}\}^{\frac{1}{\sigma}}}}\right],\right.$$

$$\left.\left[\frac{1}{\sqrt[3]{1+\{\tau_1 l_1\left(\frac{\left(\bar{k}_1^-\right)^3}{1-\left(\bar{k}_1^-\right)^3}\right)^{\sigma}\}^{\frac{1}{\sigma}}}},\frac{1}{\sqrt[3]{1+\{\tau_1 l_1\left(\frac{\left(\bar{k}_1^+\right)^3}{1-\left(\bar{k}_1^+\right)^3}\right)^{\sigma}\}^{\frac{1}{\sigma}}}}\right]\right\rangle$$

$$\oplus\left\langle\left[\sqrt[3]{1-\frac{1}{1+\{\tau_2 l_2\left(\frac{\left(\bar{\xi}_2^-\right)^3}{1-\left(\bar{\xi}_2^-\right)^3}\right)^{\sigma}\}^{\frac{1}{\sigma}}}},\sqrt[3]{1+\frac{1}{1+\{\tau_2 l_2\left(\frac{\left(\bar{\xi}_2^+\right)^3}{1-\left(\bar{\xi}_2^+\right)^3}\right)^{\sigma}\}^{\frac{1}{\sigma}}}}\right]\right.$$

$$\left[\frac{1}{\sqrt[3]{1+\{\tau_2 l_2\left(\frac{\left(\bar{k}_2^-\right)^3}{1-\left(\bar{k}_2^-\right)^3}\right)^{\sigma}\}^{\frac{1}{\sigma}}}},\frac{1}{\sqrt[3]{1+\{\tau_2 l_2\left(\frac{\left(\bar{k}_2^+\right)^3}{1-\left(\bar{k}_2^+\right)^3}\right)^{\sigma}\}^{\frac{1}{\sigma}}}}\right]\Bigg\rangle$$

$$=\Bigg\langle\left[\sqrt[3]{1-\frac{1}{1+\{\tau_1 l_1\left(\frac{\left(\bar{\xi}_1^-\right)^3}{1-\left(\bar{\xi}_1^-\right)^3}\right)^{\sigma}+\tau_2 l_2\left(\frac{\left(\bar{\xi}_2^-\right)^3}{1-\left(\bar{\xi}_2^-\right)^3}\right)^{\sigma}\}^{\frac{1}{\sigma}}}},\right.$$

$$\left.\sqrt[3]{1-\frac{1}{1+\{\tau_1 l_1\left(\frac{\left(\bar{\xi}_1^+\right)^3}{1-\left(\bar{\xi}_1^+\right)^3}\right)^{\sigma}+\tau_2 l_2\left(\frac{\left(\bar{\xi}_2^+\right)^3}{1-\left(\bar{\xi}_2^+\right)^3}\right)^{\sigma}\}^{\frac{1}{\sigma}}}}\right]$$

$$\left[\sqrt[3]{1-\frac{1}{1+\{\tau_1 l_1\left(\frac{\left(\bar{k}_1^-\right)^3}{1-\left(\bar{k}_1^-\right)^3}\right)^{\sigma}+\tau_2 l_2\left(\frac{\left(\bar{k}_2^-\right)^3}{1-\left(\bar{k}_2^-\right)^3}\right)^{\sigma}\}^{\frac{1}{\sigma}}}},\right.$$

$$\left.\sqrt[3]{1-\frac{1}{1+\{\tau_1 l_1\left(\frac{\left(\bar{k}_1^+\right)^3}{1-\left(\bar{k}_1^+\right)^3}\right)^{\sigma}+\tau_2 l_2\left(\frac{\left(\bar{k}_2^+\right)^3}{1-\left(\bar{k}_2^+\right)^3}\right)^{\sigma}\}^{\frac{1}{\sigma}}}}\right]$$

$$=\left\langle \left[\sqrt[3]{1-\frac{1}{1+\{\tau_i l_i\left(\frac{\left(\bar{\xi}_i^-\right)^3}{1-\left(\bar{\xi}_i^-\right)^3}\right)^{\sigma}\}^{\frac{1}{\sigma}}}},\ \sqrt[3]{1+\frac{1}{1+\{\tau_i l_i\left(\frac{\left(\bar{\xi}_i^+\right)^3}{1-\left(\bar{\xi}_i^+\right)^3}\right)^{\sigma}\}^{\frac{1}{\sigma}}}}\right]\right.$$

$$\left.\left[\frac{1}{\sqrt[3]{1+\{\tau_i l_i\left(\frac{\left(\bar{k}_i^-\right)^3}{1-\left(\bar{k}_i^-\right)^3}\right)^{\sigma}\}^{\frac{1}{\sigma}}}},\ \frac{1}{\sqrt[3]{1+\{\tau_i l_i\left(\frac{\left(\bar{k}_i^+\right)^3}{1-\left(\bar{k}_i^+\right)^3}\right)^{\sigma}\}^{\frac{1}{\sigma}}}}\right]\right\rangle.$$

Now for $k = m$, Equation (9.2) becomes:

$$CIVFFDWA\left(\tilde{f}_1,\tilde{f}_2\ldots\tilde{f}_m\right)=\left\langle \left[\sqrt[3]{1+\frac{1}{1+\{\sum_{i=1}^{m}\tau_1 l_1\left(\frac{\left(\bar{\xi}_1^-\right)^3}{1-\left(\bar{\xi}_1^-\right)^3}\right)\sigma\}^{\frac{1}{\sigma}}}},\ \sqrt[3]{1+\frac{1}{1+\{\sum_{i=1}^{m}\tau_1 l_1\left(\frac{\left(\bar{\xi}_1^+\right)^3}{1-\left(\bar{\xi}_1^+\right)^3}\right)\sigma\}^{\frac{1}{\sigma}}}}\right],\right.$$

$$\left.\left[\frac{1}{\sqrt[3]{1+\{\sum_{i=1}^{m}\tau_i l_i\left(\frac{\left(\bar{k}_i^-\right)^3}{1-\left(\bar{k}_i^-\right)^3}\right)\sigma\}^{\frac{1}{\sigma}}}},\ \frac{1}{\sqrt[3]{1+\{\sum_{i=1}^{m}\tau_i l_i\left(\frac{\left(\bar{k}_1^+\right)^3}{1-\left(\bar{k}_1^+\right)^3}\right)\sigma\}^{\frac{1}{\sigma}}}}\right]\right\rangle$$

Now,

$$CIVFFDWA\left(\tilde{f}_1, \tilde{f}_2 \ldots \tilde{f}_m, \tilde{f}_{m+1}\right) = \overset{m+1}{\underset{1}{\oplus}} \tau_i\left(l_i \tilde{f}_i\right)$$

$$\left\langle \left[\sqrt[3]{1+\frac{1}{1+\{\sum_{i=1}^{m}\tau_1 l_1\left(\frac{\left(\bar{\xi}_1^-\right)^3}{1-\left(\bar{\xi}_1^-\right)^3}\right)\sigma\}^{\frac{1}{\sigma}}}}, \sqrt[3]{1+\frac{1}{1+\{\sum_{i=1}^{m}\tau_1 l_1\left(\frac{\left(\bar{\xi}_1^+\right)^3}{1-\left(\bar{\xi}_1^+\right)^3}\right)\sigma\}^{\frac{1}{\sigma}}}} \right], \right.$$

$$\left. \left[\frac{1}{\sqrt[3]{1+\{\sum_{i=1}^{m}\tau_i l_i\left(\frac{\left(\bar{k}_i^-\right)^3}{1-\left(\bar{k}_i^-\right)^3}\right)\sigma\}^{\frac{1}{\sigma}}}}, \frac{1}{\sqrt[3]{1+\{\sum_{i=1}^{m}\tau_i l_i\left(\frac{\left(\bar{k}_1^+\right)^3}{1-\left(\bar{k}_1^+\right)^3}\right)\sigma\}^{\frac{1}{\sigma}}}} \right] \right\rangle$$

Therefore, by the method of mathematical induction, the result in Equation (9.2) is verified.

Now, we will show that the aggregated value of the IVFFNs $f_i, i = 1(1)k$ using the CIVFFDWA operator is again an IVFFN.

Since $\tilde{f}_i = \left(\left[\bar{\xi}_i^-, \bar{\xi}_i^+\right], \left[\bar{\kappa}_i^-, \bar{\kappa}_i^+\right]\right)$ $(i = 1, 2, \ldots, k)$ are all IVFFNs, so $0 \le \left(\xi_i^+\right)^3 + \left(\kappa_i^+\right)^3 \le 1, \forall i = 1(1)k.$

$$\left(\xi_i^+\right)^3 + \left(\kappa_i^+\right)^3 \le 1, \forall i = 1(1)k \Rightarrow \left(\xi_i^+\right)^3 \le 1 - \left(\kappa_i^+\right)^3, \forall i = 1(1)k.$$

$$\left(\xi_i^+\right)^3 + \left(\kappa_i^+\right)^3 \le 1, \forall i = 1(1)k \Rightarrow 1 - \left(\xi_i^+\right)^3 \ge \left(\kappa_i^+\right)^3, \forall i = 1(1)k$$

Also,
$$\Rightarrow \frac{1}{1-\left(\xi_i^+\right)^3} \le \frac{1}{\left(\kappa_i^+\right)^3}, \forall i = 1(1)k.$$

Combining the above results, we get $\frac{\left(\xi_i^+\right)^3}{1+\left(\xi_i^+\right)^3} \le \frac{\left(k_i^+\right)^3}{1+\left(k_i^+\right)^3}, \forall i = 1(1)k \Rightarrow \tau_i l_i\left(\frac{\left(\xi_i^+\right)^3}{1+\left(\xi_i^+\right)^3}\right)^\alpha \le$

$$\frac{\left(\xi_i^+\right)^3}{1+\left(\xi_i^+\right)^3} \le \frac{\left(k_i^+\right)^3}{1+\left(k_i^+\right)^3}, \forall i = 1(1)k \Rightarrow \tau_i l_i\left(\frac{\left(\xi_i^+\right)^3}{1+\left(\xi_i^+\right)^3}\right)^\alpha \le$$

$$\tau_i l_i\left(\frac{\left(1-k_i^+\right)^3}{\left(k_i^+\right)^3}\right)^\alpha, \forall i = 1(1)k \Rightarrow \sum_{i=1}^{k}\tau_i l_i\left(\frac{\left(\xi_i^+\right)^3}{1-\left(\xi_i^+\right)^3}\right)^\alpha \le \sum_{i=1}^{k}\tau_i l_i\left(\frac{\left(1-k_i^+\right)^3}{\left(k_i^+\right)^3}\right)^\alpha \Rightarrow$$

$$1+\left(\sum_{i=1}^{k}\tau_i l_i\left(\frac{\left(\xi_i^+\right)^3}{1-\left(\xi_i^+\right)^3}\right)^{\alpha}\right)^{1/\alpha}\leq 1+\left(\sum_{i=1}^{k}\tau_i l_i\left(\frac{\left(1-k_i^+\right)^3}{\left(k_i^+\right)^3}\right)^{\alpha}\right)^{1/\alpha}\Rightarrow$$

$$\frac{1}{1+\left(\sum_{i=1}^{k}\tau_i l_i\left(\frac{\left(\xi_i^+\right)^3}{1-\left(\xi_i^+\right)^3}\right)^{\alpha}\right)^{1/\alpha}}\geq\frac{1}{1+\left(\sum_{i=1}^{k}\tau_i l_i\left(\frac{\left(1-k_i^+\right)^3}{\left(k_i^+\right)^3}\right)^{\alpha}\right)^{1/\alpha}}\Rightarrow$$

$$1-\frac{1}{1+\left(\sum_{i=1}^{k}\tau_i l_i\left(\frac{\left(\xi_i^+\right)^3}{1-\left(\xi_i^+\right)^3}\right)^{\alpha}\right)^{1/\alpha}}\leq 1-\frac{1}{1+\left(\sum_{i=1}^{k}\tau_i l_i\left(\frac{\left(1-k_i^+\right)^3}{\left(k_i^+\right)^3}\right)^{\alpha}\right)^{1/\alpha}}\Rightarrow$$

$$\left(1-\frac{1}{1+\left(\sum_{i=1}^{k}\tau_i l_i\left(\frac{\left(\xi_i^+\right)^3}{1-\left(\xi_i^+\right)^3}\right)^{\alpha}\right)^{1/\alpha}}\right)+\left(\frac{1}{1+\left(\sum_{i=1}^{k}\tau_i l_i\left(\frac{\left(1-k_i^+\right)^3}{\left(k_i^+\right)^3}\right)^{\alpha}\right)^{1/\alpha}}\right)\leq 1.$$

Therefore, $CIVFFDWA\left(\tilde{f}_1,\tilde{f}_2,\ldots,\tilde{f}_k\right)$ is an IVFFN.

The remaining portion of the theorem can be demonstrated similarly. □

For the sets of IVFFNs $\{\tilde{f}_i | i=1(1)k\}$ and $\{\tilde{f}_i^{'} | i=1(1)k\}$, our proposed CIVFFDWA and CIVFFDWG operators satisfy the following three properties.

1. **Idempotency property:** If all the IVFFNs $f_i\left(i=1,2,\ldots,k\right)$ are equal, i.e., if $\tilde{f}_i=\tilde{f},\forall i=1,2,\ldots,k,$ then $CIVFFDWA\left(\tilde{f}_1,\tilde{f}_2,\ldots,\tilde{f}_k\right)=\tilde{f}$ and $CIVFFDWG\left(\tilde{f}_1,\tilde{f}_2,\ldots,\tilde{f}_k\right)=\tilde{f}$.
2. **Monotonicity property**: If $\tilde{f}_i\preceq\tilde{f}_i^{'}$, $\forall\ i$, then $CIVFFDWA\left(\tilde{f}_1,\tilde{f}_2,\ldots,\tilde{f}_k\right)$ $\preceq CIVFFDWA\left(\tilde{f}_1^{'},\tilde{f}_2^{'},\ldots,\tilde{f}_k^{'}\right)$ and $CIVFFDWG\left(\tilde{f}_1,\tilde{f}_2,\ldots,\tilde{f}_k\right)\preceq CIVFFDWG$ $\left(\tilde{f}_1^{'},\tilde{f}_2^{'},\ldots,\tilde{f}_k^{'}\right)$.
3. **Boundedness property**: If $\tilde{f}^-=\min_i\left\{\tilde{f}_i\right\}$ and $\tilde{f}^+=\max_i\left\{\tilde{f}_i\right\}$, then $\tilde{f}^-\preceq CIVFFDWA\left(\tilde{f}_1,\tilde{f}_2,\ldots,\tilde{f}_k\right)\preceq\tilde{f}^+$ and $\tilde{f}^-\preceq CIVFFDWG\left(\tilde{f}_1,\tilde{f}_2,\ldots,\tilde{f}_k\right)\preceq\tilde{f}^+$.

9.4 MAGDM FRAMEWORK USING CIVFFDWA AND CIVFFDWG OPERATORS

This section presents a MAGDM approach that employs the proposed CIVFFDWA and CIVFFDWG operators. This approach involves three key stages. The first step involves collecting data from multiple experts for each available alternative related to the attributes, which is organized into matrices. The next phase involves combining the decision matrices gathered from various experts. Finally, alternatives are evaluated and prioritized according to the score values derived from the aggregated data. Consider the set $T=\{T_1,T_2,\dots,T_l\}$ representing l alternatives and the set $L=\{L_1,L_2,\dots,L_q\}$ representing q criteria. The weights corresponding to these criteria are denoted as $W_{L_1},W_{L_2},\dots,W_{L_q}$, which fulfill the conditions $\sum_{n=1}^{q} W_{L_n}=1$ and $0\le W_{L_n}\le 1$ for each n. Suppose there are i decision makers. $D_1,D_2,\dots,D_i$ with weights $W_{D_1},W_{D_2},\dots,W_{D_i}$, respectively, so that $\sum_{r=1}^{i} w_{D_r}=1$ and $0\le W_{D_r}\le 1$. Let the confidence level of each expert about each criterion be $cl_t^{\,r}$ such that $0\le cl_t^{\,r}\le 1$. We have chosen the value of $\xi=2$ for this problem.

Here, we introduce a MAGDM framework that utilizes the proposed CIVFFDWA and CIVFFDWG operators within the context of the IVFF environment.

Step 1.

Identify the problem and select the experts with their confidence level.

Step 2.

Define the alternatives and criteria associated with the given problem.

Step 3.

Gather the decision matrices expressed in linguistic terms from each decision expert.

Step 4.

Transform the linguistic variables into IVFFNs. This results in the decision matrix $M_r=\left(\left(\sigma_{ab}^{r}\right)\right)_{l\times q}$ for $r=1,2,\dots,i$, where $a=1,2,\dots,l$ and $b=1,2,\dots,q$.

Step 5.

Using the confidence level and weights of DEs and CIVFFDWA aggregation operator, compute the aggregated decision matrix. Let $\bar{M}=\left(\bar{\sigma}_{ab}\right)_{l\times q}$, where $\bar{\sigma}_{ab}=\left(\left[\bar{\xi}_{ab}^{-},\bar{\xi}_{ab}^{+}\right],\left[\bar{\kappa}_{ab}^{-},\bar{\kappa}_{ab}^{+}\right]\right)$ be the aggregated decision matrix.

Step 6.

Calculate the combined value α_t for the alternative T_t using Equation (9.2) or Equation (9.3).

Step 7.

Determine the score value $\mathcal{S}(\alpha_t)\left(t=1(1)l\right)$ utilizing Equation (9.1) for each combined value α_t.

Step 8.

If one specific alternative has a higher score than the others, it is identified as the most preferable option. Thus, the most appropriate alternative is T_c if $\mathcal{S}(\alpha_c)=\max_{1\le t\le l}\{\mathcal{S}(\alpha_t)\}$. If no single alternative has the highest score, proceed to the next step.

Step 9.

If multiple alternatives have the same score value, use Definition (2.3) to compute the accuracy function Φ for these alternatives. Now,

- When multiple alternatives yield the highest score values, the optimal choice is determined by selecting the alternative with the greatest Φ-functional value.
- If multiple alternatives have identical Φ−functional values, any of them can be considered as a suitable option.

Figure 9.1 shows the step of the proposed MAGDM approach.

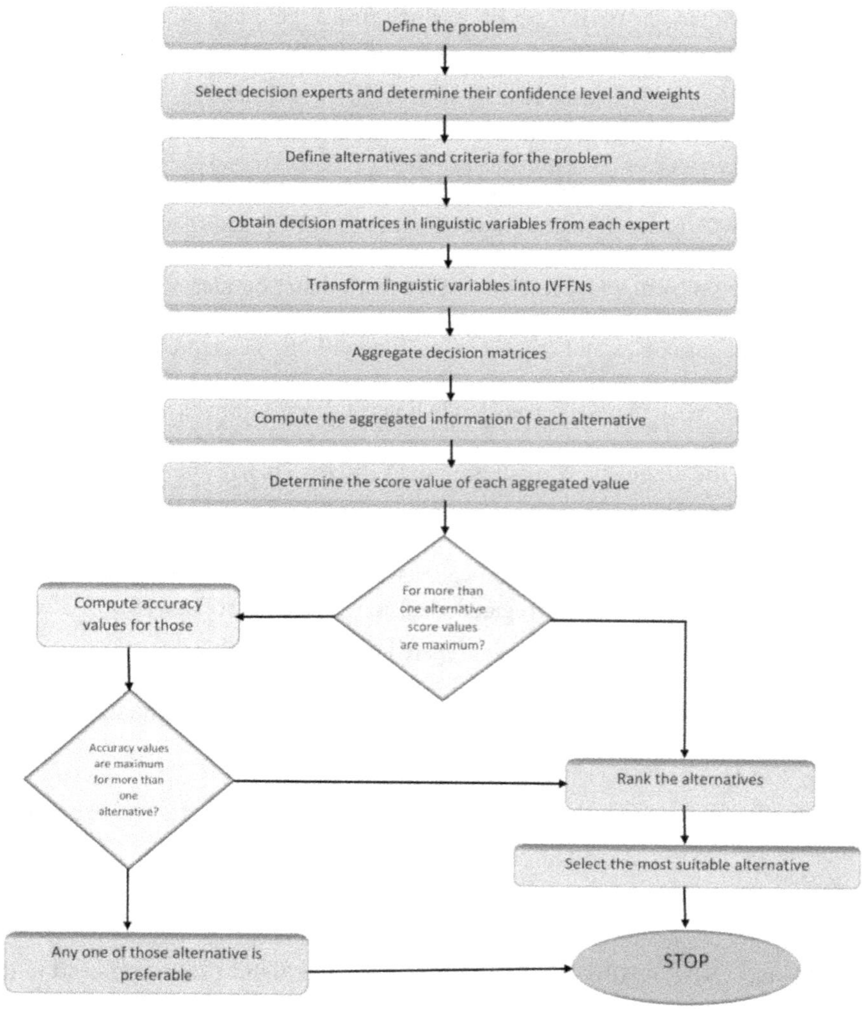

FIGURE 9.1 Framework of the proposed model.

9.5 ILLUSTRATIVE EXAMPLE

This section illustrates the proposed MAGDM technique by solving a numerical problem about the selection of the most effective plastic waste management method.

Plastic waste management has become a critical issue globally due to the harmful environmental impacts of plastic pollution. Several alternative methods have been proposed to manage plastic waste effectively. However, each method has different implications across various dimensions, such as environmental sustainability, economic cost, social acceptance, and operational effectiveness. Decision experts from different fields will assess these methods based on multiple criteria to make an informed decision.

Four decision experts, each from a different field, provide diverse perspectives on the problem. Table 9.1 displays the demographic details of the decision experts.

The five alternative plastic waste management methods under consideration are:

1. **Recycling** (T_1) (Sharma et al., 2024; Patria et al., 2024): Recycling involves collecting, sorting, and processing plastics into new materials, typically through mechanical or chemical methods. Mechanical recycling melts and remolds plastics but may reduce their quality over time. Chemical recycling, however, breaks down polymers into monomers, allowing for new plastics without quality loss. Recycling helps reduce waste and conserve resources, but it faces challenges like inefficient sorting, economic constraints, and limited infrastructure, particularly for chemical recycling.
2. **Incineration with energy recovery** (T_2) (Lombardi et al., 2024; Patria et al., 2024): This method burns plastic waste at high temperatures to generate energy, reducing landfill waste and providing heat or electricity. While it helps manage nonrecyclable plastics, the process can emit harmful gases and produce toxic ash, necessitating strict controls. Though incineration offers a practical solution for energy recovery, its environmental risks and potential air pollution make it a controversial choice in waste management.

TABLE 9.1
Demographic information of decision experts

Expert	Expertise	Experience	Field of expertise
D_1	Environmental scientist	15 years	Environmental impact and policy
D_2	Economist	12 years	Economic feasibility and cost-benefit analysis
D_3	Social scientist	10 years	Social behavior and public acceptance
D_4	Waste management engineer	18 years	Waste treatment and recycling technologies

3. **Landfilling** (T_3) (Patria et al., 2024; Sharma et al., 2024): Landfilling is the most common method for disposing of plastic waste, especially for non-recyclable materials. While it is cost-effective, it contributes to long-term environmental issues, such as soil and water contamination and the slow decomposition of plastics. Moreover, landfill sites are becoming scarcer, and their management generates greenhouse gases, like methane, which further exacerbates climate change concerns.
4. **Pyrolysis** (T_4) (Lombardi et al., 2024; Praveenkumar et al., 2024): Pyrolysis is a chemical process that converts plastic waste into fuel through high-temperature decomposition in the absence of oxygen. It can manage non-recyclable plastics and reduce landfill waste. However, the process requires substantial energy and remains in the early stages of development, with challenges in economic feasibility and large-scale implementation. Despite these hurdles, it shows promise as an innovative waste management technology.
5. **Biodegradable and compostable plastics** (T_5) (Lombardi et al., 2024; Patria et al., 2024): These plastics are designed to decompose in natural environments or composting facilities, reducing long-term pollution. Made from renewable sources, they offer an eco-friendly alternative to traditional plastics. However, their effectiveness depends on specialized composting infrastructure, and they may not degrade properly in landfills. Additionally, concerns about their production resources and potential contamination of recycling streams make their environmental benefits debatable.

The experts will assess the aforementioned options according to four primary criteria and their corresponding 10 subcriteria:

- Environmental impact $\left(L_1\right)$: This criterion measures the level of harm or benefit each method causes to the environment. Lower emissions, resource efficiency, and minimal pollution are preferred.
  - Reduction in environmental pollution $\left(L_{11}\right)$ (Pandey et al., 2023): This subcriterion assesses how well a waste management method minimizes air, water, and soil pollution. It focuses on reducing harmful emissions, such as toxic chemicals and microplastics, throughout the waste process. Methods with lower pollution footprints are rated higher.
  - Enhancement of biodiversity and ecosystem $\left(L_{12}\right)$ (Kibria et al., 2023): This subcriterion evaluates the method's contribution to conserving biodiversity and improving ecosystem health. It considers the method's ability to prevent habitat destruction and support the restoration of natural environments. Methods promoting long-term sustainability and species preservation score higher.
- Economic feasibility $\left(L_2\right)$: This criterion assesses the financial viability of implementing each method, including initial costs, operational costs,

and potential revenue generation. There are three subcriteria of economic feasibility.

- Return on investment $\left(L_{21}\right)$ (Ghodrat et al., 2019): Evaluates the financial returns relative to the initial investment. Higher ROI indicates better financial sustainability and attractiveness for stakeholders.
- Operational efficiency $\left(L_{22}\right)$ (Rafey and Siddiqui, 2023): Assesses how effectively resources are utilized during operation. Methods with higher efficiency reduce energy and resource waste, lowering overall costs.
- Cost-efficiency $\left(L_{23}\right)$ (Sarra et al., 2017): Measures the overall cost-effectiveness of a method, including setup and maintenance. A more cost-efficient method achieves the desired outcomes with lower financial expenditure.

- Social acceptance $\left(L_3\right)$: This criterion evaluates how well society will likely accept the method. It includes public perception, willingness to participate, and effects on communities.
  - Public affordability $\left(L_{31}\right)$ (Mandal and Seikh, 2023): This subcriterion evaluates whether the waste management method is financially accessible to the general public. It considers the costs imposed on consumers and the overall economic feasibility of widespread implementation.
  - Public acceptability $\left(L_{32}\right)$ (Mandal and Seikh, 2023): This subcriterion measures the community's willingness to support and engage with the waste management method. It looks at public perception and the method's integration into daily life.
  - Job creation $\left(L_{33}\right)$ (Mandal and Seikh, 2023): This subcriterion assesses the method's potential to generate employment opportunities. It considers creating jobs in areas such as waste management, recycling facilities, and associated sectors.
- Effectiveness $\left(L_4\right)$: This criterion measures how effective the method is at reducing plastic waste volume and whether it provides long-term sustainability in managing waste.
  - Long-term success $\left(L_{41}\right)$ (Mandal and Seikh, 2023): This subcriterion evaluates how well a method provides lasting benefits over time. It measures the method's effectiveness in continuously reducing plastic waste and maintaining environmental improvements. Methods with proven long-term reliability receive higher scores.
  - Time to get effective results $\left(L_{42}\right)$ (Mandal and Seikh, 2023): This subcriterion assesses how quickly a method achieves noticeable improvements in waste management. It focuses on the speed at which the method shows significant reductions in plastic waste. Faster results are favored for their immediate impact.

We have assumed the importance of the linguistic variables of each decision expert in calculating their weights. In addition, we assumed each decision expert's confidence level in subcriteria based on their degree, experience, and field of competence.

Here are 10 sub-criteria, and we are assuming their weights as: $W_{L_{11}} = 0.104$, $W_{L_{12}} = 0.124$, $W_{L_{13}} = 0.103$, $W_{L_{21}} = 0.124$, $W_{L_{22}} = 0.112$, $W_{L_{31}} = 0.109$, $W_{L_{32}} = 0.110$, $W_{L_{33}} = 0.117$, $W_{L_{41}} = 0.123$, and $W_{L_{42}} = 0.074$. The addition of all subcriteria weights is equal to 1. Also, the weight of the decision experts D_1, D_2, D_3, and D_4 are chosen as $0.34, 0.13, 0.21$, and 0.32, respectively. The linguistic assessment for evaluating DEs, criteria, and alternatives is displayed in Table 9.2. The confidence level of the experts is given in Table 9.3.

By seeing this, we can find the alternative one, that is, recycling, is the best alternative.

9.5.1 Comparative Analysis

In a comparative analysis, we compare the proposed methods CIVFFDWA and CIVFFDWG with two existing aggregation operators: confidence level IVFF weighted averaging operators (CIVFFWA) and confidence level IVFF weighted geometric operators (CIVFFWG), which are given by Seikh and Chatterjee (2024), and confidence level IVFF Einstein weighted averaging operators (CIVFFEWA) and confidence level IVFF Einstein weighted geometric operators (CIVFFEWG), which

TABLE 9.2
Linguistic assessment for evaluating DEs, criteria, and alternatives

LVs	IVFFN
Exceptionally high (EH)	([0.9, 0.95], [0.05, 0.1])
Very high (VH)	([0.8, 0.9], [0.2, 0.35])
High (H)	([0.65, 0.8], [0.4,0.5])
Medium (M)	([0.5,0.5], [0.5, 0.5])
Low (L)	([0.4, 0.5], [0.6, 0.7])
Very low (VL)	([0.2, 0.3], [0.8, 0.85])
Exceptionally low (EL)	([0.1, 0.2], [0.85, 0.9])

TABLE 9.3
DEs' confidence level for each attribute

Subcriteria	D_1	D_2	D_3	D_4
L_1	0.95	0.80	0.80	0.85
L_2	0.80	0.90	0.7	0.80
L_3	0.7	0.85	0.85	0.75
L_4	0.85	0.7	0.75	0.90

TABLE 9.4
Decision matrix in LVs

Decision maker	Alternatives	Subcriteria									
		L_{11}	L_{12}	L_{13}	L_{21}	L_{22}	L_{31}	L_{32}	L_{33}	L_{41}	L_{42}
D_1	T_1	VH	VL	H	M	L	VH	EL	H	M	EH
	T_2	M	H	VH	EL	H	EH	M	L	VL	M
	T_3	H	VH	EL	VL	M	VH	H	M	EH	L
	T_4	VL	VH	M	M	L	VL	EH	M	H	EL
	T_5	L	VH	EH	VH	M	L	H	EL	H	VL
D_2	T_1	M	VH	M	EL	H	EH	M	L	VL	H
	T_2	VH	L	EH	M	L	EL	H	VL	M	VH
	T_3	EH	EH	M	EL	VH	H	L	M	H	VL
	T_4	L	M	EH	VH	H	VL	L	EL	VH	M
	T_5	VL	H	EL	VH	L	M	H	EL	M	EH
D_3	T_1	M	H	EL	L	VH	EH	L	M	H	VL
	T_2	L	EH	M	EL	L	H	VL	H	M	VH
	T_3	EH	M	H	EL	H	L	VL	EL	VH	M
	T_4	EL	VH	H	M	EH	VL	H	VH	H	EH
	T_5	VL	H	M	EL	EL	VH	M	H	EH	L
D_4	T_1	EH	H	VH	M	L	VL	EL	EH	M	H
	T_2	H	L	EH	EL	M	VH	H	M	VL	EL
	T_3	VH	H	VH	EH	L	M	M	EL	EH	VL
	T_4	M	VL	M	H	VH	EH	EL	H	VH	L
	T_5	L	VH	H	M	EH	EL	L	VL	VH	H

are given by Rani et al. (2022). Table 9.8 presents the final ranking outcomes using various aggregation operators.

Because confidence level-based operators for IVFF numbers are not available, we compared the suggested method to the Dombi weighted aggregation operator. First, we compare the proposed method of averaging aggregation operator with CIVFFWA and CIVFFEWA. By comparing this, we get the ranking of alternatives as $T_1 \succ T_4 \succ T_3 \succ T_5 \succ T_2$. So in this case, the first alternative is the best. Also, we have compared the proposed method of geometric aggregation operator

TABLE 9.5
Decision matrix

Decision maker	Alternatives	L_{11}	L_{12}	L_{13}	L_{21}	L_{22}
D1	T_1	([0.8, 0.9], [0.2, 0.35])	([0.2, 0.3], [0.8, 0.85])	([0.65, 0.8], [0.4, 0.5])	([0.5, 0.5], [0.5, 0.5])	([0.4, 0.5], [0.6, 0.7])
	T_2	([0.5, 0.5], [0.5, 0.5])	([0.65, 0.8], [0.4, 0.5])	([0.8, 0.9], [0.2, 0.35])	([0.1, 0.2], [0.85, 0.9])	([0.65, 0.8], [0.4, 0.5])
	T_3	([0.65, 0.8], [0.4, 0.5])	([0.8, 0.9], [0.2, 0.35])	([0.1, 0.2], [0.85, 0.9])	([0.2, 0.3], [0.8, 0.85])	([0.5, 0.5], [0.5, 0.5])
	T_4	([0.2, 0.3], [0.8, 0.85])	([0.8, 0.9], [0.2, 0.35])	([0.5, 0.5], [0.5, 0.5])	([0.5, 0.5], [0.5, 0.5])	([0.4, 0.5], [0.6, 0.7])
	T_5	([0.4, 0.5], [0.6, 0.7])	([0.8, 0.9], [0.2, 0.35])	([0.9, 0.95], [0.05, 0.1])	([0.8, 0.9], [0.2, 0.35])	([0.5, 0.5], [0.5, 0.5])
D2	T_1	([0.5, 0.5], [0.5, 0.5])	([0.8, 0.9], [0.2, 0.35])	([0.5, 0.5], [0.5, 0.5])	([0.1, 0.2], [0.85, 0.9])	([0.65, 0.8], [0.4, 0.5])
	T_2	([0.8, 0.9], [0.2, 0.35])	([0.4, 0.5], [0.6, 0.7])	([0.9, 0.95], [0.05, 0.1])	([0.5, 0.5], [0.5, 0.5])	([0.4, 0.5], [0.6, 0.7])
	T_3	([0.9, 0.95], [0.05, 0.1])	([0.9, 0.95], [0.05, 0.1])	([0.5, 0.5], [0.5, 0.5])	([0.1, 0.2], [0.85, 0.9])	([0.8, 0.9], [0.2, 0.35])
	T_4	([0.4, 0.5], [0.6, 0.7])	([0.5, 0.5], [0.5, 0.5])	([0.9, 0.95], [0.05, 0.1])	([0.8, 0.9], [0.2, 0.35])	([0.65, 0.8], [0.4, 0.5])
	T_5	([0.2, 0.3], [0.8, 0.85])	([0.65, 0.8], [0.4, 0.5])	([0.1, 0.2], [0.85, 0.9])	([0.8, 0.9], [0.2, 0.35])	([0.4, 0.5], [0.6, 0.7])
D3	T_2	([0.5, 0.5], [0.5, 0.5])	([0.65, 0.8], [0.4, 0.5])	([0.1, 0.2], [0.85, 0.9])	([0.4, 0.5], [0.6, 0.7])	([0.8, 0.9], [0.2, 0.35])
	T_2	([0.4, 0.5], [0.6, 0.7])	([0.9, 0.95], [0.05, 0.1])	([0.5, 0.5], [0.5, 0.5])	([0.1, 0.2], [0.85, 0.9])	([0.4, 0.5], [0.6, 0.7])
	T_3	([0.9, 0.95], [0.05, 0.1])	([0.5, 0.5], [0.5, 0.5])	([0.65, 0.8], [0.4, 0.5])	([0.1, 0.2], [0.85, 0.9])	([0.65, 0.8], [0.4, 0.5])
	T_4	([0.1, 0.2], [0.85, 0.9])	([0.8, 0.9], [0.2, 0.35])	([0.65, 0.8], [0.4, 0.5])	([0.5, 0.5], [0.5, 0.5])	([0.9, 0.95], [0.05, 0.1])

(Continued)

TABLE 9.5 (CONTINUED)
Decision matrix

Decision maker	Alternatives	L_{11}	L_{12}	L_{13}	L_{21}	L_{22}
	T_5	([0.2, 0.3], [0.8, 0.85])	([0.65, 0.8], [0.4, 0.5])	([0.5, 0.5], [0.5, 0.5])	([0.1, 0.2], [0.85, 0.9])	([0.1, 0.2], [0.85, 0.9])
D4	T_1	([0.9, 0.95], [0.05, 0.1])	([0.65, 0.8], [0.4, 0.5])	([0.8, 0.9], [0.2, 0.35])	([0.5, 0.5], [0.5, 0.5])	([0.4, 0.5], [0.6, 0.7])
	T_2	([0.65, 0.8], [0.4, 0.5])	([0.4, 0.5], [0.6, 0.7])	([0.9, 0.95], [0.05, 0.1])	([0.1, 0.2], [0.85, 0.9])	([0.5, 0.5], [0.5, 0.5])
	T_3	([0.8, 0.9], [0.2, 0.35])	([0.65, 0.8], [0.4, 0.5])	([0.8, 0.9], [0.2, 0.35])	([0.9, 0.95], [0.05, 0.1])	([0.4, 0.5], [0.6, 0.7])
	T_4	([0.5, 0.5], [0.5, 0.5])	([0.2, 0.3], [0.8, 0.85])	([0.5, 0.5], [0.5, 0.5])	([0.65, 0.8], [0.4, 0.5])	([0.8, 0.9], [0.2, 0.35])
	T_5	([0.4, 0.5], [0.6, 0.7])	([0.8, 0.9], [0.2, 0.35])	([0.65, 0.8], [0.4, 0.5])	([0.5, 0.5], [0.5, 0.5])	([0.9, 0.95], [0.05, 0.1])
Decision maker	**Alternatives**	L_{31}	L_{32}	L_{33}	L_{41}	L_{42}
D1	T_1	([0.8, 0.9], [0.2, 0.35])	([0.1, 0.2], [0.85, 0.9])	([0.65, 0.8], [0.4, 0.5])	([0.5, 0.5], [0.5, 0.5])	([0.9, 0.95], [0.05, 0.1])
	T_2	([0.9, 0.95], [0.05, 0.1])	([0.5, 0.5], [0.5, 0.5])	([0.4, 0.5], [0.6, 0.7])	([0.2, 0.3], [0.8, 0.85])	([0.5, 0.5], [0.5, 0.5])
	T_3	([0.8, 0.9], [0.2, 0.35])	([0.65, 0.8], [0.4, 0.5])	([0.5, 0.5], [0.5, 0.5])	([0.9, 0.95], [0.05, 0.1])	([0.4, 0.5], [0.6, 0.7])
	T_4	([0.2, 0.3], [0.8, 0.85])	([0.9, 0.95], [0.05, 0.1])	([0.5, 0.5], [0.5, 0.5])	([0.65, 0.8], [0.4, 0.5])	([0.1, 0.2], [0.85, 0.9])
	T_5	([0.4, 0.5], [0.6, 0.7])	([0.65, 0.8], [0.4, 0.5])	([0.1, 0.2], [0.85, 0.9])	([0.65, 0.8], [0.4, 0.5])	([0.2, 0.3], [0.8, 0.85])
D2	T_1	([0.9, 0.95], [0.05, 0.1])	([0.5, 0.5], [0.5, 0.5])	([0.4, 0.5], [0.6, 0.7])	([0.2, 0.3], [0.8, 0.85])	([0.65, 0.8], [0.4, 0.5])
	T_2	([0.1, 0.2], [0.85, 0.9])	([0.65, 0.8], [0.4, 0.5])	([0.2, 0.3], [0.8, 0.85])	([0.5, 0.5], [0.5, 0.5])	([0.8, 0.9], [0.2, 0.35])

(Continued)

TABLE 9.5 (CONTINUED)
Decision matrix

Decision maker	Alternatives	L_{31}	L_{32}	L_{33}	L_{41}	L_{42}
	T_3	([0.65, 0.8], [0.4, 0.5])	([0.4, 0.5], [0.6, 0.7])	([0.5, 0.5], [0.5, 0.5])	([0.65, 0.8], [0.4, 0.5])	([0.2, 0.3], [0.8, 0.85])
	T_4	([0.2, 0.3], [0.8, 0.85])	([0.4, 0.5], [0.6, 0.7])	([0.1, 0.2], [0.85, 0.9])	([0.8, 0.9], [0.2, 0.35])	([0.5, 0.5], [0.5, 0.5])
	T_5	([0.5, 0.5], [0.5, 0.5])	([0.65, 0.8], [0.4, 0.5])	([0.1, 0.2], [0.85, 0.9])	([0.5, 0.5], [0.5, 0.5])	([0.9, 0.95], [0.05, 0.1])
D3	T_1	([0.9, 0.95], [0.05, 0.1])	([0.4, 0.5], [0.6, 0.7])	([0.5, 0.5], [0.5, 0.5])	([0.65, 0.8], [0.4, 0.5])	([0.2, 0.3], [0.8, 0.85])
	T_2	([0.65, 0.8], [0.4, 0.5])	([0.2, 0.3], [0.8, 0.85])	([0.65, 0.8], [0.4, 0.5])	([0.5, 0.5], [0.5, 0.5])	([0.8, 0.9], [0.2, 0.35])
	T_3	([0.4, 0.5], [0.6, 0.7])	([0.2, 0.3], [0.8, 0.85])	([0.1, 0.2], [0.85, 0.9])	([0.8, 0.9], [0.2, 0.35])	([0.5, 0.5], [0.5, 0.5])
	T_4	([0.2, 0.3], [0.8, 0.85])	([0.65, 0.8], [0.4, 0.5])	([0.8, 0.9], [0.2, 0.35])	([0.65, 0.8], [0.4, 0.5])	([0.9, 0.95], [0.05, 0.1])
	T_5	([0.8, 0.9], [0.2, 0.35])	([0.5, 0.5], [0.5, 0.5])	([0.65, 0.8], [0.4, 0.5])	([0.9, 0.95], [0.05, 0.1])	([0.4, 0.5], [0.6, 0.7])
D4	T_1	([0.2, 0.3], [0.8, 0.85])	([0.1, 0.2], [0.85, 0.9])	([0.9, 0.95], [0.05, 0.1])	([0.5, 0.5], [0.5, 0.5])	([0.65, 0.8], [0.4, 0.5])
	T_2	([0.8, 0.9], [0.2, 0.35])	([0.65, 0.8], [0.4, 0.5])	([0.5, 0.5], [0.5, 0.5])	([0.2, 0.3], [0.8, 0.85])	([0.1, 0.2], [0.85, 0.9])
	T_3	([0.5, 0.5], [0.5, 0.5])	([0.5, 0.5], [0.5, 0.5])	([0.1, 0.2], [0.85, 0.9])	([0.9, 0.95], [0.05, 0.1])	([0.2, 0.3], [0.8, 0.85])
	T_4	([0.9, 0.95], [0.05, 0.1])	([0.1, 0.2], [0.85, 0.9])	([0.65, 0.8], [0.4, 0.5])	([0.8, 0.9], [0.2, 0.35])	([0.2, 0.3], [0.8, 0.85])
	T_5	([0.1, 0.2], [0.85, 0.9])	([0.4, 0.5], [0.6, 0.7])	([0.2, 0.3], [0.8, 0.85])	([0.8, 0.9], [0.2, 0.35])	([0.65, 0.8], [0.4, 0.5])

TABLE 9.6
Aggregated decision matrix

T_i	L_{11}	L_{12}	L_{13}	L_{21}	L_{22}
T_1	([0.807, 0.895], [0.077, 0.153])	([0.602, 0.754], [0.365, 0.533])	([0.667, 0.805], [0.292, 0.462])	([0.428, 0.446], [0.573, 0.584])	([0.575, 0.720], [0.352, 0.547])
T_2	([0.597, 0.731], [0.364, 0.504])	([0.721, 0.836], [0.090, 0.180])	([0.833, 0.912], [0.069, 0.137])	([0.258, 0.279], [0.764, 0.785])	([0.514, 0.642], [0.506, 0.577])
T_3	([0.811, 0.900], [0.076, 0.153])	([0.753, 0.863], [0.104, 0.206])	([0.644, 0.783], [0.297, 0.479])	([0.742, 0.850], [0.079, 0.157])	([0.575, 0.706], [0.365, 0.522])
T_4	([0.360, 0.392], [0.647, 0.667])	([0.703, 0.831], [0.251, 0.427])	([0.669, 0.777], [0.106, 0.208])	([0.602, 0.732], [0.356, 0.493])	([0.750, 0.860], [0.094, 0.186])
T_5	([0.345, 0.440], [0.664, 0.756])	([0.750, 0.868], [0.233, 0.395])	([0.793, 0.884], [0.073, 0.145])	([0.675, 0.804], [0.270, 0.443])	([0.751, 0.850], [0.078, 0.157])
	L_{31}	L_{32}	L_{33}	L_{41}	L_{42}
T_1	([0.798, 0.890], [0.075, 0.150])	([0.303, 0.351], [0.716, 0.758])	([0.757, 0.859], [0.080, 0.160])	([0.499, 0.585], [0.512, 0.544])	([0.783, 0.880], [0.075, 0.151])
T_2	([0.788, 0.885], [0.080, 0.159])	([0.524, 0.661], [0.501, 0.576])	([0.474, 0.590], [0.543, 0.608])	([0.338, 0.366], [0.672, 0.686])	([0.615, 0.746], [0.308, 0.487])
T_3	([0.632, 0.765], [0.307, 0.479])	([0.490, 0.606], [0.530, 0.598])	([0.363, 0.369], [0.646, 0.653])	([0.860, 0.928], [0.059, 0.120])	([0.352, 0.410], [0.658, 0.702])
T_4	([0.733, 0.840], [0.080, 0.161])	([0.747, 0.853], [0.081, 0.161])	([0.619, 0.757], [0.328, 0.492])	([0.712, 0.842], [0.264, 0.430])	([0.679, 0.794], [0.093, 0.184])
T_5	([0.565, 0.703], [0.342, 0.541])	([0.529, 0.668], [0.496, 0.579])	([0.402, 0.547], [0.627, 0.734])	([0.771, 0.875], [0.091, 0.181])	([0.645, 0.776], [0.111, 0.220])

TABLE 9.7
Ranking outcomes from the proposed method

Alternatives	IVFFN	Score	Rank
T_1	([0.708, 0.825], [0.100, 0.199])	1.452	1
T_2	([0.673, 0.796], [0.113, 0.222])	1.399	5
T_3	([0.695, 0.815], [0.106, 0.211])	1.4340	3
T_4	([0.698, 0.817], [0.106, 0.210])	1.4476	2
T_5	([0.688, 0.811], [0.112, 0.222])	1.424	4

TABLE 9.8
Ranking outcomes using different aggregation operator-based methods

Aggregation operators	Score values					Ranking
	$\mathcal{S}(T_1)$	$\mathcal{S}(T_2)$	$\mathcal{S}(T_3)$	$\mathcal{S}(T_4)$	$\mathcal{S}(T_5)$	
CIVFFDWA (proposed)	1.4540	1.3987	1.4339	1.4384	1.4236	$T_1 \succ T_4 \succ T_3 \succ T_5 \succ T_2$
CIVFFDWG (proposed)	0.9464	0.8359	0.8712	0.9510	0.9546	$T_5 \succ T_4 \succ T_1 \succ T_3 \succ T_2$
CIVFFWA (Seikh and Chatterjee, 2024)	0.2966	0.1846	0.2658	0.2885	0.2472	$T_1 \succ T_4 \succ T_3 \succ T_5 \succ T_2$
CIVFFWG (Seikh and Chatterjee, 2024)	0.4616	0.3790	0.4542	0.4283	0.4975	$T_5 \succ T_4 \succ T_1 \succ T_3 \succ T_2$
CIVFFEWA (Rani et al., 2022)	0.3292	0.2647	0.3002	0.3208	0.2790	$T_1 \succ T_4 \succ T_3 \succ T_5 \succ T_2$
CIVFFEWG (Rani et al., 2022)	0.3901	0.2751	0.3849	0.3504	0.4197	$T_5 \succ T_4 \succ T_1 \succ T_3 \succ T_2$

with CIVFFWG and CIVFFEWG. By comparing this, we get that the ranking of CIVFFDWG and CIVFFWG is the same, that is, $T_5 \succ T_4 \succ T_1 \succ T_3 \succ T_2$, and the ranking of CIVFFEWG is $T_5 \succ T_4 \succ T_1 \succ T_3 \succ T_2$. So we can say that, in this case, the fifth alternative is the best. This establishes that our proposed method is practically applicable.

Figure 9.2 displays the ranking outcomes using different aggregation operator-based MAGDM methods.

The novelty of the proposed aggregation operators, CIVFFDWA and CIVFFDWG, lies in their enhanced capability to handle uncertainty with greater precision through the incorporation of confidence levels in an IVFF framework. The comparative analysis with existing operators, CIVFFWA, CIVFFWG, CIVFFEWA, and CIVFFEWG, demonstrates several key points of distinction and advancement.

- By integrating the DTN and DTCN, the proposed methods introduce a parameterized approach that adjusts the aggregation process based on confidence levels, leading to more reliable decision-making outcomes. This flexibility is absent in the traditional operators such as CIVFFWA and CIVFFEWA, which use simpler averaging techniques without considering varying degrees of confidence.
- Unlike the existing methods, the proposed operators explicitly incorporate confidence levels, which adds an additional layer of reliability to the decision-making process. This ensures that the ranking of alternatives not only

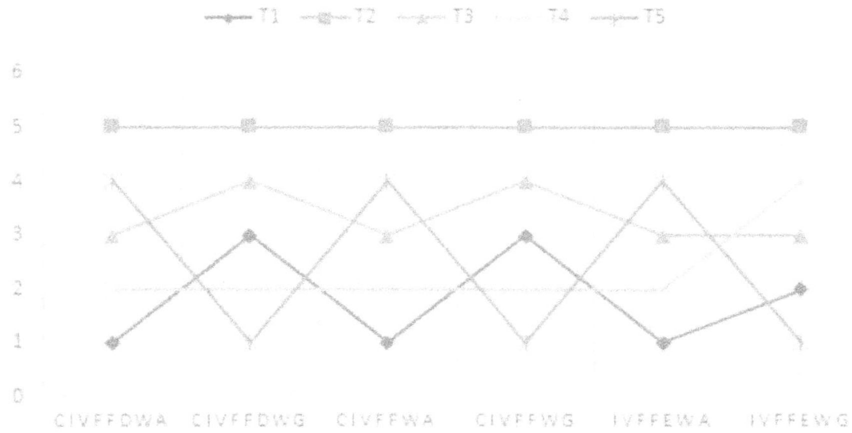

FIGURE 9.2 Comparative analysis of proposed MAGDM process

reflects their inherent performance but also the certainty associated with the input data.

- The proposed CIVFFDWA and CIVFFDWG operators provide more refined score values, resulting in a clear and consistent ranking of alternatives across different scenarios. For instance, the rankings produced by CIVFFDWA and CIVFFWA are the same ($T_1 \succ T_4 \succ T_3 \succ T_5 \succ T_2$), but CIVFFDWA achieves higher score values, indicating a more precise differentiation of alternatives.

9.5.2 Sensitivity Analysis

In this study, we performed a comprehensive sensitivity analysis of the proposed approach by systematically varying the parameter values. Specifically, we focused on a single parameter σ, integral to the Dombi aggregation operators. Initially, we set $\sigma = 2$ for the numerical problem, which served as our baseline configuration. To assess the robustness of our method, we subsequently modified the value of σ across a range from 1 to 10, observing how these adjustments influenced the outcomes.

Table 9.9 presents the scores of the alternatives across the ranking sequence defined by $\sigma = 1(1)10$. Table 9.9 details the ranking results corresponding to each parameter value, illustrating how changes in σ affect the overall rankings of the alternatives. Through this sensitivity analysis, we not only evaluate the performance of the proposed method under varying conditions but also reinforce its reliability in producing consistent results.

Here, we find that the rank for the parameter $\sigma = 1,2,3,4$ is same, that is, $T_1 \succ T_4 \succ T_3 \succ T_5 \succ T_2$, and the rank for the parameter $\sigma = 5,6,7,8,9$ is same, that is, $T_1 \succ T_4 \succ T_3 \succ T_2 \succ T_5$, and the rank for the parameter $\sigma = 10$ is $T_1 \succ T_3 \succ T_2 \succ T_5 \succ T_4$. So we can see that the first alternative, that is, recycling, is the most effective PWM

TABLE 9.9
Sensitivity of proposed method

σ	$\mathcal{S}(T_1)$	$\mathcal{S}(T_2)$	$\mathcal{S}(T_3)$	$\mathcal{S}(T_4)$	$\mathcal{S}(T_5)$	Rank
1	1.454	1.398	1.433	1.438	1.423	$T_1 \succ T_4 \succ T_3 \succ T_5 \succ T_2$
2	1.600	1.563	1.584	1.588	1.572	$T_1 \succ T_4 \succ T_3 \succ T_5 \succ T_2$
3	1.663	1.638	1.651	1.653	1.641	$T_1 \succ T_4 \succ T_3 \succ T_5 \succ T_2$
4	1.697	1.679	1.687	1.689	1.679	$T_1 \succ T_4 \succ T_3 \succ T_5 \succ T_2$
5	1.717	1.703	1.709	1.711	1.703	$T_1 \succ T_4 \succ T_3 \succ T_2 \succ T_5$
6	1.712	1.699	1.705	1.707	1.699	$T_1 \succ T_4 \succ T_3 \succ T_2 \succ T_5$
7	1.740	1.730	1.734	1.735	1.730	$T_1 \succ T_4 \succ T_3 \succ T_2 \succ T_5$
8	1.747	1.738	1.742	1.743	1.738	$T_1 \succ T_4 \succ T_3 \succ T_2 \succ T_5$
9	1.752	1.745	1.748	1.749	1.745	$T_1 \succ T_4 \succ T_3 \succ T_2 \succ T_5$
10	1.756	1.750	1.753	1.732	1.750	$T_1 \succ T_3 \succ T_2 \succ T_5 \succ T_4$

method. Figure 9.3 illustrates the ranking outcomes of the alternatives across various values of σ, providing insight into how changes in σ affect the relative positioning of each alternative within the decision-making framework.

The findings indicate that the proposed method is robust and adaptable, demonstrating a strong capacity to maintain reliable performance across different settings of the parameter σ. This sensitivity analysis not only validates our approach but also provides insights into the critical influence of parameter selection in decision-making processes.

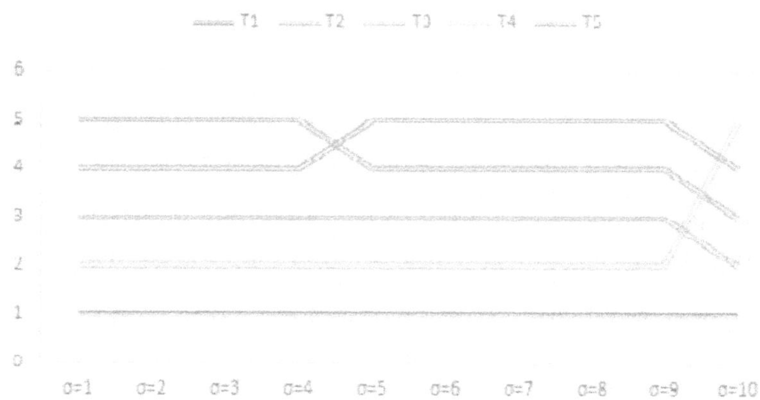

FIGURE 9.3 Ranking outcomes for various σ

9.5.3 Practical Implications

The study presents practical implications for decision-making in plastic waste management by integrating DTN- and DTCN-based aggregation operators within an IVFF framework. The incorporation of confidence levels enhances the ability to evaluate and prioritize waste management alternatives based on both performance and the reliability of the information. This approach allows for more accurate modeling of uncertainty in decision-making processes, offering a robust tool for assessing sustainability initiatives and resource allocation. Additionally, it can be applied to optimize waste collection, recycling processes, and policymaking, contributing to more efficient and effective plastic waste management strategies.

9.6 CONCLUSION

In this chapter, we introduced a MAGDM model utilizing confidence level-based Dombi aggregation operators tailored for IVFFSs. We began by developing confidence level-based Dombi weighted averaging (geometric) aggregation operators for IVFFSs. Following this, a MAGDM methodology was proposed using these operators to address decision-making problems within the IVFF framework. A numerical example utilizing IVFFSs was presented to demonstrate the practical relevance and efficiency of the model in evaluating PWM methods. Furthermore, a comparison with current models underscored the strengths and practical relevance of our approach. Finally, a sensitivity analysis was performed to evaluate the stability of the results obtained.

While this research offers valuable insights, it is important to acknowledge several limitations. Firstly, the study exclusively utilized the DTN and DTCN for developing the aggregation operators, which limits the scope of the analysis. Additionally, the criteria weights, decision expert weights, and confidence levels were assumed to be crisp values, and comparisons were limited to aggregation operator-based MAGDM methods. Moreover, only one decision expert was considered per criterion, which may restrict the diversity of perspectives. Future research will address these limitations. First, weight-determining methods such as SWARA and entropy will be explored to more accurately determine the weights of criteria and decision experts. Second, comparisons with methods like TOPSIS and VIKOR will be conducted to better validate the practical applicability of the proposed model. Finally, incorporating multiple decision experts across various criteria will offer a broader perspective on diverse viewpoints, resulting in more effective decision-making outcomes.

In the future, we will develop some more confidence level-based aggregation operators for the extended versions of fuzzy sets using different TNs and TCNs. Numerous MAGDM difficulties, such as including electric vehicle selection (Golui et al., 2024), economic condition analysis of a country (Jana et al., 2024), and many more, can be solved with the help of the suggested approach. Furthermore, the proposed framework can be applied within various uncertainty sets, such as hesitant fuzzy sets (Gupta and Kumar, 2024), picture fuzzy sets (Bani-Doumi et al., 2024), etc.

CONFLICT OF INTEREST

The authors declare that they have no conflict of interest.

REFERENCES

K. T. Atanassov. Intuitionistic fuzzy sets. *Fuzzy Sets and Systems*, 20:87–96, 1986.

S. B. Aydemir and S. Y. Gunduz. Fermatean fuzzy TOPSIS method with Dombi aggregation operators and its application in multi-criteria decision making. *Journal of Intelligent & Fuzzy Systems*, 39(1):851–869, 2020.

M. Bani-Doumi, J. Serrano-Guerrero, F. Chiclana, F. Romero, and J. A. Olivas. A picture fuzzy set multi criteria decision-making approach to customize hospital recommendations based on patient feedback. *Applied Soft Computing*, 153:111331, 2024.

J. Dombi. A general class of fuzzy operators, the demorgan class of fuzzy operators and fuzziness measures induced by fuzzy operators. *Fuzzy Sets and Systems*, 8(2):149–163, 1982.

M. A. A. J. Ghodrat, J. A. Alonso, D. Hagare, R. Yang, and B. Samali. Economic feasibility of energy recovery from waste plastic using pyrolysis technology: An Australian perspective. *International Journal of Environmental Science and Technology*, 16:3721–3734, 2019.

S. Golui, B. S. Mahapatra, and G. S. Mahapatra. A new correlation-based measure on Fermatean fuzzy applied on multi-criteria decision making for electric vehicle selection. *Expert Systems with Applications*, 237:121605, 2024.

Ö. F. Görçün, A. Ulutaş, A. Topal, and F. Ecer. Telescopic forklift selection through a novel interval-valued Fermatean fuzzy PIPRECIA–WISP approach. *Expert Systems with Applications*, 255:124674, 2024.

R. Gupta and S. Kumar. Novel similarity measure between hesitant fuzzy set and their applications in pattern recognition and clustering analysis. *Journal of Engineering and Applied Science*, 71(1):5, 2024.

C. Jana, T. Senapati, and M. Pal. Pythagorean fuzzy Dombi aggregation operators and its applications in multiple attribute decision-making. *International Journal of Intelligent Systems*, 34:2019–2038, 2019.

C. Jana, V. Simic, M. Pal, B. Sarkar, and D. Pamucar. Hybrid multi-criteria decision-making method with a bipolar fuzzy approach and its applications to economic condition analysis. *Engineering Applications of Artificial Intelligence*, 132:107837, 2024.

S. Jeevaraj. Ordering of interval-valued Fermatean fuzzy sets and its applications. *Expert Systems with Applications*, 185:115613, 2021. https://doi.org/10.1016/j.eswa.2021.115613.

M. G. Kibria, N. I. Masuk, R. Safayet, H. Q. Nguyen, and M. Mourshed. Plastic waste: Challenges and opportunities to mitigate pollution and effective management. *International Journal of Environmental Research*, 17(1):20, 2023.

M. Kirişci. Interval-valued Fermatean fuzzy based risk assessment for self-driving vehicles. *Applied Soft Computing*, 152:111265, 2024.

S. Korucuk, A. Aytekin, Ö. Görçün, V. Simic, and Ö. F. Görçün. Warehouse site selection for humanitarian relief organizations using an interval-valued Fermatean fuzzy LOPCOW-RAFSI model. *Computers & Industrial Engineering*, 192:110160, 2024.

F. Lombardi, L. Bartolucci, S. Cordiner, G. Costa, A. Falsetti, P. Mele, M. Mercurio, V. Mulone, and D. Sorino. Chemical-physical characterization of bio-based biodegradable plastics in view of identifying suitable recycling/recovery strategies and numerical modeling of PLA pyrolysis. *Waste and Biomass Valorization*, 15(3):1653–1670, 2024.

X. Ma, H. Sun, H. Qin, Y. Wang, and Y. Zheng. A novel multi-attribute decision making method based on interval-valued Fermatean fuzzy Bonferroni mean operators. *Journal of Intelligent & Fuzzy Systems*, (Preprint):1–21, 2024.

U. Mandal and M. R. Seikh. Interval-valued spherical fuzzy MABAC method based on Dombi aggregation operators with unknown attribute weights to select plastic waste management process. *Applied Soft Computing*, 145(2):110516, 2023. https://doi.org/10.1016/j.asoc.2023.110516.

P. Pandey, M. Dhiman, A. Kansal, and S. P. Subudhi. Plastic waste management for sustainable environment: Techniques and approaches. *Waste Disposal & Sustainable Energy*, 5(2):205–222, 2023.

R. D. Patria, S. Rehman, C. B. Yuen, D. J. Lee, A. K. Vuppaladadiyam, and S. Y. Leu. Energy-environment-economic (3e) hub for sustainable plastic management–upgraded recycling, chemical valorization, and bioplastics. *Applied Energy*, 357:122543, 2024.

T. R. Praveenkumar, M. Sekar, R. R. Pasupuleti, B. Gavurová, G. A. Kumar, and M. V. Kumar. Current technologies for plastic waste treatment for energy recovery, its effects on poly aromatic hydrocarbons emission and recycling strategies. *Fuel*, 357:129379, 2024.

A. Rafey and F. Z. Siddiqui. A review of plastic waste management in India—challenges and opportunities. *International Journal of Environmental Analytical Chemistry*, 103(16):3971–3987, 2023.

P. Rani, A. R. Mishra, M. Deveci, and J. Antucheviciene. New complex proportional assessment approach using Einstein aggregation operators and improved score function for interval-valued Fermatean fuzzy sets. *Computers & Industrial Engineering*, 169:108165, 2022.

A. Sarra, M. Mazzocchitti, and A. Rapposelli. Evaluating joint environmental and cost performance in municipal waste management systems through data envelopment analysis: Scale effects and policy implications. *Ecological Indicators*, 73:756–771, 2017.

M. R. Seikh and P. Chatterjee. Identifying sustainable strategies for electronic waste management utilizing confidence-based group decision-making method in interval-valued fermatean fuzzy environment. *Engineering Applications of Artificial Intelligence*, 135:108701, 2024.

M. R. Seikh and U. Mandal. Intuitionistic fuzzy dombi aggregation operators and their application to multiple attribute decision-making. *Granular Computing*, 6:473–488, 2021.

M. R. Seikh and U. Mandal. Interval-valued Fermatean fuzzy Dombi aggregation operators and SWARA based PROMETHEE ii method to bio-medical waste management. *Expert Systems with Applications*, 226, 2023.

T. Senapati and R. R. Yager. Fermatean fuzzy sets. *Journal of Ambient Intelligence and Humanized Computing*, 11:663–674, 2020.

D. Sharma, N. I. Moondra, R. K. Bharatee, A. Nema, K. Sweta, M. K. Yadav, and N. S. Maurya. Processing and recycling of plastic wastes for sustainable material management. In J. Parameswaranpillai & K. Deshmukh (Eds.), *Plastic Waste Management: Methods and Applications*, pages 89–116, Wiley-VCH GmbH, 2024.

R. R. Yager. Pythagorean membership grades in multicriteria decision making. *IEEE Transactions on Fuzzy Systems*, 22:958–965, 2014. https://doi.org/10.1109/TFUZZ.2013.2278989.

L. A. Zadeh. Fuzzy sets. *Information and Control*, 8(3):338–353, 1965.

10 Evaluation of Heavy Forest Fire Helicopters Using q-Rung Orthopair Fuzzy Set-Based TOPSIS Decision-Making Model

Hakan Ayhan Dağıstanlı and
Cevriye Temel Gencer

10.1 INTRODUCTION

Forest fires are a type of fire that is difficult to control, primarily occurring in natural areas. Their spread and intensity can be influenced by factors such as wind, weather conditions, and vegetation, potentially threatening agricultural lands, residential areas, and human lives (Dhall et al., 2020). Unlike other natural disasters, forest fires can have a range of causes. These include natural factors such as lightning strikes, sparks from rolling rocks, and volcanic activity, as well as human-induced factors such as unattended campfires, stubble burning, unextinguished cigarette butts, and electrical faults where flammable materials come into contact with a fire source (Bouguettaya et al., 2022; Enoh et al., 2021).

Forest fires have numerous short- and long-term negative effects. In the short term, they can disrupt essential services such as transportation, communications, electricity, natural gas, and water supplies. Additionally, they can degrade air quality (Sivrikaya and Küçük, 2022) and cause significant loss of property, crops, and human lives (Naderpour et al., 2019). In the long term, forest fires release substantial amounts of carbon dioxide, carbon monoxide, and fine particulate matter into the atmosphere, which can influence weather patterns, contribute to climate change, and exacerbate air pollution. The resulting pollution can lead to a variety of physical and mental health issues, including respiratory and cardiovascular problems (Chaturvedi et al., 2022).

Today, regional and global climate changes are contributing to the increasing frequency and intensity of forest fires (Kilic and Ozkan, 2019). Rising average temperatures and prolonged periods of drought facilitate the ignition and spread of these

 DOI: 10.1201/9781003529088-10

fires. Given the severe consequences of forest fires, the need for effective prevention and early warning mechanisms is becoming increasingly critical. Technological advancements offer new possibilities for developing systems that can predict fires in advance (Ozkan and Kilic, 2023). However, even with predictive systems in place, challenges may still arise in the detection and prevention of potential fires. In such instances, early-stage fire detection becomes paramount. Academic research has increasingly focused on early detection methods (Dağıstanlı and Gencer, 2023). Nevertheless, despite advances in prediction and detection technologies, forest fires continue to devastate vast areas globally, largely due to factors like ecological changes. Consequently, governments have tasked official institutions with extinguishing forest fires and implementing various measures such as inventory reinforcement, workforce expansion, training, and public awareness campaigns, which intensify each year. In this context, it is crucial that academic studies on combating forest fires be supported by analytical approaches.

This study conducts an analysis of helicopters, which are among the most critical tools for combating forest fires. Drawing on data provided by the General Directorate of Forestry, as well as statements made by relevant authorities to the press and strategic reports issued by government bodies, efforts are underway to establish a forest fire helicopter fleet. At its core, this issue presents a helicopter selection problem. In the literature, the FMCDM approach is commonly proposed as a solution for such problems. In this study, we propose a solution utilizing q-ROFS, one of the most advanced and robust fuzzy sets available. The primary motivation for this choice is that q-ROFS enables significantly more sensitive analyses compared to other fuzzy sets (Erdebilli et al., 2023). Furthermore, while helicopter selection has been addressed in previous studies, no research to date has specifically focused on selecting helicopters for forest fire management, nor has q-ROFS been applied to this type of problem. In this study, the q-ROFS method is integrated with the TOPSIS method to generate rankings for the results. Additionally, a sensitivity analysis for different q levels has been conducted to validate the robustness of the findings, and the results are compared using various MCDM methods.

The remainder of this chapter is structured as follows: Section 10.2 provides a review of the relevant literature. Section 10.3 delineates the problem, criteria, and alternatives. Section 10.4 presents preliminary information on q-ROFS. Section 10.5 demonstrates the application of the methodology to the identified problem. Section 10.6 encompasses a sensitivity analysis and presents comparative results. Finally, the last section includes conclusions and suggestions for future work.

10.2 LITERATURE REVIEW

The literature review is conducted from multiple perspectives. In the first stage, studies related to forest fire management are examined. In the second stage, research on helicopter selection is explored. Finally, in the third stage, studies involving q-ROFS are summarized in tabular form.

10.2.1 Literature Review for Fighting Forest Fires

Regardless of the reasons for their occurrence, forest fires present a significant threat to a sustainable future worldwide. Consequently, the academic literature, much like state institutions, has not remained indifferent to this challenge. Various studies are being conducted to address the issue of forest fire management from multiple perspectives.

Temiz and Tecim (2009) developed a spatial approach to combating forest fires by integrating geographic components with MCDM methods. The combined use of geographic information systems (GISs) and MCDM enables forest managers to visualize solutions and gain a deeper understanding of the problems they face.

Pappis and Rachaniotis (2010) focused on a single-source scheduling plan for managing multiple fires requiring control, considering the deterioration of fire extinguishing times. They derived the processing time, the duration needed to control the fire in the affected area, by accounting for preparation times. Akay et al. (2012) created a GIS-based decision support system aimed at identifying the shortest and safest routes to forest fires, featuring a case study in Turkey's Mediterranean region.

Yuan et al. (2015) highlighted the significant potential of unmanned aerial vehicles (UAVs), equipped with vision-based systems, for monitoring, detecting, and combating forest fires. Their study began with a brief overview of UAV system development and architecture for fire management. It further reviews the technological capabilities of UAVs related to fire monitoring, detection, and intervention, including fire detection and the mitigation of image vibrations. Finally, the study summarized the challenges and potential solutions associated with using UAVs in fire management.

Akay and Aziz (2015) proposed a GIS-based model for forest road networks to enhance forest protection in their study conducted in the Erbil region of Iraq. Their goal was to ensure that fire trucks and ground crews could reach the forest fire areas in the shortest possible time. Sakellariou et al. (2017) conducted a comprehensive review of decision support systems for forest fire prevention and suppression. This included database management systems, mathematical and economic algorithms for the spatial optimization of firefighting resources, forest fire simulators, and satellite technology for real-time detection and prediction of fire development, as well as GIS platforms equipped with various tools for processing and analyzing geographic data to develop strategic and operational plans.

Ghamry et al. (2017) presented a study based on the assumption that fire locations would be detected in advance and their coordinates transmitted to firefighting UAV teams. The objective was to minimize the distance traveled between each UAV's initial location and the designated fire spot by allocating UAVs according to their relative distances. The particle swarm optimization algorithm was employed to optimize the path planning to the assigned fire spot for the UAVs. Zhang et al. (2018) developed a path planning algorithm capable of learning about the terrain environment and generating movement policies to determine the most suitable paths for UAVs engaged in forest surveillance for firefighting missions.

Bal et al. (2019) explored collaboration opportunities between Turkey and Greece in the Aegean Region concerning aerial firefighting. They collected data on the fire history of forest assets in both countries, fire risk and intensity maps, aerial firefighting vehicles and related infrastructure (including runways and water sources), and vehicle allocations. The study examined mathematical programming models for optimal resource reallocation in both decentralized and centralized decision-making scenarios, proposing a game theory-based approach to identify potential cooperative allocation strategies. The potential benefits and drawbacks of cooperation were also discussed.

Caetano and Morgado (2019) analyzed the Portuguese Air Force's Research, Development, and Innovation (RD&I) program aimed at developing unmanned aircraft systems and applying these technologies for societal benefit. The program emphasized UAV activities, particularly in maritime surveillance and forest fire detection and control. In collaboration with the National Civil Protection Corps, the Portuguese Air Force focused on UAV studies regarding their integrated roles in forest fire detection, monitoring, and management, with plans for a comprehensive approach involving the entire air fleet in subsequent phases.

Jain et al. (2020) provided an overview of machine learning applications in fire science and management, specifically discussing algorithms for forest fire detection. They emphasized that helicopter route determination studies could enhance fire management efforts. Özkan (2021) focused on UAV guidance for forest fire detection, developing an algorithm that hybridizes simulated annealing and local search metaheuristics with an integer linear programming model. The effectiveness of this methodology was validated through a real-life case study in Turkey and compared with a genetic algorithm.

Abdo et al. (2022) employed GIS-based frequency ratio and analytical hierarchy process (AHP) methods for forest fire susceptibility mapping in the western region of Syria. The study aimed to compare the performance of the frequency ratio and AHP techniques in determining the spatial distribution of forest fire susceptibility in the Al-Draikich region. An inventory map of historical forest fire events was created by spatially digitizing 32 incidents that occurred during the summer months of 2019, 2020, and 2021. These fire events were divided into a 70% training dataset (22 events) and a 30% test dataset (10 events). Analyses conducted using both techniques with data from independent variables revealed that AHP, as a multi-criteria decision-making technique, yielded higher values than the frequency ratio. The findings provided valuable spatial information to inform forest management strategies in the study area.

Maraş et al. (2023) developed a mathematical model based on cluster coverage for GIS-based identification of optimal heliport and water source locations for forest fire extinguishment, employing multi-objective programming. The primary objective of the study was to determine heliport locations to enable rapid intervention in the event of a fire. The secondary aim was to propose new water source locations based on a comprehensive fire scenario to evaluate the existing water sources.

Ozkan and Kilic (2023) addressed the UAV routing problem through simulation-based optimization approaches to mitigate forest fire risks. Their study introduced a novel heuristic distribution rule and a simulation-based optimization algorithm for creating routes for UAVs, particularly for detecting forest fires in remote areas.

A notable aspect of the proposed algorithms was their focus on minimizing the probability of fire occurrence. Experimental results were analyzed across various scenarios.

In conclusion, numerous studies within the academic field address the issue of forest fire management from diverse perspectives, utilizing various analytical methods. The research is predominantly concentrated on terrain analysis using GIS, detection methodologies employing UAV technologies, and strategy development. Additionally, the integration of advanced technological tools, software analysis systems, artificial intelligence, MCDM, and optimization methods is evident in the literature.

10.2.2 Literature Review on Helicopter Selection Problem

The helicopter selection problem is not a novel area of research. However, upon reviewing the literature, it becomes apparent that it is not extensively studied. Over the years, research has focused on various helicopter types associated with different mission groups. Table 10.1 presents a summary of the literature concerning helicopter selection studies.

TABLE 10.1
Literature summary for helicopter selection problem

Year	References	Subject	Fuzzy sets	Method
1999	Cheng et al.	Attack helicopter selection	✓	AHP
2018	Qian and Wang	Training helicopter selection	X	AHP
2019	Üsküdar et al.	Cargo helicopter selection	✓	AHP
2020	Costa et al.	Naval attack helicopter selection	✓	AHP-TOPSIS
2021	Maeda et al.	Naval attack helicopter selection	X	AHP-TOPSIS
2021	Moreira et al.	Attack helicopter selection	X	PROMETHEESAPEVO-M1
2022	Ardil	Attack helicopter selection	X	Variance weight procedure
2023	Günaltili et al.	Light class archetypal helicopterelections	X	Competitor analysis
2023	Assis et al.	Police helicopter selection	X	WASPAS
2024	Rodrigues et al.	Maritime logistics helicopterelections	X	AHP-TOPSIS- THOR2

Helicopters are generally classified into light, medium, and heavy categories based on their weight, carrying capacity, and intended purpose. Heavy-class helicopters, in particular, play a crucial role in firefighting operations due to their ability to transport substantial amounts of water or fire suppression chemicals, thanks to their high carrying capacity. In forested and rural areas where fires can spread rapidly, the quick intervention capability of these helicopters is of paramount importance. Consequently, this study focuses on heavy-class helicopters designed for forest fire suppression.

A review of the literature on helicopter selection studies reveals a predominant reliance on classical methods. Although fuzzy sets are frequently employed in literature concerning problems that necessitate expert opinions, such as the selection of weapons, equipment, and tools, this approach appears to be relatively scarce in the context of helicopter selection studies. Moreover, it can be inferred that existing research primarily utilizes criteria based solely on technical specifications, resulting in analyses that inadequately address the experiential criteria associated with the mission definitions of the helicopters in question.

In this study, we aim to address this significant gap by employing q-ROFSs, which have been demonstrated to offer superior performance compared to intuitionistic fuzzy sets (IFSs) and Pythagorean fuzzy sets (PFSs) in the existing literature.

10.2.3 Literature Review on q-ROFS

q-ROFS, which is favored for addressing the helicopter selection problem, has gained significant traction across various fields in the literature, with its applications in relevant areas continually expanding. Table 10.2 provides a summary of the current research domains utilizing q-ROFS.

TABLE 10.2
Literature summary for studies using q-ROFS

Year	References	Subject	Method
2024	Ghoushchi et al.	Understanding the causes of land subsidence	FMEA-SWARA
2024	Zeb et al.	Selecting electrical power system	Hamacheraggregation operator
2023	Zulqarnain et al.	Selecting cloud service provider	TOPSIS
2024	Javeed et al.	Selecting medical device manufacturer	Yager aggregation operator
2024	Naz et al.	Transportation investment project selection	DEMATEL
2024	Erdebilli and Sıcakyüz	Sustainable and efficient supply chain management	VIKOR
2024	Cao et al.	Evaluation of food waste treatment techniques	TODIM
2024	Li et al.	Evaluation of mobile medical application service quality	BWM-SEMMARCOS

An examination of Table 10.2 reveals that q-ROFS has been integrated with various methods to address real-life problems; however, no applications of q-ROFS in helicopter selection problems are found in the literature. In this study, weights derived from q-ROFS are integrated with the TOPSIS method to rank heavy helicopters. The q-ROFS-TOPSIS approach has been explored in significant studies within the literature.

For instance, Erdebilli et al. (2023) developed a comprehensive model for selecting private health insurance policies, utilizing q-ROFS-based TOPSIS to achieve a solution. They also tested the robustness of their results using the VIšeKriterijumska Optimizacija I Kompromisno Rešenje (VIKOR) method. Similarly, Pınar et al. (2021) applied the q-ROFS-based TOPSIS method in the context of sustainable supplier selection, comparing their results with classical TOPSIS and intuitionistic TOPSIS. Their findings highlighted the superiority of the q-ROFS-based TOPSIS results over other methods and recommended further comparisons with additional multi-criteria decision-making (MCDM) methods.

Upon examining the literature related to forest fires, helicopter selection problems, and qROFS applications, the following conclusions can be drawn:

1. Forest fire combat studies are primarily limited to the diagnosis and detection stages.
2. Most detection and diagnosis studies focus on the initial phase of fire management. While there are analyses involving decision-making and optimization, such as the placement of water wells and UAV locations, there remains a significant gap in planning studies concerning the specialized equipment required. In this context, helicopters are emphasized in Turkey's efforts to purchase, lease, and strengthen reserve power sources within its firefighting inventory.
3. A review of helicopter selection studies reveals that they are conducted with a specific focus on the roles assigned to helicopters. This study specifically addresses heavy-class forest fire helicopters, which play a pivotal role in combating forest fires.
4. The literature indicates a predominant reliance on classical MCDM methods. However, expert evaluations are crucial for such specialized tasks. Therefore, a solution proposal utilizing FMCDM has been presented for the relevant problem.
5. The q-ROFS approach is selected as a fuzzy set methodology because it effectively addresses the limitations inherent in IFS and PFS, providing robust support for decision makers in complex, uncertain, and contradictory situations.
6. To assess the robustness of the solution proposed by q-ROFS, the results are validated using the TOPSIS method, which has been applied to real-world problem areas in the literature. This approach allows for more efficient and adaptable modeling of higher degrees of uncertainty.

Given these considerations, the evaluation of heavy-class fire helicopters as an effective resource in combating forest fires in a previously unexplored problem area enhances the value of this study. This research may also provide insights for the country to reassess its policies regarding the purchasing, leasing, and reserve capabilities of firefighting resources.

10.3 PROBLEM DESCRIPTION AND DETERMINING THE DECISION CRITERIA AND ALTERNATIVES

In Turkey, the General Directorate of Forestry publishes official statistics and data on forest fires annually. Given the frequency of these fires and the substantial loss of hectares of land they cause, it is imperative to enhance the measures currently in place. One of the strategic initiatives being implemented focuses on strengthening inventory capabilities. A critical aspect of this initiative involves the employment of qualified personnel alongside the use of land and aerial vehicles. Developments related to fire control aircraft and water dispensers are integral to these efforts.

In recent years, there has been a notable increase in leasing agreements for forest fire helicopters, accompanied by significant budget allocations for purchases. However, such long-term processes may not be sufficiently rapid to address forest fires, which can escalate to critical levels almost instantaneously. Consequently, deficiencies are mitigated by requesting helicopter reserves from military and civilian institutions.

In this study, three different types of heavy-class forest fire helicopters that are utilized for leasing, purchasing, and reserve purposes, each playing an active role in firefighting, have been identified as alternatives. Heavy-class helicopters from the inventories of various countries are excluded from the analysis. The reasons for this exclusion include the lack of necessary licenses among pilots due to brand-specific differences, the expertise of pilots being limited to actively used helicopters, and the absence of different alternative helicopter types in the recently shared leasing agreements.

Details regarding the alternative heavy-class helicopters are provided below.

A_1:*CH-47 (Chinook)*: The Chinook is a heavy-lift helicopter manufactured by Boeing, distinguished by its twin engines and tandem rotor configuration, setting it apart from other helicopter types. Originally designed for military applications, its primary purpose is to lift heavy loads; however, it is also utilized for troop transport and search and rescue operations. Despite its military origins, the Chinook possesses several features that make it particularly suitable for aerial firefighting. Key attributes include its capability to navigate challenging terrain, built-in system redundancy, high cruising speed, extended range, and the flexibility to carry water both internally and externally. Its most notable advantage is its exceptional load capacity, enabling it to transport more water than any other rotorcraft during firefighting operations.

A_2:*Mi-26 (Halo)*: The Mi-26 is a twin-engine heavy transport helicopter designed and manufactured in Russia for both military and civilian applications. Entering production in 1980, it is notable for being the only helicopter in the world equipped with an eight-bladed main rotor. Specifically designed for firefighting, the Mi-26TP can dispense a substantial volume of water using its VSU-15 bucket, which was first introduced in 1994. It is widely recognized as one of the foremost heavy-class fire helicopters globally.

A_3:*Sikorsky S-64 (Skycrane)*: The Sikorsky S-64 Skycrane is an American twin-engine heavy-lift helicopter designed for various applications, including firefighting. Equipped with specialized tank and water intake systems, the helicopter can effectively intervene in forest fires, boasting a remarkably high water capacity. It features two Pratt & Whitney JFTD12-4A engines, each capable of producing 4,500 horsepower, enabling the helicopter to achieve speeds exceeding 200 km/h and a range surpassing 370 km.

To evaluate three different types of helicopters, a literature review is conducted to determine the requisite criteria. Expert opinions are also consulted to incorporate rental costs and water-carrying capacity as criteria specific to forest fire helicopters. Additionally, the aircraft selection problem addressed in studies by Simic and Sabic (2022) and Sánchez-Lozano et al. (2022) has provided valuable insights for criteria determination. Consequently, three main criteria are identified: cost (Simic and Sabic, 2022; Sánchez-Lozano et al., 2022), general characteristics (Qian and Wang, 2018), and performance (Üsküdar et al., 2019; Costa et al., 2020; Ardil, 2022). These main criteria encompass a total of seven subcriteria, which include flight and rental costs for the first criterion, technical specifications and water-carrying capacity for the second, and maximum speed, cruise speed, and range for the third. The first two criteria are cost related, whereas the remaining criteria are benefit oriented. This study contributes to the literature by incorporating two cost criteria that are commonly used in aircraft selection studies but have not been considered in helicopter selection analyses. Furthermore, since this is the inaugural instance of conducting a fire aircraft selection, the inclusion of water-carrying capacity in helicopter selection problems also represents a novel contribution. Additionally, the incorporation of subcriteria under the general characteristics and performance main criteria has enhanced the comprehensiveness of the selection problem, which has traditionally been approached with a limited number of main criteria in the existing literature.

10.4 q-RUNG ORTHOPAIR FUZZY SETS

Fuzzy set theory is a mathematical approach that addresses concepts of uncertainty and degrees of membership. While traditional sets distinctly determine whether an element belongs to the set, fuzzy sets allow each element to be defined by a certain degree of membership (Zadeh, 1965). This flexibility makes fuzzy set theory applicable in various fields, such as modeling complex systems, decision-making processes, and artificial intelligence, due to its capability to better represent uncertainties present in real-world situations.

q-ROFSs were introduced by Yager (2016) as generalized fuzzy sets, extending intuitionistic fuzzy sets (IFSs) introduced by Atanassov (1986) and Pythagorean fuzzy sets (PFSs) developed by Yager and Abbasov (2013). In addition to the two degrees present in IFSs and PFSs, q-ROFS incorporates a degree of hesitation. Like all fuzzy sets, q-ROFS serves as a general structure for modeling uncertainty and complexity. While traditional orthopair fuzzy sets rely on two fundamental components, q-ROFS offers a more intricate structure that accommodates multiple levels of evaluation. This structure comprises two values: the degree to which an object or alternative possesses a certain property (degree of membership) and the degree to which it lacks that property (degree of nonmembership). The q-rung structure facilitates finer-grained evaluations, thereby reflecting levels of uncertainty and ambiguity with greater accuracy.

q-ROFS is extensively utilized in decision-making processes and MCDM problems (Dutta et al., 2024; Zeb et al., 2024; Revalde et al., 2024). These sets provide the flexibility needed to assess complex situations and assist in comparing various alternatives and criteria (Rani and Kumar, 2024; Shahzadi et al., 2024). Furthermore, unlike traditional fuzzy sets, q-ROFS delivers richer information, enabling decision makers to conduct more comprehensive analyses (Kumar and Chen, 2024; Zhang and Wang, 2024). For this reason, q-ROFS has emerged as a significant tool for generating more effective and efficient solutions in uncertain contexts and has been employed in this study. The necessary mathematical definitions for q-ROFS are provided below.

Definition 4.1: Let X be a nonempty set. A q-ROFS $\hat{Q}$ in X is defined by Yager (2016):

$$\hat{Q} = \left\{\left(x, a_{\hat{Q}}(x), \beta_{\hat{Q}}(x)\right) \mid x \in X\right\} \tag{10.1}$$

where $a_{\hat{Q}}, \beta_{\hat{Q}}: \rightarrow [0,1]$ represent the degree of membership and nonmembership of the element $x \in X$ to the set $\hat{Q}$, respectively. q-ROFS in X must satisfy the condition as

$$0 = \left\{\left(a_{\hat{Q}}(x), \beta_{\hat{Q}}(x)\right) \mid x \in X\right\} \tag{10.2}$$

The degree of hesitancy is defined as

$$\gamma_{\hat{Q}}(x) = \sqrt[q]{1 - \left(a_{\hat{Q}}(x)\right)^q, \left(\beta_{\hat{Q}}(x)\right)^q} \tag{10.3}$$

where $\gamma_{\hat{Q}}(x)$ denotes the degree of hesitancy. For simplicity, $\hat{S} = (\alpha, \beta)$ is called a q-rung orthopair fuzzy number (q-ROFN).

Definition 4.2: Let $\hat{S}_1 = \alpha_{\widehat{S_1}}, \beta_{\widehat{S_1}}$ and $\hat{S}_2 = \alpha_{\widehat{S_2}}, \beta_{\widehat{S_2}}$ be two q-ROFNs, and their operations are defined as follows (Deveci et al., 2022):

$$\hat{S}_1 \cap \hat{S}_2 = \left(\min\left\{\alpha_{\widehat{S_1}}, \alpha_{\widehat{S_2}}\right\}, \max\left\{\beta_{\widehat{S_1}}, \beta_{\widehat{S_2}}\right\}\right) \tag{10.4}$$

$$\hat{S}_1 \cup \hat{S}_2 = \left(\max\left\{\alpha_{\widehat{S_1}}, \alpha_{\widehat{S_2}}\right\}, \min\left\{\beta_{\widehat{S_1}}, \beta_{\widehat{S_2}}\right\}\right) \tag{10.5}$$

$$\hat{S}_1 \oplus \hat{S}_2 = \left(\sqrt{q a^q_{\widehat{S_1}} + a^q_{\widehat{S_2}} - a^q_{\widehat{S_1}} a^q_{\widehat{S_2}}}, \beta^q_{\widehat{S_1}} \beta^q_{\widehat{S_2}}\right) \tag{10.6}$$

$$\hat{S}_1 \otimes \hat{S}_2 = a^{\widehat{S_1}} a^{\widehat{S_2}} \sqrt[q]{\beta^q_{\widehat{S_1}} + \beta^q_{\widehat{S_2}} - \beta^q_{\widehat{S_1}} \beta^q_{\widehat{S_2}}} \tag{10.7}$$

$$\omega \hat{S}_2 = \left(\sqrt{q 1 - \left(1 - a^q_{\hat{S}}\right)^{\omega}}, \beta^{\omega}_{\hat{S}}\right), \omega > 0 \tag{10.8}$$

$$\hat{S}^{\omega} = \left(\beta^{\omega}_{\hat{S}} \sqrt[q]{1 - \left(1 - \beta^q_{\hat{S}}\right)^{\omega}}\right), \omega > 0 \tag{10.9}$$

Definition 4.3: Let $\hat{S} = (\alpha, \beta)$ be a q-ROFN, then the score function ($\hat{S}$) and accuracy function ($\hat{S}$) of $\hat{S}$ can be expressed as in Equations (10.10) (Wei et al., 2018) and (10.11) (Liu and Wang, 2018), respectively:

$$A\left(\hat{S}\right) = \frac{1}{2}\left(1 + a^q_{\hat{S}} - \beta^q_{\hat{S}}\right) \tag{10.10}$$

$$A\left(\hat{S}\right) = a^q_{\hat{S}} + \beta^q_{\hat{S}} \tag{10.11}$$

Definition 4.4: Let $\hat{S}_i = \left(\alpha_{\widehat{S_i}}, \beta_{\widehat{S_i}}\right)(i = 1, 2, \cdots, e)$ be a set of q-ROFNs, and $w = (w_1, w_2, \ldots, w_e)^{T}$ be weight vector of $\hat{S}_i$ with $\sum_{i=1}^{e} w_i = 1$. A q-rung orthopair fuzzy weighted average (q − ROFWA) operator can be defined as follows (Deveci et al., 2022):

$$q - ROFWA\left(\hat{S}_1, \hat{S}_2, \cdots, \hat{S}_e\right) = \left(\left(1 - \prod_{i=1}^{e}\left(1 - a^q_{\hat{S}_1}\right)^{wi}\right)^{\frac{1}{q}}, \prod_{i=1}^{e} \beta^{w_i}_{\widehat{S_i}}\right) \tag{10.12}$$

Definition 4.5: Let $\hat{S}_i = \left(\alpha_{\widehat{S_i}}, \beta_{\widehat{S_i}}\right)(i = 1, 2, \cdots, e)$ be a set of q-ROFNs.

And $w = (w_1, w_2, \ldots, w_e)^T$ be weight vector of SC_* with $\sum_{i=1}^{e} w_i = 1$. A q-rung orthopair fuzzy weighted geometric (q − ROFWG) operator can be defined as follows (Deveci et al., 2022):

$$q - ROFWG\left(\hat{S}_1, \hat{S}_2, \cdots, \hat{S}_e\right) = \left(\prod_{i=1}^{e} a^{w_i}_{\widehat{S_i}}, \left(1 - \prod_{i=1}^{e}\left(1 - \beta^q_{\widehat{S_i}}\right)^{w_i}\right)^{\frac{1}{q}}\right) \tag{10.13}$$

Definition 4.6: Let $\hat{S}_1 = \alpha_{\widehat{S_1}}, \beta_{\hat{1}}$ and $\hat{S}_2 = \alpha_{\widehat{S_2}}, \beta_{\widehat{S_2}}$ be the q-ROFNs. Then the distance measure between ̂S1 and ̂S2 can be defined as follows (Liu and Liu, 2019):

$$D\left(\hat{S}_1, \hat{S}_2\right) = \frac{1}{2}\left(\left|a^q_{\widehat{S_1}} - a^q_{\widehat{S_2}}\right| + \left|\beta^q_{\widehat{S_1}} - \beta^q_{\widehat{S_2}}\right| + \left|\gamma^q_{\widehat{S_1}} - \gamma^q_{\widehat{S_2}}\right|\right) \tag{10.14}$$

In this study, the q-ROF methodology is integrated with the TOPSIS. The TOPSIS method is one of the MCDM approaches utilized for ranking alternatives in decision problems, first developed by Yoon and Hwang (1995). As with all decision-making processes, the identification of necessary criteria is conducted with meticulous attention. Alternatives are evaluated by following the prescribed steps of the method. The objective of TOPSIS is to select the alternative that is closest to the positive ideal solution and farthest from the negative ideal solution. The positive ideal solution represents the option that maximizes the benefit criterion, while the negative ideal solution corresponds to the option that minimizes it. TOPSIS reveals suitable solutions by presenting both positive and negative ideal solutions. This method, known for its ease of use, can be applied across various fields due to its clarity in evaluating and interpreting results. The steps of the TOPSIS method are explained below.

Step 1. The decision matrix A is constructed by determining the values associated with each criterion. The evaluated alternatives are represented in the rows, while the criteria are displayed in the columns.

$$A_{ij} = \begin{bmatrix} a_{11} & a_{12} & \cdots & a_{1n} \\ a_{21} & a_{22} & & a_{2n} \\ \vdots & \vdots & \ddots & \vdots \\ a_{m1} & a_{m2} & \cdots & a_{mn} \end{bmatrix} \tag{10.15}$$

Step 2. The normalized decision matrix R is created by applying vector normalization techniques.

$$r_{ij} = \frac{a_{ij}}{\sqrt{\sum_{i=1}^{m} a_{ij}^2}} \tag{10.16}$$

$$R_{ij} = \begin{bmatrix} r_{11} & r_{12} & \cdots & r_{1n} \\ r_{21} & r_{22} & & r_{2n} \\ \vdots & \vdots & \ddots & \vdots \\ r_{m1} & r_{m2} & \cdots & r_{mn} \end{bmatrix} \tag{10.17}$$

Step 3. If criterion weights (w_i) have been established for the problem at hand, they are multiplied by the normalized decision matrix R to obtain the weighted normalized decision matrix V.

$$V_{ij} = \begin{bmatrix} w_1 r_{11} & w_2 r_{12} & \cdots & w_n r_{1n} \\ w_1 r_{21} & w_2 r_{22} & & w_n r_{2n} \\ \vdots & \vdots & \ddots & \vdots \\ w_1 r_{m1} & w_2 r_{m2} & \cdots & w_n r_{mn} \end{bmatrix} \tag{10.18}$$

Step 4. The ideal positive solution (A^*) and the ideal negative solution (A^-) are determined for all alternatives.

$$A^* = \left\{\left(\max_i V_{ij}\right) j \in J\right\}, \left\{\left(\min_i V_{ij}\right) j \in J'\right\}, A^- = \left\{\left(\min_i V_{ij}\right) j \in J\right\}, \left\{\left(\max_i V_{ij}\right) j \in J'\right\} \tag{10.19}$$

Step 5. The distances for each alternative are calculated, yielding the Ideal Positive Discrimination (S_i^*) and Ideal Negative Discrimination (S_i^-) measures.

$$S_i^* = \sqrt{\sum_{j=1}^{n} \left(v_{ij} - v_i^*\right)^2}, S_i^- = \sqrt{\sum_{j=1}^{n} \left(v_{ij} - v_i^-\right)^2} \tag{10.20}$$

Step 6. The ranking process is conducted by calculating the relative distances (C_i^*) of the evaluated alternatives.

$$C_i^* = \frac{S_i^-}{S_i^- + S_i^*} \tag{10.21}$$

The C_i^* value ranges from 0 to 1 ($0 \leq C_i^* \leq 1$). A value closer to 0 indicates a greater distance from the ideal solution, while a value closer to 1 signifies proximity to the ideal solution.

10.5 APPLICATION OF q-ROF-BASED TOPSIS METHOD FOR HEAVY-CLASS HELICOPTER SELECTION

In this section, q-ROF-based TOPSIS is applied to the heavy-class forest fire helicopter selection problem. In this study, the decision-making methodology used for the first time for helicopter selection is applied to rank 3 types of heavy class helicopters based on three main and seven subcriteria. For this purpose, the aim of working with experienced helicopter pilots and also technical experts in forest fire is to define the decision criteria and alternative helicopters, and then the calculations are made step by step. Four experts who are involved in determining the alternatives evaluated the helicopters according to the decision criteria using the language scale proposed by Pinar and Boran (2020) and are given in Table 10.3.

Then, the criteria are prioritized by four experts using the language scale given in the Bozanic et al. (2022) study, and the weight coefficients of the criteria are calculated. Linguistic evaluations for the criteria are given in Table 10.4.

TABLE 10.3
Linguistic assessments for helicopters

		Criteria						
Alternatives	**Expert**	C_1	C_2	C_3	C_4	C_5	C_6	C_7
	E_1	EH	H	MH	MH	M	MH	H
	E_2	H	H	M	M	M	MH	H
	E_3	H	VH	ML	M	M	MH	H
A_1	E_4	VH	MH	ML	MH	MH	VH	VH
	E_1	H	VH	MH	MH	M	MH	H
	E_2	VH	VH	M	M	M	MH	VH
	E_3	MH	VH	M	M	ML	M	MH
A_2	E_4	M	MH	M	MH	M	MH	H
	E_1	EH	H	M	ML	ML	MH	MH
	E_2	H	VH	ML	ML	ML	ML	MH
	E_3	VH	VH	ML	L	ML	ML	MH
A_3	E_4	VH	MH	ML	ML	M	L	MH

TABLE 10.4
Linguistic assessments for the decision criteria

	Expert			
Criteria	**Expert 1**	**Expert 2**	**Expert 3**	**Expert 4**
MC_1: cost	AH	EH	MH	AH
MC_2: general characteristics	H	MH	E	H
MC_3: performance	EH	EH	H	AH
MC_1: cost				
C_1	AH	AH	EH	AH
C_2	AH	EH	AH	H
MC_2: general characteristics				
C_3	E	MH	H	EH
C_4	AH	AH	EH	EH
MC_3: performance				
C_5	MH	H	E	H
C_6	H	EH	MH	EH
C_7	AH	EH	H	AH

As a result of the calculations, the relationship vector is obtained. The relationship vector data are analyzed, and the weight coefficients of the criteria are found, and the total vector of these coefficients is calculated and presented in Table 10.5. All these steps are performed using equation 2731 given in the study by Deveci et al. (2022).

The three main criteria considered for helicopter selection, along with the scores of the corresponding subcriteria, are presented in Table 10.5. When the main criteria are taken into consideration, the criteria priorities are performance, cost, and general characteristics, respectively. There is a minimal difference between performance and cost criteria. However, when the criteria are detailed, the situation becomes different. Upon examining the subcriteria in Table 10.5, it is evident that the cost criteria have the highest weights, followed by criteria related to general characteristics. The performance criteria rank lowest. This prioritization stems from the need for cost-effective interventions in forest fires, adhering to a safety-first principle. Experts emphasize the importance of selecting helicopters that offer suitable costs, minimum flight time, and compliance with technical specifications.

Following the cost criteria, the next most critical factor is water-carrying capacity, as this is a key feature of heavy-class forest fire helicopters, directly influencing their ability to extinguish fires. The technical competence of the helicopters ranks fourth; this criterion, which encompasses all technical features, should not be overlooked. Regarding performance criteria, the range for intervening in a fire is prioritized, followed by speed-related criteria. All these results show the importance of detailing the criteria.

The decision matrix is constructed by aggregating evaluations from four experts for the alternatives listed in Table 10.3. A weighted normalized matrix is then generated by normalizing the decision matrix and applying the weights calculated in Table 10.5.

The score values of alternatives are found by Equations (10.15)–(10.19) using weighted normalized matrix and are presented in Table 10.6.

TABLE 10.5
Values of criterion weights

Criteria	Local	Global	Crisp values
MC_1: cost	(0.258, 0.344, 0.457)	—	
C_1	(0.395, 0.506, 0.647)	(0.102, 0.174, 0.296)	0.1781
C_2	(0.383, 0.494, 0.640)	(0.099, 0.170, 0.292)	0.1740
MC_2: general characteristics	(0.225, 0.310, 0.430)	—	
C_3	(0.343, 0.471, 0.648)	(0.077, 0.164, 0.300)	0.1514
C_4	(0.402, 0.530, 0.697)	(0.090, 0.170, 0.200)	0.1696
MC_3: performance	(0.260, 0.346, 0.463)	—	
C_5	(0.225, 0.313, 0.435)	(0.058, 0.108, 0.202)	0.1119
C_6	(0.246, 0.334, 0.457)	(0.064, 0.115, 0.212)	0.1193
C_7	(0.264, 0.353, 0.468)	(0.069, 0.122, 0.217)	0.1254

TABLE 10.6
Score matrix

	C_1	C_2	C_3	C_4	C_5	C_6	C_7
A_1	0.1292	0.1040	0.0871	0.1080	0.0712	0.0785	0.0797
A_2	0.1049	0.0969	0.0963	0.1080	0.0644	0.0722	0.0746
A_3	0.1251	0.1004	0.0779	0.0736	0.0576	0.0534	0.0617
A+	0.1049	0.0969	0.0963	0.1080	0.0712	0.0785	0.0797
A-	0.1292	0.1040	0.0779	0.0736	0.0576	0.0534	0.0617

TABLE 10.7
The final score values of alternatives

	S+	S-	C	Rank
A_1	0.0269	0.0490	0.6460	2
A_2	0.0106	0.0522	0.8315	1
A_3	0.0555	0.0054	0.0887	3

The rankings created using Equations (10.20) and (10.21) for the ranking process with TOPSIS are given in Table 10.7. The order of alternatives is A2 > A1 > A3. According to this ranking, it can be concluded that the A2 alternative is the best alternative and the A1 alternative is the worst alternative.

10.6 SENSITIVITY ANALYSIS FOR q LEVELS AND COMPARISON WITH MCDM METHODS

In this section, sensitivity analyses are conducted to verify the robustness of the results of the proposed models. Firstly, the sensitivity analysis of the proposed q-ROF-based TOPSIS model for q levels is given in Figure 10.1.

Secondly, in the proposed q-ROF-based TOPSIS model, sensitivity analysis is conducted to highlight the results when similar procedures are applied to different MCDM methods. The results are compared with the SAW, WASPAS, and MABAC methods from the MCDM methods and are given in Figure 10.2. In each cluster, A2 is found to be the best alternative and A3 is found to be the worst alternative.

When both Figures 10.1 and 10.2 are examined, it can be seen that the ranking order of the best alternatives is A2 > A1 > A3.

10.7 DISCUSSION

This study assesses heavy-class forest fire helicopters, recognized as the most formidable assets in combating the growing threat of forest fires in recent years.

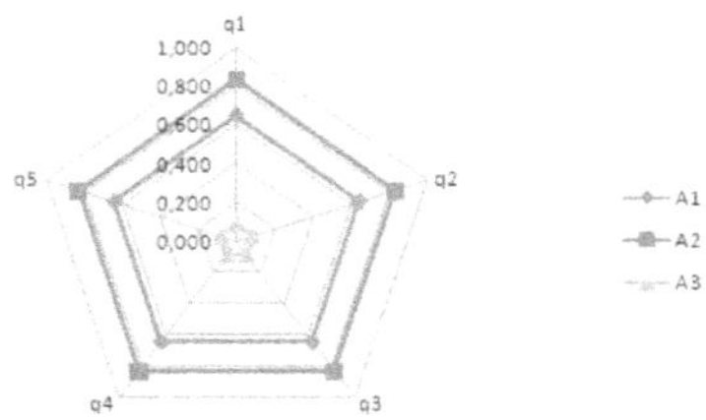

FIGURE 10.1 q-value analysis for q-ROF TOPSIS

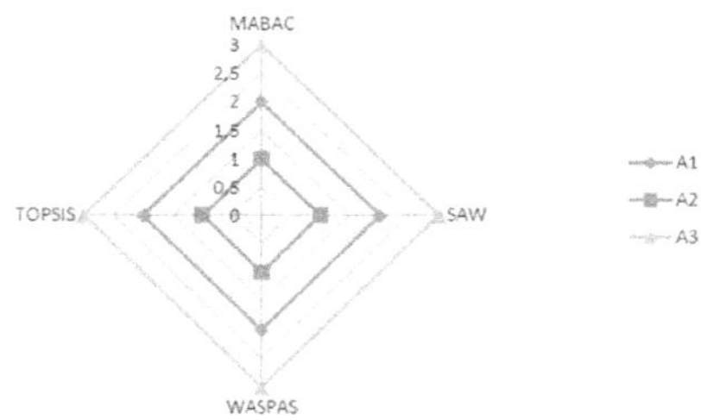

FIGURE 10.2 Comparison among other MCDM techniques

Consequently, the proposed methodology provides valuable insights for countries regarding the rental and reserve capabilities of helicopters. The analysis reveals that the Mi-26 helicopter emerges as the best alternative. The Mi-26 Halo is one of the largest and most powerful transport helicopters globally, distinguished by its load-carrying capacity exceeding 20 tons, spacious cabin area, and powerful engines. These attributes render it exceptionally suited for a range of military and civilian tasks, particularly in firefighting operations. Equipped with large water tanks, Mi-26s can rapidly transport water, access fire sites, and distribute significant quantities of water effectively. Their rapid response capability and substantial carrying capacity make them vital support elements for firefighting teams. Additionally, their ability to operate in challenging weather conditions and rugged terrains enhances their effectiveness in fire situations.

However, it is essential to acknowledge several critical disadvantages associated with this helicopter type. The Mi-26 is in limited inventory across many countries, and issues related to spare parts support can complicate its deployment. Moreover, its considerable size may present challenges in confined operational areas. As a result, the heavy-lift helicopter market is primarily dominated by two U.S. designs: the CH-47 Chinook and the S-64 Skycrane.

Upon examining the study results, the CH-47 Chinook emerges as the second best alternative. Chinook helicopters are twin-engine heavy transport aircraft renowned for their high payload capacity and dual-rotor design. These features enable them to carry over 10 tons of cargo while maintaining stability in challenging weather conditions. Their rapid mobility and spacious cabin make them suitable for a variety of missions, including military operations, humanitarian aid, and logistical support.

In the context of combating forest fires, Chinooks play a significant role. Equipped with large water tanks, they facilitate effective interventions by swiftly transporting water or fire extinguishing agents to the affected areas. Moreover, they provide crucial support in firefighting efforts by enabling the rapid transport of personnel and equipment to fire sites. Consequently, Chinooks are considered reliable and powerful assets in firefighting operations, and their versatility further contributes to their widespread preference. Although their carrying capacity, cabin volume, and performance in adverse weather and rough terrain do not match those of the Mi-26, they are favored over the third alternative, the S-64 Skycrane helicopter.

The final alternative, the S-64 Skycrane, presents several disadvantages. In contrast to the Chinook's modular structure, which offers enhanced flexibility for various missions, the Skycrane design, while effective for specific tasks, lacks adaptability across a broader range of operations. Additionally, the CH-47's twin-rotor configuration enhances flight stability, whereas the S-64 may exhibit less reliability in challenging weather conditions. Designed primarily for military operations, the CH-47 is equipped with numerous integrated systems that bolster its operational capabilities. Although the Skycrane can be employed in military missions, its compatibility with military standards is comparatively limited, which hinders its effectiveness in reserve power and multipurpose roles. Furthermore, the S-64's less widespread deployment complicates maintenance and the supply of spare parts, potentially increasing operational costs. These factors contribute to the S-64 Skycrane being less favored than the CH-47 Chinook and Mi-26 Halo.

10.8 CONCLUSIONS

The military rotorcraft sector has traditionally been divided into three primary roles: scout, lift, and attack. This categorization encompasses platforms such as dedicated attack helicopters (AHs), light utility helicopters (LUHs), medium support helicopters (MSHs), and heavy-lift machines. In this study, the focus is on heavy-class helicopters, which are crucial for critical operational tasks. Heavy-class fire helicopters intended for forest fire suppression were evaluated using the q-ROFS-based TOPSIS method. To test the model and validate the robustness of the results, comparisons were made with three different MCDM methods, as suggested by Pınar et al. (2021).

In this study, the advantages of q-ROFS, which encompasses significantly more data to convey information pertinent to the critical issue of helicopter selection, were leveraged. Moreover, the integration of q-ROFS with the widely used and straightforward calculation method known as TOPSIS facilitated convenience in the decision-making process. This integration addresses the challenges associated with multistage and fuzzy decision-making problems, particularly the limitations posed by the scarcity of experts and the lack of mathematical infrastructure. The expert opinions utilized in this study, along with the number of heavy-class helicopter alternatives and the variations in criteria employed in forest fire vehicle selection as reported in the literature, are acknowledged as limitations.

Future research could explore the combination of q-ROFS with novel MCDM methodologies that have yet to be applied, thereby paving the way for the development of new hybrid approaches. Additionally, the problem could be adapted to include medium-class helicopters, which represent a more unpredictable segment of the market. The TOPSIS method, when integrated with q-ROFS and validated for robustness in this study, can be employed in selection problems that have not been previously addressed, particularly in military contexts, thereby supporting the attainment of robust results in strategic-level decision-making processes.

REFERENCES

Abdo, H. G., Almohamad, H., Al Dughairi, A. A., & Al-Mutiry, M. (2022). GIS-based frequency ratio and analytic hierarchy process for forest fire susceptibility mapping in the western region of Syria. *Sustainability*, 14(8), 4668.

Akay, A., & Aziz, B. (2015, July). GIS-Based forest road network model for forest protection purposes. In *38th Annual COFE Meeting. Engineering Solutions for Non-Industrial Private Forest Operations* (pp. 266–281).

Akay, A. E., Wing, M. G., Sivrikaya, F., & Sakar, D. (2012). A GIS-based decision support system for determining the shortest and safest route to forest fires: A case study in Mediterranean Region of Turkey. *Environmental Monitoring and Assessment*, 184, 1391–1407.

Ardil, C. (2022). Military attack helicopter selection using distance function measures in multiple criteria decision making analysis. *International Journal of Aerospace and Mechanical Engineering*, 16(2), 15–22.

de Assis, G. S., dos Santos, M., & Basilio, M. P. (2023). Use of the WASPAS method to select suitable helicopters for aerial activity carried out by the military police of the state of Rio de Janeiro. *Axioms*, 12(1), 77.

Atanassov, K. T. (1986). Intuitionistic fuzzy sets. *Fuzzy Sets and Systems*, 20, 87–96. https://doi.org/10.1016/S0165-0114(86)80034-3.

Bal, M., Ozpeynirci, O., & Yurt, O. (2019). International cooperation on airborne forest firefighting: Opportunities for Turkey and Greece. *International Journal of Ecology & Development*, 34(1), 61–74.

Bouguettaya, A., Zarzour, H., Taberkit, A. M., & Kechida, A. (2022). A review on early wildfire detection from unmanned aerial vehicles using deep learning-based computer vision algorithms. *Signal Processing*, 190, 108309.

Božanić, D., Pamučar, D., Milić, A., Marinković, D., & Komazec, N. (2022). Modification of the logarithm methodology of additive weights (LMAW) by a triangular fuzzy number and its application in multi-criteria decision making. *Axioms*, 11(3), 89.

Caetano, J. V., & Morgado, J. P. (2019). Portuguese air force UAV development, certification and operations in maritime environment. Unmanned air vehicles: Technological challenges, concepts of operations and regulatory issues. https://publications.sto.nato.int/publications/STO%20Educational%20Notes/STO-EN-AVT-274/EN-AVT-274-11.pdf.

Cao, Y., Wu, X., Ding, L., & Wang, W. (2024). A hybrid framework for selecting food waste treatment techniques using q-Rung orthopair fuzzy CRITIC-Generalized TODIM method. *International Journal of Fuzzy Systems*, 1–20.

Chaturvedi, S., Khanna, P., & Ojha, A. (2022). A survey on vision-based outdoor smoke detection techniques for environmental safety. *ISPRS Journal of Photogrammetry and Remote Sensing*, 185, 158–187.

Cheng, C. H., Yang, K. L., & Hwang, C. L. (1999). Evaluating attack helicopters by AHP based on linguistic variable weight. *European Journal of Operational Research*, 116(2), 423–435.

Costa, I., Junior, M., Maêda, S., Fávero, L., Costa, A., Corriça, J., Gomes, C. F. e Dos Santos, M. (2020). Application of the AHP-TOPSIS-2N hybrid method for selection of an attack helicopter to be acquired by the Brazilian Navy. In *International Symposium on the Analytic Hierarchy Process (ISAHP 2020)*. doi: 10.13033/isahp.y2020.018.

Dağıstanlı, H. A., & Gencer, C. (2023). Hibrit Tehdit Perspektifinden Orman Yangınları ve Türkiye'nin Mücadele Politikası. *SAVSAD Savunma ve Savaş Araştırmaları Dergisi*, (1), 35–70.

Dhall, A., Dhasade, A., Nalwade, A., VK, M. R., & Kulkarni, V. (2020). A survey on systematic approaches in managing forest fires. *Applied Geography*, 121, 102266.

Deveci, M., Gokasar, I., & Brito-Parada, P. R. (2022). A comprehensive model for socially responsible rehabilitation of mining sites using Q-rung orthopair fuzzy sets and combinative distance-based assessment. *Expert Systems with Applications*, 200, 117155.

Dutta, P., Saikia, B., & Banik, A. K. (2024). Novel distance measure for q-rung orthopair fuzzy sets with application to transportation problem. *International Journal of Information Technology*, 1–18.

Enoh, M. A., Okeke, U. C., & Narinua, N. Y. (2021). Identification and modelling of forest fire severity and risk zones in the Cross–Niger transition forest with remotely sensed satellite data. *The Egyptian Journal of Remote Sensing and Space Science*, 24(3), 879–887.

Erdebilli, B., Gecer, E., Yılmaz, İ., Aksoy, T., Hacıoglu, U., Dinçer, H., & Yüksel, S. (2023). Q-ROF fuzzy TOPSIS and VIKOR methods for the selection of sustainable private health insurance policies. *Sustainability*, 15(12), 9229.

Erdebilli, B., & Sıcakyüz, Ç. (2024). An Integrated Q-Rung Orthopair Fuzzy (Q-ROF) for the selection of supply-chain management. *Sustainability*, 16(12), 4901.

Ghamry, K. A., Kamel, M. A., & Zhang, Y. (2017, June). Multiple UAVs in forest fire fighting mission using particle swarm optimization. In *2017 International Conference on Unmanned Aircraft Systems (ICUAS)* (pp. 1404–1409). IEEE.

Ghoushchi, S. J., Vahabzadeh, S., & Pamucar, D. (2024). Applying hesitant q-rung orthopair fuzzy sets to evaluate uncertainty in subsidence causes factors. *Heliyon*, 10(8), 1–15.

Gunaltili, E., Ekici, S., Kalkan, A., Gocmen, F. E., Kale, U., Yilmazoglu, Z., & Karakoc, T. H. (2023). Conceptual design and optimization of a sustainable and environmentally friendly archetypal helicopter within the selection criteria and limitations. *Heliyon*, 9(6), 1–28.

Jain, P., Coogan, S. C., Subramanian, S. G., Crowley, M., Taylor, S., & Flannigan, M. D. (2020). A review of machine learning applications in wildfire science and management. *Environmental Reviews*, 28(4), 478–505.

Javeed, S., Javed, M., Shafique, I., Shoaib, M., Khan, M. S., Cui, L., ... & Alshamrani, A. M. (2024). Complex q-rung orthopair fuzzy Yager aggregation operators and their application to evaluate the best medical manufacturer. *Applied Soft Computing*, 157, 111532.

Kilic, S., & Ozkan, O. (2019, July). A self-adaptive UAV routing for forest fire risk mitigation: A conceptual model. In *Proceedings of the 2019 Summer Simulation Conference* (pp. 1–12).

Kumar, K., & Chen, S. M. (2024). Multiattribute decision making based on qrung orthopair fuzzy Yager prioritized weighted arithmetic aggregation operator of qrung orthopair fuzzy numbers. *Information Sciences*, 657, 119984.

Li, Z., Xing, Y., & Dong, P. (2024). A novel q-rung orthopair fuzzy best-worst method, Shannon entropy and MARCOS method for mobile medical app service quality evaluation. *Applied Soft Computing*, 155, 111417.

Liu, P., & Liu, W. (2019). Multiple-attribute group decision-making method of linguistic q-rung orthopair fuzzy power Muirhead mean operators based on entropy weight. *International Journal of Intelligent Systems*, 34(8), 1755–1794. https://doi.org/10.1002/int.22114.

Liu, P., & Wang, P. (2018). Some q-rung orthopair fuzzy aggregation operators and their applications to multiple-attribute decision making. *International Journal of Intelligent Systems*, 33(2), 259–280. https://doi.org/10.1002/int.21927.

Maêda, S. M. D. N., Costa, I. P. D. A., Castro Junior, M. A. P. D., Fávero, L. P., Costa, A. P. D. A., Corriça, J. V. D. P., ... & Santos, M. D. (2021). Multi-criteria analysis applied to aircraft selection by Brazilian Navy. *Production*, 31, e20210011.

Maraş, E. E., Dönmez, K., & Emecen, Y. (2023). GIS-based determination of the optimal heliport and water source locations for forest fire suppression using multi-objective programming. *Aerospace*, 10(3), 305.

Moreira, M. Â. L., de Araújo Costa, I. P., Pereira, M. T., Dos Santos, M., Gomes, C. F. S., & Muradas, F. M. (2021). PROMETHEE-SAPEVO-M1 a Hybrid approach based on ordinal and cardinal inputs: Multi-Criteria evaluation of helicopters to support Brazilian navy operations. *Algorithms*, 14(5), 140.

Naderpour, M., Rizeei, H. M., Khakzad, N., & Pradhan, B. (2019). Forest fire induced Natech risk assessment: A survey of geospatial technologies. *Reliability Engineering & System Safety*, 191, 106558.

Naz, S., Saeed, M. R., & Butt, S. A. (2024). Multi-attribute group decision-making based on 2-tuple linguistic cubic q-rung orthopair fuzzy DEMATEL analysis. *Granular Computing*, 9(1), 12.

Ozkan, O. (2021). Optimization of the distance-constrained multi-based multi-UAV routing problem with simulated annealing and local search-based matheuristic to detect forest fires: The case of Turkey. *Applied Soft Computing*, 113, 108015.

Ozkan, O., & Kilic, S. (2023). UAV routing by simulation-based optimization approaches for forest fire risk mitigation. *Annals of Operations Research*, 320(2), 937973.

Pappis, C. P., & Rachaniotis, N. P. (2010). Scheduling a single fire fighting resource with deteriorating fire suppression times and set-up times. *Operational Research*, 10, 27–42.

Pinar, A., & Boran, F. E. (2020). A q-rung orthopair fuzzy multi-criteria group decision making method for supplier selection based on a novel distance measure. *International Journal of Machine Learning and Cybernetics*, 11, 1749–1780.

Pınar, A., Babak Daneshvar, R., & Özdemir, Y. S. (2021). q-Rung orthopair fuzzy TOPSIS method for green supplier selection problem. *Sustainability*, 13(2), 985.

Qian, Y., & Wang, H. (2018, May). Training helicopter selection based on Analytic Hierarchy Process. In *2018 7th International Conference on Energy, Environment and Sustainable Development (ICEESD 2018)* (pp. 622–625). Atlantis Press.

Rani, V., & Kumar, S. (2024). An innovative distance measure for quantifying the dissimilarity between Q-Rung orthopair fuzzy sets. *Decision Analytics Journal*, 11, 100440.

Revalde, H., Almerino Jr, P., Rodriguez, M., Hortelano, P., Revalde, M., Chan, C., ... & Ocampo, L. (2024). A q-rung orthopair fuzzy set extension of the DEMATEL and its application in the education sector. *Advances in Fuzzy Systems*, 2024(1), 5593966.

Rodrigues, M. V. G., dos Santos, M., & Gomes, C. F. S. (2024). Selection of helicopters for offshore service using three multi-criteria decision analysis methods: AHP-TOPSIS-2N, THOR 2 and Gaussian AHP-TOPSIS-2N. *Journal of Control and Decision*, 12(3), 434–448.

Sakellariou, S., Tampekis, S., Samara, F., Sfougaris, A., & Christopoulou, O. (2017). Review of state-of-the-art decision support systems (DSSs) for prevention and suppression of forest fires. *Journal of Forestry Research*, 28(6), 1107–1117.

Sánchez-Lozano, J. M., Correa-Rubio, J. C., & Fernández-Martínez, M. (2022). A double fuzzy multi-criteria analysis to evaluate international high-performance aircrafts for defense purposes. *Engineering Applications of Artificial Intelligence*, 115, 105339.

Shahzadi, G., Siddique, S., Shehzadi, H., & Deveci, M. (2024). Multi-attribute decision-making using (p, q)-rung orthopair fuzzy Hamacher interactive aggregation operators. *Granular Computing*, 9(3), 1–20.

Simic, E., & Sabic, M. (2022). Aircraft selection for aerial firefighting operations. In *Conference: 20th International Conference on Transport Science ICTS 2022 At: Slovene Association of Transport Sciences*. Ljubljana, Slovenia: University of Ljubljana, Faculty of Maritime Studies and Transport.

Sivrikaya, F., & Küçük, Ö. (2022). Modeling forest fire risk based on GIS-based analytical hierarchy process and statistical analysis in Mediterranean region. *Ecological Informatics*, 68, 101537.

Temiz, N., & Tecim, V. (2009). The use of GIS and multi-criteria decision-making as a decision tool in forestry. *OR Insight*, 22(2), 105–123.

Üsküdar, A., Türkan, Y. S., Özdemir, Y. S., & Öz, A. H. (2019, March). Fuzzy AHP-center of gravity method helicopter selection and application. In *2019 8th International Conference on Industrial Technology and Management (ICITM)* (pp. 170174). IEEE.

Wei, G., Gao, H., & Wei, Y. (2018). Some q-rung orthopair fuzzy Heronian mean operators in multiple attribute decision making. *International Journal of Intelligent Systems*, 33(7), 1426–1458. https://doi.org/10.1002/int.21985.

Yager, R. R. (2016). Generalized orthopair fuzzy sets. *IEEE Transactions on Fuzzy Systems*, 25(5), 1222–1230. https://doi.org/10.1109/TFUZZ.2016.2604005.

Yager, R. R., & Abbasov, A. M. (2013). Pythagorean membership grades, complex numbers, and decision making. *International Journal of Intelligent Systems*, 28(5), 436–452. https://doi.org/10.1002/int.21584.

Yoon, K. P., & Hwang, C. L. (1995). *Multiple Attribute Decision Making: An Introduction*. Sage publications.

Yuan, C., Liu, Z., & Zhang, Y. (2015, June). UAV-based forest fire detection and tracking using image processing techniques. In *2015 International Conference on Unmanned Aircraft Systems (ICUAS)* (pp. 639–643). IEEE.

Zadeh, L. A. (1965). Fuzzy sets. *Information and Control*, 8(3), 338–353. https://doi.org/10.1016/S0019-9958(65)90241-X.

Zeb, A., Ahmad, W., Asif, M., Simic, V., Senapati, T., & Hou, M. (2024). Optimizing decision-making in electric power system selection: A generalized approach based on Hamacher aggregation operators for q-rung orthopair fuzzy soft sets. *Applied Energy*, 367, 123405.

Zhang, D., & Wang, G. (2024). Q-rung orthopair fuzzy decision-making method of multi-source information based on the compression mapping and inverse score function. *Expert Systems with Applications*, 242, 122574.

Zhang, L., Liu, Z., Zhang, Y., & Ai, J. (2018, August). Intelligent path planning and following for UAVs in forest surveillance and fire fighting missions. In *2018 IEEE CSAA Guidance, Navigation and Control Conference (CGNCC)* (pp. 1–6). IEEE.

Zulqarnain, R. M., Siddique, I., Mahboob, A., Ahmad, H., Askar, S., & Gurmani, S. H. (2023). Optimizing construction company selection using Einstein weighted aggregation operators for q-rung orthopair fuzzy hypersoft set. *Scientific Reports*, 13(1), 6511.

11 Some Interval-Valued Intuitionistic Fuzzy Confidence Level-Based Aggregation Operators Using Frank t-Norm and t-Conorm

Prayosi Chatterjee and Mijanur Rahaman Seikh

11.1 INTRODUCTION

"Multi-attribute group decision-making" (MAGDM) is a widely used method for integrating the expertise and insights of multiple decision makers. It aims to identify a collaborative solution to a problem by allowing various decision makers to express their preferences regarding different options and criteria. These strategies can be advantageous across various fields, including business, engineering, healthcare, and many others (Dong et al., 2016). Zadeh (1965) presented the notion of fuzzy sets, which had only membership degrees. Later, Atanassov (1986) introduced "intuitionistic fuzzy sets" (IFSs), where the single membership degree in a fuzzy set is divided into three distinct components: membership and nonmembership degrees (MND) and hesitation degrees. However, due to society's rapid evolution, the IFSs are unable to accurately define uncertain information. Then the concept of "interval-valued intuitionistic fuzzy set (IVIFS)" was subsequently put forward by Atanassov (1989) as an extension of the IFS that can handle more complicated scenarios.

A crucial step in decision-making problems, where several attributes and alternatives are involved, is aggregation. For aggregating IVIFNs, the most popular method is the usage of different aggregation operators (AOs). Xu (2007) developed the weighted averaging and geometric (WA&G) AOs for IVIFSs. Xu and Jian (2007) defined ordered weighted averaging and hybrid averaging AOs for IVIFSs. Later, they developed some geometric AOs for IVIFSs (Xu & Chen, 2007). Wang and Liu (2013a) defined several Einstein norm-based averaging AOs for IVIFNs. Similarly,

 DOI: 10.1201/9781003529088-11

some geometric AOs using the Einstein norm were defined for IVIFSs by Wang and Liu (2013b).

Frank norms represent an important class of t-norms and t-conorms that satisfies compatibility conditions, offering greater flexibility compared to other t-norms and t-conorms (Liu, 2017). Frank norms have been extended to several fuzzy environments like IVIFSs (Liu, 2017; Zhang, 2017), picture fuzzy numbers (Seikh & Mandal, 2021), interval-valued picture fuzzy numbers (Abolfathi et al., 2023), q-rung orthopair fuzzy numbers (Seikh & Mandal, 2022), and many others to develop AOs. Recently, they have been extended to several advanced fuzzy environments and combined with other norms as well (Petchimuthu et al., 2024; Yang et al., 2023). Aggregation operators that are based on confidence levels are a crucial tool for data fusion and decision-making procedures. These operators expand on traditional aggregation techniques by incorporating the idea of confidence or dependability linked with each input source or piece of information. Decision makers can more easily assess each input's relative relevance while accounting for varying degrees of reliability when they consider the confidence levels assigned to different data sources (Rahim et al., 2024). Confidence level (CL)-based AOs have been developed for various fuzzy sets like IFS (Seikh & Chatterjee, 2024a; Rahman et al., 2021), "intuitionistic fuzzy rough sets" (Mahmood et al., 2023), "Pythagorean fuzzy sets" (Garg, 2017), "interval-valued Fermatean fuzzy sets" (Seikh & Chatterjee, 2024), and many more.

One of the most competitive issues faced by modern businesses is the supplier selection dilemma (Konys, 2019). To increase sustainability, businesses need to improve their processes for managing a chain of green supply that is environmentally friendly and upholds ecological principles and technology. Choosing suppliers that are committed to environmental sustainability is just as important as traditional factors like price, quality, and dependability when it comes to selecting green providers. A complex, multifaceted problem, choosing the best green supplier, is essential to managing the green supply chain. Recently, academics and industry specialists have become more interested in green supplier selection problems (Zhu & Wang, 2024; Nafei et al., 2024; Wang et al., 2013; Zeng et al., 2023; Verma et al., 2022). This chapter solves a green supplier selection MAGDM problem using a novel group decision-making method utilizing newly proposed CL-based Frank AOs in an interval-valued intuitionistic fuzzy (IVIF) environment.

The subsequent gaps in research identified by the literature review motivate the contribution of this chapter:

- The absence of CL-based AOs for IVIFSs drives the need to develop "CL-based IVIF Frank WA&G AOs".
- A new MAGDM procedure is proposed that makes use of the newly developed CL-based AOs for IVIFSs.
- The suggested approach is used to solve a practical MAGDM problem of assessing and selecting green suppliers.

The chapter's novelty lies in introducing CL-based Frank WA&G AOs for IVIFSs. By incorporating decision makers' confidence levels, these operators enhance the robustness and reliability of decision-making under uncertainty. They are used to form a novel group decision-making framework, which is further validated through its application to a green supplier selection problem, demonstrating its practical relevance and effectiveness.

The following is the arrangement of the chapter: Section 11.2 comprises the preliminary information. The recently proposed "IVIF Frank weighted AOs" with confidence levels are presented and discussed in Section 11.3. A novel MAGDM method is proposed using the newly developed AOs in Section 11.4. In Section 11.5, the suggested approach is used to solve a green supplier selection problem. Subsection 11.5.1 provides a comparative study of the suggested approach, and a study of the suggested method's sensitivity is conducted in Subsection 11.5.2. Section 11.6 contains the conclusion, limitations, and perspectives for further studies.

11.2 PRELIMINARIES

A few foundational definitions and opening remarks are discussed in this section. The universal set is referred to in this work as X.

Definition 2.1 (Atanassov, 1989) The IVIFS, I in X is defined as: $I=\{\langle \varrho, ([e(\varrho), f(\varrho)], [s(\varrho), t(\varrho)])\rangle | \varrho \in X\}$, where $[e(\varrho), f(\varrho)] \subseteq [0,1]$ and $[s(\varrho), t(\varrho)] \subseteq [0,1]$ are the MND interval of each element ϱ in the set I with the conditions that $0 \leq f(\varrho)+t(\varrho) \leq 1$, $\forall \varrho \in X$. For any given $\varrho \in X$, the pair $([e,f], [s,t])$ in I is called an IVIFN.

Definition 2.2 (Xu, 2007; Xu & Jian, 2007) The score function $s(\varrho)$ and accuracy degree $h(\varrho)$ of IVIFN $\varrho=([e,f], [s,t])$ are as follows:

$$s(\varrho)=\frac{(e-s+f-t)}{2}$$

$$h(\varrho)=\frac{(e+f+s+t)}{2}. \tag{11.1}$$

Let ϱ_i (i = 1, 2) be any two IVIFNs, $s(\varrho_i)$ (i = 1, 2) and $h(\varrho_i)$ (i = 1, 2) be the score function and accuracy degrees of ϱ_i (i=1, 2), respectively, then:

- If $s(\varrho_1)>s(\varrho_2)$, then $\varrho_1 > \varrho_2$.
- If $s(\varrho_1)=s(\varrho_2)$, then:
  - - if $h(\varrho_1)>h(\varrho_2)$, then $\varrho_1 > \varrho_2$,
  - - if $h(\varrho_1)<(\varrho_2)$, then $\varrho_1 < \varrho_2$, and
  - - if $h(\varrho_1)=(\varrho_2)$, then $\varrho_1 = \varrho_2$.

Definition 2.3 (Klement & Mesiar, 1997) The Frank t-norm $\left(F_t\right)$ and t-conorm $\left(F_c\right)$ are defined for all $\varpi > 1$ by:

$$F_t\left(x,\ y\right) = \log_{\varpi}\left(\ 1 + \frac{\left(\varpi^{x} - 1\right)\left(\varpi^{y} - 1\right)}{\varpi - 1}\right),$$

$$F_c\left(x,\ y\right) = 1 - \log_{\varpi}\left(\ 1 + \frac{\left(\varpi^{1-x} - 1\right)\left(\varpi^{1-x} - 1\right)}{\varpi - 1}\right).$$

Definition 2.4 (Oraki et al., 2021) Assume that $\varrho_1 = \left(\left[e_1, f_1\right], \left[s_1,\ t_1\right]\right)$ and $\varrho_2 = \left(\left[e_2, f_2\right], \left[s_2, t_2\right]\right)$ be any two IVIFNs, and if $\varpi > 1$ and n is any real number greater than 0, then the following are the defined operational rules based on the Frank t-norm and t-conorm:

- $$\varrho_1 \oplus \varrho_2 = \left(\left[1 - \log_{\varpi}(1 + \frac{\left(\varpi^{1-e_1} - 1\right)\left(\varpi^{1-e_2} - 1\right)}{\varpi - 1}), 1 - \log_{\varpi}(1 + \frac{\left(\varpi^{1-f_1} - 1\right)\left(\varpi^{1-f_2} - 1\right)}{\varpi - 1})\right], \left[\log_{\varpi}(1 + \frac{\left(\varpi^{s_1} - 1\right)\left(\varpi^{s_2} - 1\right)}{\varpi - 1}),\ \log_{\varpi}(1 + \frac{\left(\varpi^{t_1} - 1\right)\left(\varpi^{t_2} - 1\right)}{\varpi - 1})\right]\right).$$

- $$\varrho_1 \otimes \varrho_2 = \left(\left[\log_{\varpi}(1 + \frac{\left(\varpi^{e_1} - 1\right)\left(\varpi^{e_2} - 1\right)}{\varpi - 1}),\ \log_{\varpi}(1 + \frac{\left(\varpi^{f_1} - 1\right)\left(\varpi^{f_2} - 1\right)}{\varpi - 1})\right], \left[1 - \log_{\varpi}(1 + \frac{\left(\varpi^{1-s_1} - 1\right)\left(\varpi^{1-s_2} - 1\right)}{\varpi - 1}),\ 1 - \log_{\varpi}(1 + \frac{\left(\varpi^{1-t_1} - 1\right)\left(\varpi^{1-t_2} - 1\right)}{\varpi - 1})\right]\right).$$

- $$\varrho_1^n = \left(\begin{array}{l}\left[\log_{\varpi}\left(\ 1 + \frac{\left(\varpi^{e_1} - 1\right)^n}{\left(\varpi - 1\right)^{n-1}}\right), \log_{\varpi}\left(\ 1 + \frac{\left(\varpi^{f_1} - 1\right)^n}{\left(\varpi - 1\right)^{n-1}}\right)\right], \\ \left[1 - \log_{\varpi}\left(\ 1 + \frac{\left(\varpi^{1-s_1} - 1\right)^n}{\left(\varpi - 1\right)^{n-1}}\right), 1 - \log_{\varpi}\left(\ 1 + \frac{\left(\varpi^{1-t_1} - 1\right)^n}{\left(\varpi - 1\right)^{n-1}}\right)\right]\end{array}\right).$$

- $$n\varrho_1 = \left(\begin{array}{l}\left[1 - \log_{\varpi}\left(\ 1 + \frac{\left(\varpi^{1-e_1} - 1\right)^n}{\left(\varpi - 1\right)^{n-1}}\right),\ 1 - \log_{\varpi}\left(\ 1 + \frac{\left(\varpi^{1-f_1} - 1\right)^n}{\left(\varpi - 1\right)^{n-1}}\right)\right], \\ \left[\log_{\varpi}\left(\ 1 + \frac{\left(\varpi^{s_1} - 1\right)^n}{\left(\varpi - 1\right)^{n-1}}\right), \log_{\varpi}\left(\ 1 + \frac{\left(\varpi^{t_1} - 1\right)^n}{\left(\varpi - 1\right)^{n-1}}\right)\right]\end{array}\right).$$

Definition 2.5 (Oraki et al., 2021): Assume that $\varrho_z = \left(\left[e_z, f_z\right], \left[s_z, t_z\right]\right)$ $(z = 1,2,3...,n)$ are IVIFNs. Then, the "IVIF Frank weighted averaging" (IVIFFWA) AO, IVIFFWA: $\gamma^n \to \gamma$ is

$$IVIFFWA\left(\varrho_1, \varrho_2, \ldots, \varrho_n\right) = \bigoplus_{z=1}^{n}\left(\check{S}_z \varrho_z\right)$$

$$= \left(\left[1 - \log_{\varpi}\left(1 + \frac{\prod_{z=1}^{n}\left(\varpi^{1-e_z} - 1\right)^{\check{S}_z}}{(\varpi - 1)^{\sum_{z=1}^{n}(\check{S}_z)-1}}\right), 1 - \log_{\varpi}\left(1 + \frac{\prod_{z=1}^{n}\left(\varpi^{1-f_z} - 1\right)^{\check{S}_z}}{(\varpi - 1)^{\sum_{z=1}^{n}(\check{S}_z)-1}}\right)\right],\right.$$

$$\left.\left[\log_{\varpi}\left(1 + \frac{\prod_{z=1}^{n}\left(\varpi^{s_z} - 1\right)^{\check{S}_z}}{(\varpi - 1)^{\sum_{z=1}^{n}(\check{S}_z)-1}}\right), \log_{\varpi}\left(1 + \frac{\prod_{z=1}^{n}\left(\varpi^{t_z} - 1\right)^{\check{S}_z}}{(\varpi - 1)^{\sum_{z=1}^{n}(\check{S}_z)-1}}\right)\right]\right).$$

The "interval-valued intuitionistic fuzzy Frank weighted geometric" (IVIFFWG) AO, IVIFFWG: $\gamma^n \to \gamma$ is

$$\text{IVIFFWG}\left(\varrho_1, \varrho_2, \ldots, \varrho_n\right) = \bigoplus_{z=1}^{n}\left(\varrho_z^{\check{S}_z}\right)$$

$$= \left(\left[\log_{\varpi}\left(1 + \frac{\prod_{z=1}^{n}\left(\varpi^{e_z} - 1\right)^{\check{S}_z}}{(\varpi - 1)^{\sum_{z=1}^{n}(\check{S}_z)-1}}\right), \log_{\varpi}\left(1 + \frac{\prod_{z=1}^{n}\left(\varpi^{f_z} - 1\right)^{\check{S}_z}}{(\varpi - 1)^{\sum_{z=1}^{n}(T_z)-1}}\right)\right],\right.$$

$$\left.\left[1 - \log_{\varpi}\left(1 + \frac{\prod_{z=1}^{n}\left(\varpi^{1-s_z} - 1\right)^{\check{S}_z}}{(\varpi - 1)^{\sum_{z=1}^{n}(\check{S}_z)-1}}\right), 1 - \log_{\varpi}\left(1 + \frac{\prod_{z=1}^{n}\left(\varpi^{1-t_z} - 1\right)^{\check{S}_z}}{(\varpi - 1)^{\sum_{z=1}^{n}(\check{S}_z)-1}}\right)\right]\right).$$

Here, the weight vectors of $\left(\varrho_1, \varrho_2, \ldots, \varrho_n\right)$ are $\mp\left(\daleth_1, \daleth_2 \ldots, \daleth_n\right)^T$, and γ represents the set of all IVIFNs. It is ensured that $\daleth_z > 0$ and $\sum_{z=1}^{n} \daleth_z = 1$.

11.3 CL-BASED IVIF FRANK WEIGHTED AOS

The CL-based Frank WA&G AOs for IVIFNs are proposed and their properties are discussed in this section.

Definition 3.1 Consider the set of IVIFNs $\varrho_z = \left(\left[e_z, f_z\right], \left[s_z, t_z\right]\right)(z = 1, 2, 3 \ldots, n)$. Let η_z represent the IVIFNs' CL, then CIVIFFWA: $\gamma^n \to \gamma$ is:

$$CIVIFFWA\left(\varrho_1, \varrho_2, \ldots, \varrho_n\right) = \mathop{\oplus}_{z=1}^{n} \daleth_z\left(\eta_z \varrho_z\right)$$

where γ is the set of all IVIFNs and $\daleth = \left(\daleth_1, \daleth_2, \ldots, \daleth_n\right)^T$ is weight vector of $\left(\varrho_1, \varrho_2, \ldots, \varrho_n\right)$ such that $\daleth_z > 0$ and $\sum_{z=1}^{n} \daleth_z = 1$. Also, the confidence levels, η_z, lie in [0,1].

Theorem 3.1 *Let* $\varrho_z = \left(\left[e_z, f_z\right], \left[s_z, t_z\right]\right)(z = 1, 2, 3 \ldots, n)$ *be a collection of the IVIFNs and* η_z *be the CL of IVIFNs, then the result aggregated from the above definition is an IVIFN, and:*

$$CIVIFFWA\left(\varrho_1, \varrho_2, \ldots, \varrho_n\right) = \left(\left[\begin{array}{l} 1 - \log_\varpi \left(1 + \dfrac{\prod_{z=1}^{n} \left(\varpi^{1-e_z} - 1\right)^{\daleth_z \eta_z}}{(\varpi - 1)^{\sum_{z=1}^{n} (\daleth_z \eta_z) - 1}} \right), 1 \\ - \log_\varpi \left(1 + \dfrac{\prod_{z=1}^{n} \left(\varpi^{1-f_z} - 1\right)^{\daleth_z \eta_z}}{(\varpi - 1)^{\sum_{z=1}^{n} (\daleth_z \eta_z) - 1}} \right) \end{array} \right], \right. \tag{11.2}$$

$$\left. \left[\log_\varpi \left(1 + \frac{\prod_{z=1}^{n} \left(\varpi^{s_z} - 1\right)^{\daleth_z \eta_z}}{(\varpi - 1)^{\sum_{z=1}^{n} (\daleth_z \eta_z) - 1}} \right), \log_\varpi \left(1 + \frac{\prod_{z=1}^{n} \left(\varpi^{t_z} - 1\right)^{\daleth_z \eta_z}}{(\varpi - 1)^{\sum_{z=1}^{n} (\daleth_z \eta_z) - 1}} \right) \right] \right).$$

Proof: First, we apply the Frank aggregation procedure for two IVIFNs, ϱ_1 and ϱ_2, in order to demonstrate this theorem using the mathematical induction. When $n = 1$ we get

$$CIVIFFWA\left(\varrho_1\right) = \left(\left[1 - \log_\varpi \left(1 + \frac{\left(\varpi^{1-e_1} - 1\right)^{\eta_1 \daleth}}{(\varpi - 1)^{\eta_1 \daleth - 1}} \right), 1 - \log_\varpi \left(1 + \frac{\left(\varpi^{1-f_1} - 1\right)^{\eta_1 \daleth}}{(\varpi - 1)^{\eta_1 \daleth - 1}} \right) \right]. \right.$$

$$\left. \left[\log_\varpi \left(1 + \frac{\left(\varpi^{s_1} - 1\right)^{\eta_1 \daleth}}{(\varpi - 1)^{\eta_1 \daleth - 1}} \right), \log_\varpi \left(1 + \frac{\left(\varpi^{t_1} - 1\right)^{\eta_1 \daleth}}{(\varpi - 1)^{\eta_1 \daleth - 1}} \right) \right] \right)$$

Thus, it is true for $n = 1$. Now for $n = 2$

$$CIVIFFWA(\varrho_2)=\left(\left[1-\log_{\varpi}\left(1+\frac{\left(\varpi^{1-e_2}-1\right)^{\eta_2\daleth_2}}{\left(\varpi-1\right)^{\eta_2\daleth_2-1}}\right),1-\log_{\varpi}\left(1+\frac{\left(\varpi^{1-f_2}-1\right)^{\eta_2\daleth_2}}{\left(\varpi-1\right)^{\eta_2\daleth_2-1}}\right)\right],\right.$$

$$\left.\left[\log_{\varpi}\left(1+\frac{\left(\varpi^{s_2}-1\right)^{\eta_2\daleth_2}}{\left(\varpi-1\right)^{\eta_2\daleth_2-1}}\right),\ \log_{\varpi}\left(1+\frac{\left(\varpi^{t_2}-1\right)^{\eta_2\daleth_2}}{\left(\varpi-1\right)^{\eta_2\daleth_2-1}}\right)\right]\right)$$

Then,

$$CIVIFFWA(\varrho_1,\varrho_2)=\ \underset{z=1}{\overset{2}{\oplus}}\check{S}_z(\eta_z,\varrho_z)$$

$$=\left(\left[1-\log_{\varpi}\left(1+\frac{\prod_{z=1}^{2}\left(\varpi^{1-e_z}-1\right)^{\check{S}_z\eta_z}}{(\varpi-1)^{\sum_{z=1}^{2}(\check{S}_z\eta_z)-1}}\right),\ 1-\log_{\varpi}\left(1+\frac{\prod_{z=1}^{2}\left(\varpi^{1-f_z}-1\right)^{\check{S}_z\eta_z}}{(\varpi-1)^{\sum_{z=1}^{2}(\check{S}_z\eta_z)-1}}\right)\right],\right.$$

$$\left.\left[\log_{\varpi}\left(1+\frac{\prod_{z=1}^{2}\left(\varpi^{s_z}-1\right)^{\check{S}_z\eta_z}}{(\varpi-1)^{\sum_{z=1}^{2}(\check{S}_z\eta_z)-1}}\right),\ \log_{\varpi}\left(1+\frac{\prod_{z=1}^{2}\left(\varpi^{t_z}-1\right)^{\check{S}_z\eta_z}}{(\varpi-1)^{\sum_{z=1}^{2}(\check{S}_z\eta_z)-1}}\right)\right]\right)$$

Consequently, for n = 2, the result holds true. Let this be true for $n=m$, then

$$CIVIFFWA\left(\tilde{n}_1,\tilde{n}_2,\ldots,\tilde{n}_m\right)=\underset{z=1}{\overset{m}{\oplus}}\daleth_m\left(\eta_m,\tilde{n}_m\right)$$

$$=\left(\left[1-\log_{\varpi}\left(1+\frac{\prod_{z=1}^{m}\left(\varpi^{1-e_z}-1\right)^{\daleth_z\eta_z}}{\left(\varpi-1\right)^{\sum_{z=1}^{m}(\daleth_z\eta_z)-1}}\right),1-\log_{\varpi}\left(1+\frac{\prod_{z=1}^{m}\left(\varpi^{1-f_z}-1\right)^{\daleth_z\eta_z}}{\left(\varpi-1\right)^{\sum_{z=1}^{m}(\daleth_z\eta_z)-1}}\right)\right].\right.$$

$$\left.\left[\log_{\varpi}\left(1+\frac{\prod_{z=1}^{m}\left(\varpi^{s_z}-1\right)^{\daleth_z\eta_z}}{\left(\varpi-1\right)^{\sum_{z=1}^{m}(\daleth_z\eta_z)-1}}\right),\log_{\varpi}\left(1+\frac{\prod_{z=1}^{m}\left(\varpi^{t_z}-1\right)^{\daleth_z\eta_z}}{\left(\varpi-1\right)^{\sum_{z=1}^{m}(\daleth_z\eta_z)-1}}\right)\right]\right)$$

Now, for $n=m+1$

$$CIVIFFWA\left(\varrho_1,\varrho_2,\ldots,\varrho_{m+1}\right)=\overset{m}{\underset{z=1}{\oplus}}\check{S}_m\left(\eta_m,\varrho_m\right)\oplus\check{S}_{m+1}\left(\eta_{m+1}\varrho_{m+1}\right)$$

$$=\left(\left[1-\log_{\varpi}\left(1+\frac{\prod_{z=1}^{m}\left(\varpi^{1-e_z}-1\right)^{\check{S}_z\eta_z}}{(\varpi-1)^{\sum_{z=1}^{m}(\check{S}_z\eta_z)-1}}\right),1-\log_{\varpi}\left(1+\frac{\prod_{z=1}^{m}\left(\varpi^{1-f_z}-1\right)^{\check{S}_z\eta_z}}{(\varpi-1)^{\sum_{z=1}^{m}(\check{S}_z\eta_z)-1}}\right)\right],\right.$$

$$\left.\left[\log_{\varpi}\left(1+\frac{\prod_{z=1}^{m}\left(\varpi^{s_z}-1\right)^{\check{S}_z\eta_z}}{(\varpi-1)^{\sum_{z=1}^{m}(\check{S}_z\eta_z)-1}}\right),\log_{\varpi}\left(1+\frac{\prod_{z=1}^{m}\left(\varpi^{t_z}-1\right)^{\check{S}_z\eta_z}}{(\varpi-1)^{\sum_{z=1}^{m}(\check{S}_z\eta_z)-1}}\right)\right]\right)$$

$$\oplus\left(\left[1-\log_{\varpi}\left(1+\frac{\left(\varpi^{1-e_{m+1}}-1\right)^{\eta_{m+1}\check{S}_{m+1}}}{(\varpi-1)^{\eta_{m+1}\check{S}_{m+1}-1}}\right),1-\log_{\varpi}\left(1+\frac{\left(\varpi^{1-f_{m+1}}-1\right)^{\eta_{m+1}\check{S}_{m+1}}}{(\varpi-1)^{\eta_{m+1}\check{S}_{m+1}-1}}\right)\right],\right.$$

$$\left.\left[\log_{\varpi}\left(1+\frac{\left(\varpi^{s_{m+1}}-1\right)^{\eta_{m+1}\check{S}_{m+1}}}{(\varpi-1)^{\eta_{m+1}\check{S}_{m+1}-1}}\right),\log_{\varpi}\left(1+\frac{\left(\varpi^{t_{m+1}}-1\right)^{\eta_{m+1}\check{S}_{m+1}}}{(\varpi-1)^{\eta_{m+1}\check{S}_{m+1}-1}}\right)\right]\right)$$

$$=\left(\left[1-\log_{\varpi}\left(1+\frac{\prod_{z=1}^{m+1}\left(\varpi^{1-e_z}-1\right)^{\check{S}_z\eta_z}}{(\varpi-1)^{\sum_{z=1}^{m+1}(\check{S}_z\eta_z)-1}}\right),1-\log_{\varpi}\left(1+\frac{\prod_{z=1}^{m+1}\left(\varpi^{1-f_z}-1\right)^{\check{S}_z\eta_z}}{(\varpi-1)^{\sum_{z=1}^{m+1}(\check{S}_z\eta_z)-1}}\right)\right],\right.$$

$$\left.\left[\log_{\varpi}\left(1+\frac{\prod_{z=1}^{m+1}\left(\varpi^{s_z}-1\right)^{\check{S}_z\eta_z}}{(\varpi-1)^{\sum_{z=1}^{m+1}(\check{S}_z\eta_z)-1}}\right),\log_{\varpi}\left(1+\frac{\prod_{z=1}^{m+1}\left(\varpi^{tf_z}-1\right)^{\check{S}_z\eta_z}}{(\varpi-1)^{\sum_{z=1}^{m+1}(\check{S}_z\eta_z)-1}}\right)\right]\right)$$

Hence, for $n=m+1$, the result is true. Hence, the theorem is proved. □

Definition 3.2 Consider the set of IVIFNs $\varrho_z=\left(\left[e_z,f_z\right],\left[s_z,\ t_z\right]\right)(z=1,2,3\ldots,n)$. Let η_z represent the IVIFNs' CL, then CIVIFFWA: $\gamma^n\to\gamma$ is:

$$CIVIFFWG\left(\varrho_1,\varrho_2,\ldots,\varrho_n\right)=\overset{n}{\underset{z=1}{\otimes}}\left(\varrho_z^{\eta_z}\right)^{\daleth_z},$$

where γ is the set of all IVIFNs and $\daleth=\left(\daleth_1,\daleth_2,\ldots,\daleth_n\right)^T$ is weight vector of $\left(\varrho_1,\varrho_2,\ldots,\varrho_n\right)$ such that $_z\daleth>0$ and $\sum_{z=1}^{n}{}_z\daleth=1$. Also, the confidence levels, η_z, lie in [0,1].

Theorem 3.2 *Let* $\varrho_z=\left(\left[e_z,f_z\right],\left[s_z,t_z\right]\right)(z=1,2,3\ldots,n)$ *be a collection of the IVIFNs and* η_z *be the CL of IVIFNs, then the result aggregated from the above definition is an IVIFN, and:*

$$CIVIFFWG(\tilde{n}_1,\tilde{n}_2,\ldots,\tilde{n}_n)=\bigotimes_{z=1}^{n}\left(\tilde{n}_z^{\eta_z}\right)^{\daleth_z}$$

$$=\left(\left[\log_{\varpi}\left(1+\frac{\prod_{z=1}^{n}\left(\varpi^{e_z}-1\right)^{\daleth_z\eta_z}}{(\varpi-1)^{\sum_{z=1}^{n}(\daleth_z\eta_z)-1}}\right),\log_{\varpi}\left(1+\frac{\prod_{z=1}^{n}\left(\varpi^{f_z}-1\right)^{\daleth_z\eta_z}}{(\varpi-1)^{\sum_{z=1}^{n}(\daleth_z\eta_z)-1}}\right)\right],\right.$$

$$\left.\left[1-\log_{\varpi}\left(1+\frac{\prod_{z=1}^{n}\left(\varpi^{1-s_z}-1\right)^{\daleth_z\eta_z}}{(\varpi-1)^{\sum_{z=1}^{n}(\daleth_z\eta_z)-1}}\right),1-\log_{\varpi}\left(1+\frac{\prod_{z=1}^{n}\left(\varpi^{1-t_z}-1\right)^{\daleth_z\eta_z}}{(\varpi-1)^{\sum_{z=1}^{n}(\daleth_z\eta_z)-1}}\right)\right]\right). \quad (11.3)$$

Proof: The proof is similar as above. □

Following are some properties of CIVIFFWA and CIVIFFWG AOs.

- Monotonicity

  Let $(\varrho_1^{'},\varrho_2^{'},\ldots,\varrho_n^{'})$ and $(\varrho_1,\varrho_2,\ldots,\varrho_n)$ be two collections of IVIFFNs such that $\varrho_z^{'}\le\varrho_z,\forall\ z=1,2,\ldots,n$ and η_z be the confidence levels of IVIFNs, then $CIVIFFWA(\varrho_1^{'},\varrho_2^{'},\ldots,\varrho_n^{'})\le CIVIFFWA(\varrho_1,\varrho_2,\ldots,\varrho_n)$.
- Idempotency

  Let $\varrho_z=\varrho\forall\ z=1,2,\ \ldots,\ n$ and $\eta_z=\eta\forall\ z=1,2,\ \ldots,\ n$ be the CL of IVIFNs, then $CIVIFFWA(\varrho_1,\varrho_2,\ldots,\varrho_n)=\eta\varrho$.
- Boundedness

  $\min(\varrho_1,\varrho_2,\ldots,\varrho_n)\le CIVIFFWA(\varrho_1,\varrho_2,\ldots,\varrho_n)\le\max(\varrho_1,\varrho_2,\ldots,\varrho_n)$.

11.4 A MAGDM PROCESS USING CIVIFFWA AO

Here, a new MAGDM technique is proposed using the CL-based Frank weighted AOs for IVIFNs. This method consists of three primary steps. The first phase involves combining the decision matrices that were acquired from several decision specialists using AOs. In the second phase, figure out each alternative's value aggregated for each criterion. The third and final step is to compute the scores of the alternatives and rank them accordingly. Here, $R=\{R_1,R_2,\ldots,R_n\}$ is the set of alternatives. Let the set of subcriteria be $G=\{G_1,G_2,\ldots,G_m\}$ having weights of $H_{G_1},H_{G_2},\ldots,H_{G_m}$, respectively. Here the sum of H_{c_j} for $j=1,2,\ldots,m$ is 1 and $0\le H_{G_j}\le 1$. Let $L_{Q_1},L_{Q_2},\ldots,L_{Q_v}$ be the weights of the v decision experts $Q_1,Q_2,\ldots,Q_v$ such that $0\le L_{Q_k}\le 1$ for $k=1,2,\ldots,v$ and their sum is 1. Lastly, for $k=1,2,\ldots,\ v$ and $j=1,2,\ldots,m$, let each DE's CL on each criterion be η_{kj} such that $0\le\eta_{kj}\le 1$. The weights of criteria, decision experts, and their confidence levels have been assumed in this problem. Then, the steps of the suggested MAGDM approach are discussed in detail.

Step 1:

Determine the combined decision matrix (DM) by utilizing CIVIFFWA AOs.

Step 1.1:
Acquire from every DE the DMs in the linguistic variables (LVs).
Step 1.2:
Make the DMs normal and convert the LVs to IVIFNs. Next, we obtain the decision matrix $M_k = \left[\varrho_{ij} \right]_{n \times m}$, in which $j = 1, 2, ..., m$ and $i = 1, 2, ..., n$.
Step 1.3:
Combine M_k according to Equation (11.2) using CIVIFFWA for $k = 1 : v$.
Step 2:
The value for each alternative is aggregated using Definition 2.5.
Step 3:
Find scores and accuracy of the aggregated value of each alternative using Equation (11.1) and rank the alternatives accordingly.

Figure 11.1 shows the steps of the MAGDM process using the CL-based Frank operator in IVIFS.

11.5 ILLUSTRATIVE EXAMPLE

This section illustrates the proposed MAGDM technique by solving a numerical problem. Here we will assume that the alternatives are R_1, R_2, R_3, and R_4. There are four decision experts available, say Q_1, Q_2, Q_3, and Q_4. These four decision experts will assess the alternative designs or models based on the following four criteria: P_1 is financial criteria; P_2 is service quality; P_3 is product quality; P_4 is environmental criteria. Furthermore, based on each decision expert's background, education, and

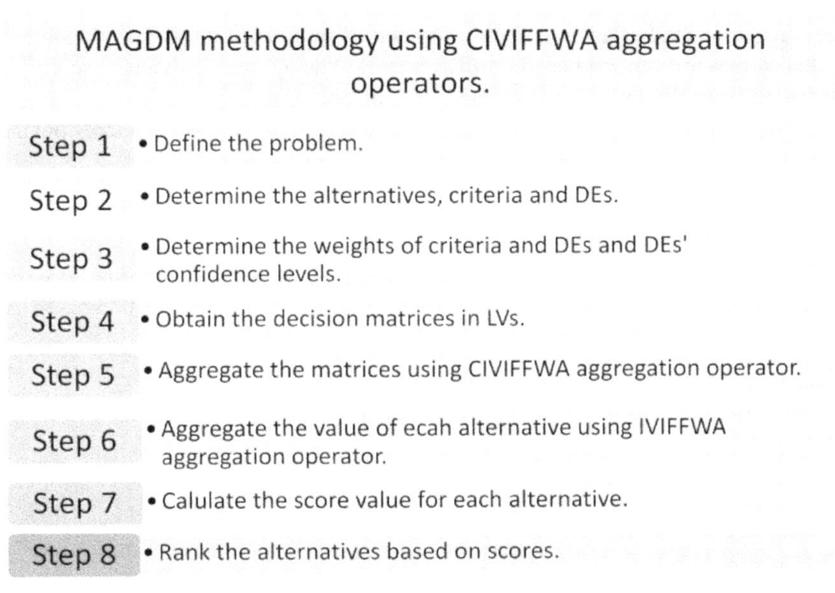

FIGURE 11.1 MAGDM process using the CIVIFFWA AO-based method

TABLE 11.1
Demographic information of decision experts

Decision experts	Aspects	Qualification	Experience (in years)
DE_1	Financial criteria	Senior financial head	8
DE_2	Service quality	Marketing manager	6
DE_3	Product quality	Executive officer	5
DE_4	Environmental criteria	Environmental expert	12

field of expertise, we have inferred assumptions about their level of confidence in specific criteria. The demographic details of the decision experts are presented in Table 11.1, while their corresponding confidence levels for each criterion are outlined in Table 11.2. Here are 12 sub-criteria, and we are assuming their weights as $H_{G_1}=0.018$, $H_{G_2}=0.012$, $H_{G_3}=0.121$, $H_{G_4}=0.23$, $H_{G_5}=0.05$, $H_{G_6}=0.123$, $H_{G_7}=0.06$, $H_{G_8}=0.022$, $H_{G_9}=0.033$, $H_{G_{10}}=0.011$, $H_{G_{11}}=0.2$, and $H_{G_{12}}=0.12$. The sum of all subcriteria weights is equal to 1. Also, the weight of the decision experts is chosen as $L_{Q_1}=0.242$, $L_{Q_2}=0.235$, $L_{Q_3}=0.269$, and $L_{Q_4}=0.254$, respectively.

Figure 11.2 gives the hierarchy of green supplier selection MAGDM problem.

The linguistic variables utilized for evaluating the alternatives are listed in Table 11.3.

The DMs obtained from the DEs using LVs from Table 11.3 are given in Table 11.4.

The DMs in LVs are then converted to IVIFNs and then aggregated using Equation (11.2). The aggregated DM is given in Table 11.5. Finally, each alternative's value is aggregated according to Definition 2.5 based on its value in IVIFNs for each criterion. The score for each alternative is then computed using Equation (11.1). Following this, the alternatives are ranked based on their scores. The aggregated values, scores, and corresponding rankings of the alternatives are displayed in Table 11.6.

It is found that R_4 is the best alternative followed by R_2 and R_3. R_4 turns out to be the least favorable alternative.

11.5.1 Comparative Analysis

To evaluate the proposed CIVIFFWA-based approach, several existing AOs have been utilized for comparison. However, there is a significant lack of CL-based AOs

TABLE 11.2
DEs' degree of confidence for every attribute

Subcriteria	DE_1	DE_2	DE_3	DE_4
P_1	0.95	0.83	0.75	0.82
P_2	0.85	0.90	0.89	0.81
P_3	0.83	0.80	0.93	0.79
P_4	0.75	0.74	0.91	0.92

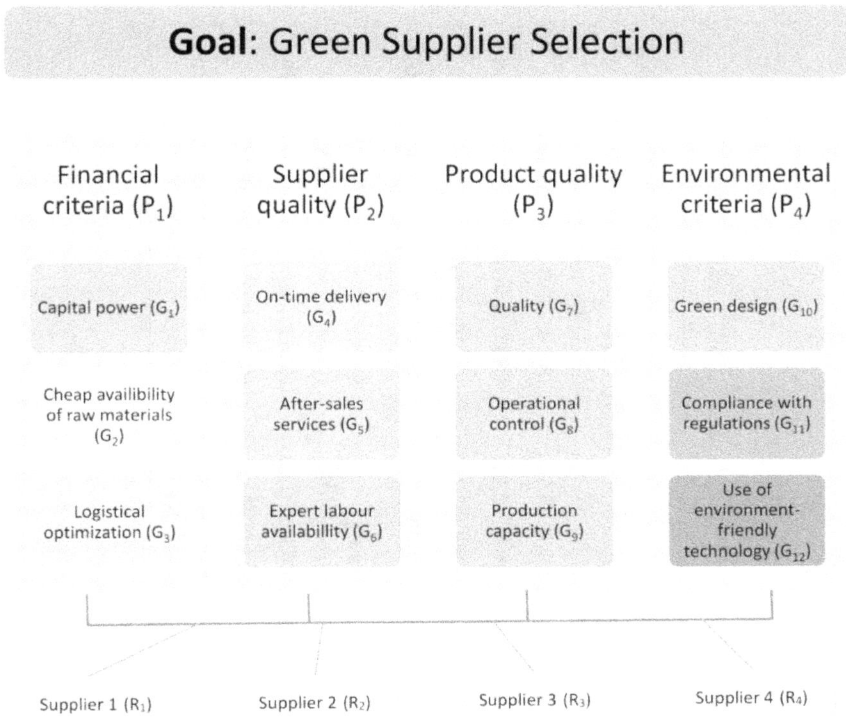

FIGURE 11.2 Hierarchy of green supplier selection MAGDM problem

TABLE 11.3
LVs used for assessing alternatives

LVs	IVIFNs
Especially good (EG)	([0.9, 0.95], [0.005, 0.01])
Very good (VG)	([0.778, 0.9], [0.01, 0.05])
Good (G)	([0.667, 0.778], [0.1, 0.172])
Relatively good (RG)	([0.556, 0.6675], [0.2, 0.283])
Medium (M)	([0.445, 0.556], [0.3, 0.394])
Relatively bad (RB)	([0.334, 0.445], [0.4, 0.505])
Bad (B)	([0.223, 0.334], [0.5, 0.616])
Very bad (VB)	([0.1, 0.223], [0.6, 0.72])
Especially bad (EB)	([0, 0.1], [0.72, 0.9])

TABLE 11.4
Decision matrix in LVs

Alternatives	G_1	G_2	G_3	G_4	G_5	G_6
R_1	(EG, VG,VG,M)	(VG,G,G,EG)	(RG,B,B,B)	(G,VB,VB,G)	(B,RB,EG,M)	(RB,M,M,RB)
R_2	(RG,G,G,G)	(G,B,EG,VG)	(B,VG,G,EG)	(RB,M,VB,B)	(M,RG,M,VG)	(VB,EG,VG,M)
R_3	(G,EG,RB,VB)	(RB,VG,M,B)	(M,M,RG,RG)	(VB,B,EG,VG)	(RG,VB,B,EG)	(VG,G,VB,G)
R_4	(VG,VG,M,RG)	(VB,G,EB,EB)	(EB,VB,VB,M)	(G,B,RG,EG)	(B,RB,M,RB)	(M,RG,EG,VG)
Alternatives	G_7	G_8	G_9	G_{10}	G_{11}	G_{12}
R_1	(M,EB,RG,EG)	(VB,RG,EB,G)	(EB,EG,EG,B)	(G,M,M,RB)	(EG,G,B,VB)	(M,VG,G,EB)
R_2	(VG,M,B,RG)	(G,G,G,RB)	(RG,VB,RB,RG)	(EB,RB,EG,EB)	(B,M,RG,M)	(EG,EB,VG,VB)
R_3	(B,M,G,RB)	(MRB,M,G)	(EG,EG,VG,RB)	(G,RG,VB,EB)	(B,EB,EB,G)	(EB,G,B,M)
R_4	(EG,EG,RB,M)	(RB,M,VG,B)	(VB,EB,B,G)	(RG,G,M,VB)	(M,VB,B,EG)	(RG,M,EG,RG)

TABLE 11.5
Aggregated decision matrix

Alternatives	G_1	G_2	G_3	G_4
R_1	([0.702, 0.821], [0.039, 0.093])	([0.708, 0.819], [0.049, 0.101])	([0.282, 0.385], [0.463, 0.566])	([0.379, 0.503], [0.323, 0.434])
R_2	([0.567, 0.680], [0.177, 0.264])	([0.646, 0.766], [0.073, 0.197])	([0.636, 0.756], [0.078, 0.144])	([0.244, 0.349], [0.498, 0.602])
R_3	([0.532, 0.648], [0.171,0 .239])	([0.416, 0.550], [0.239, 0.374])	([0.437, 0.543], [0.316, 0.407])	([0.585, 0.715], [0.093, 0.1666])
R_4	([0.595, 0.736], [0.084, 0.187])	([0.197, 0.308], [0.516, 0.651])	([0.146, 0.247], [0.597, 0.713])	([0.598, 0.707], [0.131, 0.132])
Alternatives	G_5	G_6	G_7	G_8
R_1	([0.535, 0.647], [0.169, 0.233])	([0.348, 0.452], [0.403, 0.501])	([0.523, 0.634], [0.185, 0.256])	([0.311, 0.426], [0.404, 0.524])
R_2	([0.517, 0.647], [0.170, 0.283])	([0.610, 0.739], [0.083, 0.154])	([0.460, 0.594], [0.206, 0.331])	([0.536, 0.651], [0.199, 0.292])
R_3	([0.492, 0.555], [0.199, 0.270])	([0.535, 0.673], [0.140, 0.252])	([0.396, 0.507], [0.335, 0.438])	([0.424, 0.533], [0.316, 0.414])
R_4	([0.301, 0.404], [0.448, 0.549])	([0.673, 0.787], [0.064, 0.121])	([0.655, 0.756], [0.086, 0.127])	([0.449, 0.589], [0.196, 0.331])
Alternatives	G_9	G_{10}	G_{11}	G_{12}
R_1	([0.627, 0.735], [0.089, 0.133])	([0.412, 0.520], [0.330, 0.428])	([0.477, 0.595], [0.209, 0.286])	([0.455, 0.591], [0.205, 0.338])
R_2	([0.353, 0.462], [0.388, 0.487])	([0.436, 0.552], [0.228, 0.310])	([0.378, 0.483], [0.372, 0.467])	([0.529, 0.669], [0.117, 0.211])
R_3	([0.641, 0.745], [0.092, 0.135])	([0.292, 0.407], [0.424, 0.544])	([0.223, 0.332], [0.494, 0.629])	([0.310, 0.420], [0.416, 0.532])
R_4	([0.250, 0.363], [0.466, 0.588])	([0.388, 0.502], [0.342, 0.443])	([0.491, 0.605], [0.197, 0.265])	([0.620, 0.724], [0.121, 0.174])

TABLE 11.6
Aggregated value of alternatives, score, and rank

Alternatives	Aggregated IVIFN	Score	Rank
R_1	([0.433, 0.553], [0.258, 0.356])	0.186	4
R_2	([0.462, 0.588], [0.215, 0.356]	0.258	2
R_3	([0.446, 0.568], [0.232, 0.337])	0.222	3
R_4	([0.522, 0.639], [0.176, 0.231])	0.377	1

for IVIFNs; hence, "CL-based interval-valued Fermatean fuzzy weighted averaging (CIVFFWA) AO" (Seikh & Chatterjee, 2024) and "CL interval-valued Fermatean fuzzy weighted geometric (CIVFFWG) AO" (Seikh & Chatterjee, 2024) have been used. Furthermore, the method has been compared with "IVIF Einstein weighted averaging (IVIFEWA) AO" (Wang & Liu, 2013), "IVIF Einstein weighted geometric (IVIFEWG) AO" (Wang & Liu, 2013), and "IVIF Dombi weighted averaging (IVIFDWA) AO" (He, 2019). The final ranking of alternatives is displayed in Table 11.7.

A graphical representation of the comparative analysis results, highlighting the performance differences among the various AOs, is presented in Figure 11.3. This visualization aids in understanding the effectiveness of the suggested CIVIFFWA-based approach to the existing methods.

TABLE 11.7
Comparison of different methods

Reference	Aggregation operator	Scores				Ranking
		$\mathcal{S}R_1$	$\mathcal{S}R_2$	$\mathcal{S}R_3$	$\mathcal{S}R_4$	
Seikh and Chatterjee (2024)	CIVFFWA	0.247	0.28	0.262	0.379	$R_4 > R_2 > R_3 > R_1$
	CIVFFWG	−0.247	−0.264	−0.252	−0.258	$R_1 > R_3 > R_4 > R_2$
Wang and Liu (2013)	IVIFEWA	0.701	0.728	0.705	0.769	$R_4 > R_2 > R_3 > R_1$
Wang and Liu (2013)	IVIFEWG	−0.671	−0.403	−0.446	−0.381	$R_4 > R_2 > R_3 > R_1$
He (2019)	IVIFDWA	0.471	0.462	0.493	0.511	$R_4 > R_3 > R_1 > R_2$
Proposed method	CIVIFFWA	0.247	0.28	0.262	0.379	$R_4 > R_2 > R_3 > R_1$

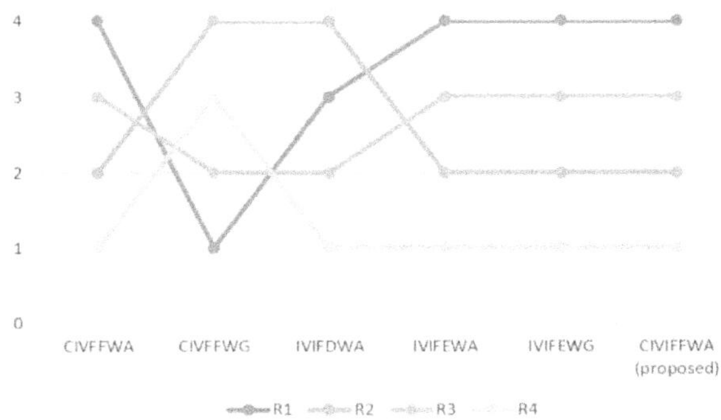

FIGURE 11.3 Evaluation of the suggested MAGDM approach in comparison

Due to the lack of availability of CL-based AOs for IVIFNs, the proposed method has first been compared with CIVFFWA and CIVFFWG AOs. From Table 11.7 and Figure 11.3, the ranking orders of the different methods are observed. The proposed method and CIVIFFWA give the same ranking order; however, in the case of CIVIFFWG, there is some variation. The ranking order obtained using CIVIFFWG is $R_1 > R_3 > R_4 > R_2$, where R_1 moves to first place and R_4 moves to third place. Next, the proposed method is compared with the IVIFEWA and IVIFEWG AOs, and in both instances, the ranking results remain consistent with those of the proposed method. Here, due to the absence of CLs in the AOs, all confidence levels have been taken as 1 during calculation. Similarly, a comparison has been made with IVIFDWA AOs, and then the ranking order obtained is $R_4 > R_3 > R_1 > R_2$, where, the positions of R_3, R_2, and R_1 vary but R_4 remains in the first position. Therefore, we observe that the ranking order is largely consistent across most cases, indicating that the proposed method demonstrates stability.

11.5.2 Analyzing Sensitivity

This section presents a sensitivity analysis of the suggested method. The sensitivity analysis is conducted by investigating how the results vary with changes in the parameter values. The only parameter that we have utilized is ϖ, which is found in Frank norms. During the numerical problem-solving process, we have arbitrarily chosen $\varpi = 2$. To perform sensitivity analysis, we alter ϖ and verify the outcome. The ranking order obtained for ϖ = 2 to 10. Table 11.8 displays the alternative scores and ranking.

Figure 11.4 demonstrates the results of sensitivity analysis graphically.

From Table 11.8 and Figure 11.4, it can be observed that the ranking results for $\varpi = 2$, $\varpi = 3$, $\varpi = 5$, and $\varpi = 10$ are the same and are $R_4 > R_2 > R_3 > R_1$. However, for $\varpi = 4$, $\varpi = 6$, $\varpi = 7$, and $\varpi = 8$ the ranking orders are slightly changed to $R_4 > R_2 > R_1 > R_3$. It is noted that the third and the fourth places are swapped. Lastly, for $\varpi = 9$, the ranking order significantly changes to $R_1 > R_4 > R_2 > R_3$. In this case,

TABLE 11.8
Sensitivity of proposed methods

ϖ	SR_1	SR_2	SR_3	SR_4	Ranking order
2	0.186	0.258	0.223	0.377	$R_4 > R_2 > R_3 > R_1$
3	0.220	0.291	0.249	0.382	$R_4 > R_2 > R_3 > R_1$
4	0.174	0.211	0.171	0.308	$R_4 > R_2 > R_1 > R_3$
5	0.141	0.215	0.175	0.332	$R_4 > R_2 > R_3 > R_1$
6	0.155	0.192	0.151	0.288	$R_4 > R_2 > R_1 > R_3$
7	0.148	0.185	0.143	0.281	$R_4 > R_2 > R_1 > R_3$
8	0.142	0.179	0.137	0.275	$R_4 > R_2 > R_1 > R_3$
9	0.344	0.304	0.202	0.327	$R_1 > R_4 > R_2 > R_3$
10	0.110	0.186	0.142	0.300	$R_4 > R_2 > R_3 > R_1$

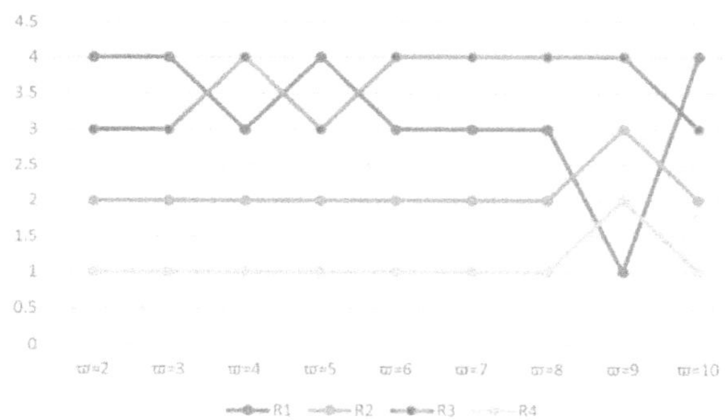

FIGURE 11.4 Sensitivity analysis of the proposed MAGDM method

a change occurs in all the positions, and R_1 gets the first place while R_4 (which was initially first) moves to the second place. Owing to the significant similarity in ranking orders in most of the cases, it can be said that the proposed method is stable.

11.6 CONCLUSION

This chapter uses Frank t-norm and t-conorm to develop CL-based AOs for IVIFNs. The chapter proposes the CL-based Frank AOs and discusses their properties as well. Using the proposed AOs, it proposes a novel MAGDM method to solve the problem of assessing and selecting optimal green supplier selection. The process is more precise and accurate owing to the MAGDM method's incorporation of decision experts' confidence levels. Moreover, Frank AOs provide a significant level of flexibility due to the inclusion of a parameter. These AOs have been employed to compile data in the decision-making process. There are several advantages of the proposed method. By incorporating decision makers' CLs, the proposed method strengthens the aggregation process, improving its robustness and reliability. This allows for more accurate and dependable decision-making, especially in situations involving uncertainty or conflicting criteria. Frank AOs outperform Dombi, Einstein, and Hamacher AOs due to their greater flexibility and robust mathematical framework. Frank t-norms and t-conorms offer significant advantages due to their adjustable parameters and smooth integration with fuzzy logic operations. This enables precise control over the aggregation process and enhances functionality. By incorporating IVIFSs, the method effectively manages uncertainty, which in turn enhances decision-making accuracy and reliability in MAGDM.

The proposed method is applied to the green supplier selection problem, where it identified the fourth supplier as the optimal choice. To assess the effectiveness and applicability of the method, a comparative analysis with existing AOs is conducted. The results of this comparison demonstrate the stability and consistency of the proposed method across various AOs. While some minor variations in the ranking were observed when compared with CIVFFWG and IVIFDWA, the overall ranking remained largely unchanged, highlighting the robustness of the approach. Additionally, a sensitivity analysis was performed by varying parameter values, which further confirmed the stability and reliability of the proposed method. The analysis revealed consistent ranking orders across most values of ϖ, with only minor shifts in specific cases. These findings collectively indicate that the proposed method is both robust and stable, delivering dependable results even under varying input parameters.

Although this research has valuable insight, it has some limitations, which have to be acknowledged. Here, we have assumed the criteria weight, decision expert weights, and CL as crisp numbers. Also, comparisons have been performed only with the AO-based MAGDM method. Additionally, a single decision expert has been chosen for each criterion, which may result in personal biases in the decision-making process.

The chapter's weaknesses can be addressed in further research. Firstly, using popular weight-determining MAGDM methods to determine criteria weight and decision experts' weights can significantly improve the methodology. Secondly, comparison with decision-making methods like TOPSIS and VIKOR can better establish the proposed method's practicality. Thirdly, incorporating several decision experts from every facet would yield a more realistic result through the comprehension of various viewpoints. Other MAGDM problems, such as the best sustainable clean energy transportation service problem, the best problem for managing plastic trash, the best problem for choosing green warehouses, the best problem for renewable energy sources, and many more, can be solved using the suggested approach to assess practical applicability in different scenarios.

CONFLICT OF INTEREST

There is no conflict of interest, according to the authors.

REFERENCES

Abolfathi, M. A., & Ebadian, A. (2023). Interval-valued picture fuzzy aggregation information based on Frank operators and their application in group decision making. *RAIRO-Operations Research*, 57(6), 3191–3222.

Atanassov, K. T. (1986). Intuitionistic fuzzy sets. *Fuzzy Sets and Systems*, 20(1), 87–96.

Atanassov, K., & Gargov, G. (1989). Interval valued intuitionistic fuzzy sets. *Fuzzy Sets and Systems*, 31(3), 343–349.

Dong, Y., Zhang, H., & Herrera-Viedma, E. (2016). Consensus reaching model in the complex and dynamic MAGDM problem. *Knowledge-Based Systems*, 106, 206–219.

Garg, H. (2017). Confidence levels based Pythagorean fuzzy aggregation operators and its application to decision-making process. *Computational and Mathematical Organization Theory*, 23(4), 546–571.

He, X. (2019). Group decision making based on Dombi operators and its application to personnel evaluation. *International Journal of Intelligent Systems*, 34(7), 1718–1731.

Klement, P. E., & Mesiar, R. (1997). Triangular norms. *Tatra Mountains Mathematical Publication*, 13(1997), 169–193.

Konys, A. (2019). Green supplier selection criteria: From a literature review to a comprehensive knowledge base. *Sustainability*, 11(15), 4208.

Liu, P. (2017). Some Frank aggregation operators for interval-valued intuitionistic fuzzy numbers and their application to group decision making. *Journal of Multiple-Valued Logic & Soft Computing*, 29, 183–223.

Mahmood, T., Ahmmad, J., Ali, Z., & Yang, M. S. (2023). Confidence level aggregation operators based on intuitionistic fuzzy rough sets with application in medical diagnosis. *IEEE Access*, 11, 8674–8688.

Nafei, A., Azizi, S. P., Edalatpanah, S. A., & Huang, C. Y. (2024). Smart TOPSIS: A neural Network-Driven TOPSIS with neutrosophic triplets for green supplier selection in sustainable manufacturing. *Expert Systems with Applications*, 255, 124744.

Oraki, M., Eshaghi Gordji, M., & Ardakani, H. (2021). Some frank aggregation operators based on the interval-valued intuitionistic fuzzy numbers. *International Journal of Nonlinear Analysis and Applications*, 12(Special Issue), 325–342.

Petchimuthu, S., Kamacı, H., & Senapati, T. (2024). Evaluation of artificial intelligence-based solid waste segregation technologies through multi-criteria decision-making and complex q-rung picture fuzzy frank aggregation operators. *Engineering Applications of Artificial Intelligence*, 133, 108154.

Rahman, K., Ayub, S., & Abdullah, S. (2021). Generalized intuitionistic fuzzy aggregation operators based on confidence levels for group decision making. *Granular Computing*, 6, 867–886.

Rahim, M., Ahmad, S., Bajri, S. A., Alharbi, R., & Khalifa, H. A. E. W. (2024). Confidence levels-based p, q, r spherical fuzzy aggregation operators and their application in selection of solar panels. *IEEE Access*, 12, 57863–57878.

Seikh, M. R., & Mandal, U. (2021). Some picture fuzzy aggregation operators based on Frank t-norm and t-conorm: Application to MADM process. *Informatica*, 45(3), 447–461.

Seikh, M. R., & Mandal, U. (2022). Q-rung orthopair fuzzy Frank aggregation operators and its application in multiple attribute decision-making with unknown attribute weights. *Granular Computing*, 7(3), 709–730.

Seikh, M. R., & Chatterjee, P. (2024a). Determination of best renewable energy sources in India using SWARA-ARAS in confidence level based interval-valued Fermatean fuzzy environment. *Applied Soft Computing*, 155, 111495.

Seikh, M. R., & Chatterjee, P. (2024b). Evaluation and selection of E-learning websites using intuitionistic fuzzy confidence level based Dombi aggregation operators with unknown weight information. *Applied Soft Computing*, 163, 111850.

Verma, M., Prem, P. R., Ren, P., Liao, H., & Xu, Z. (2022). Green supplier selection with a multiple criteria decision-making method based on thermodynamic features. *Environment, Development and Sustainability*, 27(9), 21887–21919.

Wang, W. Z., & Liu, X. W. (2013a). Interval-valued intuitionistic fuzzy hybrid weighted averaging operator based on Einstein operation and its application to decision making. *Journal of Intelligent & Fuzzy Systems,* 25, 279–290.

Wang, W., & Liu, X. (2013b). The multi-attribute decision making method based on interval-valued intuitionistic fuzzy Einstein hybrid weighted geometric operator. *Computers & Mathematics with Applications*, 66(10), 1845–1856.

Xu, Z. S. (2007). Methods for aggregating interval-valued intuitionistic fuzzy information and their application to decision making. *Control and Decision*, 22, 215–219.

Xu, Z. S., & Chen, J. (2007). On geometric aggregation over interval-valued intuitionistic fuzzy information. In: *Proceedings of the Fourth International Conference on Fuzzy Systems and Knowledge Discovery. IEEE Computer Society Press*, Washington, pp. 466–471.

Xu, Z. S., & Jian, C. (2007). Approach to group decision making based on interval-valued intuitionistic judgment matrices. *Systems Engineering-Theory & Practice*, 27(4), 126–133.

Yang, X., Mahmood, T., Ali, Z., & Hayat, K. (2023). Identification and classification of multi-attribute decision-making based on complex intuitionistic fuzzy frank aggregation operators. *Mathematics*, 11(15), 3292.

Zadeh, L. A. (1965). Fuzzy sets. *Information and Control,* 8(3), 338–3533.

Zeng, S., Chen, W., Gu, J., & Zhang, E. (2023). An integrated EDAS model for Fermatean fuzzy multi-attribute group decision making and its application in green-supplier selection. *Systems*, 11(3), 162.

Zhang, Z. (2017). Interval-valued intuitionistic fuzzy Frank aggregation operators and their applications to multiple attribute group decision making. *Neural Computing and Applications*, 28, 1471–1501.

Zhu, C., & Wang, X. (2024). A novel integrated approach based on best–worst and VIKOR methods for green supplier selection under multi-granularity extended probabilistic linguistic environment. *Complex & Intelligent Systems*, 10(2), 2029–2046.

Index

For Product Safety Concerns and Information please contact our EU representative GPSR@taylorandfrancis.com
Taylor & Francis Verlag GmbH, Kaufingerstraße 24, 80331 München, Germany

www.ingramcontent.com/pod-product-compliance
Lightning Source LLC
LaVergne TN
LVHW020642300926
849161LV00004B/778

* 9 7 8 1 0 3 2 8 6 7 6 7 0 *